# American Society of Plumbing Engineers
# Data Book

A Plumbing Engineer's Guide to System Design and Specifications

# Volume 1

# Fundamentals of Plumbing Engineering

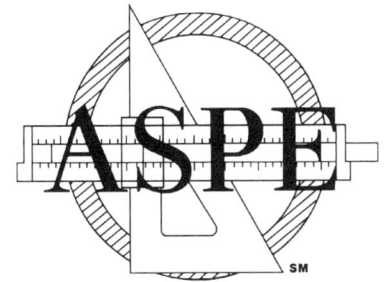

American Society of Plumbing Engineers
3617 E. Thousand Oaks Blvd., Suite 210
Westlake Village, CA 91362

The ASPE Data Book is designed to provide accurate and authoritative information for the design and specification of plumbing systems. The publisher makes no guarantees or warranties, expressed or implied, regarding the data and information contained in this publication. All data and information are provided with the understanding that the publisher is not engaged in rendering legal, consulting, engineering, or other professional services. If legal, consulting, or engineering advice or other expert assistance is required, the services of a competent professional should be engaged.

**American Society of Plumbing Engineers**
3617 E. Thousand Oaks Blvd., Suite 210
Westlake Village, CA 91362
(805) 495-7120 • Fax: (805) 495-4861
E-mail: aspchq@aol.com • Internet: www.aspe.org

**Copyright © 1999 by American Society of Plumbing Engineers**

All rights reserved, including rights of reproduction and use in any form or by any means, including the making of copies by any photographic process, or by any electronic or mechanical device, printed or written or oral, or recording for sound or visual reproduction, or for use in any knowledge or retrieval system or device, unless permission in writing is obtained from the publisher.

ISBN 1-891255-06-1
Printed in the United States of America

10  9  8  7  6  5  4  3  2  1

# Data Book
# Volume 1
# Fundamentals of Plumbing Engineering

*Data Book* Chairperson: Anthony W. Stutes, P.E., CIPE
ASPE Vice-Presidents, Technical: Patrick L. Whitworth, CIPE (1996–1998)
David Chin, P.E., CIPE (1998–2000)
Editorial Review: ASPE Technical and Research Committee
Technical and Research Committee Chairperson: Norman T. Heinig, CIPE

## CONTRIBUTORS

### Chapter 1

David A. Sealine, CIPE
Walter J. Richardson, Jr., CIPE
Saum K. Nour, P.E., PhD, CIPE

### Chapter 5

David A. Sealine, CIPE

### Chapter 6

Robert H. Evans, Jr., CIPE

### Chapter 10

Saum K. Nour, P.E., PhD, CIPE

# ABOUT ASPE

The American Society of Plumbing Engineers (ASPE) is the international organization for professionals skilled in the design and specification of plumbing systems. ASPE is dedicated to the advancement of the science of plumbing engineering, to the professional growth and advancement of its members, and to the health, welfare, and safety of the public.

The Society disseminates technical data and information, sponsors activities that facilitate interaction with fellow professionals, and, through research and education programs, expands the base of knowledge of the plumbing engineering industry. ASPE members are leaders in innovative plumbing design, effective materials and energy use, and the application of advanced techniques from around the world.

**WORLDWIDE MEMBERSHIP** — ASPE was founded in 1964 and currently has almost 7,000 members. Spanning the globe, members are located in the United States, Canada, Asia, Mexico, South America, the South Pacific, Australia, and Europe. They represent an extensive network of experienced engineers, designers, contractors, educators, code officials, and manufacturers interested in furthering their careers, their profession, and the industry. ASPE is at the forefront of technology. In addition, ASPE represents members and promotes the profession among all segments of the construction industry.

**ASPE MEMBERSHIP COMMUNICATION** — All members belong to ASPE worldwide and have the opportunity to belong and participate in one of the 56 state, provincial or local chapters throughout the U.S. and Canada. ASPE chapters provide the major communication links and the first line of services and programs for the individual member. Communications with the membership is enhanced through the Society's bimonthly newsletter, the *ASPE Report,* and the monthly magazine, *Plumbing Engineer.*

**TECHNICAL PUBLICATIONS** — The Society maintains a comprehensive publishing program, spearheaded by the profession's basic reference text, the *ASPE Data Book.* The *Data Book,* encompassing forty-five chapters in four volumes, provides comprehensive details of the accepted practices and design criteria used in the field of plumbing engineering. New additions that will shortly join ASPE's published library of professional technical manuals and handbooks include: *High-Technology Pharmaceutical Facilities Design Manual, High-Technology Electronic Facilities Design Manual, Health Care Facilities and Hospitals Design Manual,* and *Water Reuse Design Manual.*

**CONVENTION AND TECHNICAL SYMPOSIUM** — The Society hosts biennial Conventions in even-numbered years and Technical Symposia in odd-numbered years to allow professional plumbing engineers and designers to improve their skills, learn original concepts, and make important networking contacts to help them stay abreast of current trends and technologies. In conjunction with each Convention there is an Engineered Plumbing Exposition, the greatest, largest gathering of plumbing engineering and design products, equipment, and services. Everything from pipes to pumps to fixtures, from compressors to computers to consulting services is on display, giving engineers and specifiers the opportunity to view the newest and most innovative materials and equipment available to them.

**CERTIFIED IN PLUMBING ENGINEERING** — ASPE sponsors a national certification program for engineers and designers of plumbing systems, which carries the designation "Certified in Plumbing Engineering" or CIPE. The certification program provides the profession, the plumbing industry, and the general public with a single, comprehensive qualification of professional competence for engineers and designers of plumbing systems. The CIPE, designed exclusively by and for plumbing engineers, tests hundreds of engineers and designers at centers throughout the United States biennially. Created to provide a single, uniform national credential in the field of engineered plumbing systems, the CIPE program is not in any way connected to state-regulated Professional Engineer (P.E.) registration.

**ASPE RESEARCH FOUNDATION** — The ASPE Research Foundation, established in 1976, is the only independent, impartial organization involved in plumbing engineering and design research. The science of plumbing engineering affects everything . . . from the quality of our drinking water to the conservation of our water resources to the building codes for plumbing systems. Our lives are impacted daily by the advances made in plumbing engineering technology through the Foundation's research and development.

# American Society of Plumbing Engineers
# Data Book
(4 Volumes — 45 Chapters)

## Volume 2 — Plumbing Systems (Estimated date: 1999)

| Chapter | |
|---|---|
| 1 | Sanitary Drainage Systems (Chapter 1, looseleaf format) |
| 2 | Grey Water (Water Reuse) Systems |
| 3 | Vents and Venting Systems (Chapter 17, looseleaf format) |
| 4 | Storm Drainage Systems (Chapter 2, looseleaf format) |
| 5 | Cold Water Systems (Chapter 3, looseleaf format) |
| 6 | Domestic Water Heating System Fundamentals (Chapter 4, looseleaf format) |
| 7 | Fuel Gas Piping Systems (Chapter 6, looseleaf format) |
| 8 | Private Sewage Disposal Systems (Chapter 25, looseleaf format) |
| 9 | Private Water Systems (Chapter 26, looseleaf format) |
| 10 | Vacuum Systems |
| 11 | Pure Water Systems |
| 12 | Lab Waste Systems |

## Volume 3 — Special Plumbing Systems (Estimated date: 2000)

| Chapter | |
|---|---|
| 1 | Fire Protection Systems (Chapter 7, looseleaf format) |
| 2 | Plumbing Design for Health Care Facilities (Chapter 32, looseleaf format) |
| 3 | Treatment of Industrial Waste (Chapter 23, looseleaf format) |
| 4 | Irrigation Systems (Chapter 29, looseleaf format) |
| 5 | Reflecting Pools and Fountains (Chapter 30, looseleaf format) |
| 6 | Public Swimming Pools (Chapter 31, looseleaf format) |
| 7 | Gasoline and Diesel Oil Systems (Chapter 33, looseleaf format) |
| 8 | Steam and Condensate Piping (Chapter 38, looseleaf format) |
| 9 | Compressed Air Systems (Chapter 39, looseleaf format) |
| 10 | Solar Energy (Chapter 20, looseleaf format) |
| 11 | Site Utility Systems |

## Volume 4 — Plumbing Components and Equipment (Estimated revision date: 2002)

| Chapter | |
|---|---|
| 1 | Plumbing Fixtures (Chapter 8, looseleaf format) |
| 2 | Piping Systems (Chapter 10, looseleaf format) |
| 3 | Valves (Chapter 9, looseleaf format) |
| 4 | Pumps (Chapter 11, looseleaf format) |
| 5 | Piping Insulation (Chapter 12, looseleaf format) |
| 6 | Hangers and Supports (Chapter 13, looseleaf format) |
| 7 | Vibration Isolation (Chapter 14, looseleaf format) |
| 8 | Grease Interceptors (Chapter 35, looseleaf format) |
| 9 | Cross Connection Control (Chapter 24, looseleaf format) |
| 10 | Water Conditioning (Chapter 28, looseleaf format) |
| 11 | Thermal Expansion and Contractions (Chapter 5, looseleaf format) |
| 12 | Potable Water Coolers and Central Water Systems (Chapter 27, looseleaf format) |

(The chapters and subjects listed for these volume are subject to modification, adjustment and change. The contents shown for each volume are proposed and may not represent the final contents of the volume. A final listing of included chapters for each volume will appear in the actual publication.)

# Table of Contents

**CHAPTER 1   Plumbing Formulae, Symbols, and Terminology** . . . . . . . . . . . . . 1

Formulae Commonly Used in Plumbing Engineering . . . . . . . . . . . . . . . . . . . . . . . . . . . 1

    Equation 1-1, the Manning Formula . . . . . . . . . . . . . . . . . . . . . . . . . . 1

    Equation 1-2, Rate of flow . . . . . . . . . . . . . . . . . . . . . . . . . . . . . . . . . . 1

    Equation 1-3, Hydraulic radius (R) . . . . . . . . . . . . . . . . . . . . . . . . . . . . 1

    Equation 1-4, Water flow in pipes . . . . . . . . . . . . . . . . . . . . . . . . . . . . 2

    Equation 1-5, Friction head loss . . . . . . . . . . . . . . . . . . . . . . . . . . . . . 2

    Equation 1-6, Potential energy (PE) . . . . . . . . . . . . . . . . . . . . . . . . . . 3

    Equation 1-7, Kinetic energy (KE) . . . . . . . . . . . . . . . . . . . . . . . . . . . . 3

    Equation 1-8, Flow at outlet . . . . . . . . . . . . . . . . . . . . . . . . . . . . . . . . 3

    Equation 1-9, Length of vent piping . . . . . . . . . . . . . . . . . . . . . . . . . . 3

    Equation 1-11, Flow rate in fixture drain . . . . . . . . . . . . . . . . . . . . . . . 3

    Equation 1-12, Pipe expansion and contraction . . . . . . . . . . . . . . . . . . 4

    Equation 1-13, Various formulae for areas and volumes, in
        ft2 (m2) and ft3 (m3), respectively . . . . . . . . . . . . . . . . . . . . . . . . . 4

    Equation 1-14, Flow rate in outlet . . . . . . . . . . . . . . . . . . . . . . . . . . . . 6

    Equation 1-15, Gravity circulation . . . . . . . . . . . . . . . . . . . . . . . . . . . . 6

    Equation 1-16, Velocity head (h) . . . . . . . . . . . . . . . . . . . . . . . . . . . . . 6

    Equation 1-17, Bernoulli's Equation . . . . . . . . . . . . . . . . . . . . . . . . . . . 6

    Equation 1-18, Friction head (hf) . . . . . . . . . . . . . . . . . . . . . . . . . . . . 7

    Equation 1-19, Flow from outlets . . . . . . . . . . . . . . . . . . . . . . . . . . . . 7

    Equation 1-20, Hydraulic shock . . . . . . . . . . . . . . . . . . . . . . . . . . . . . 7

    Equation 1-21, Pump affinity laws . . . . . . . . . . . . . . . . . . . . . . . . . . . 7

    Equation 1-22, Pump efficiency . . . . . . . . . . . . . . . . . . . . . . . . . . . . . 8

    Equation 1-23, Rational method of storm design . . . . . . . . . . . . . . . . 8

    Equation 1-24, Spitzglass Formula . . . . . . . . . . . . . . . . . . . . . . . . . . . 8

    Equation 1-25, Weymouth Formula . . . . . . . . . . . . . . . . . . . . . . . . . . 8

Symbols . . . . . . . . . . . . . . . . . . . . . . . . . . . . . . . . . . . . . . . . . . . . . . . . . . . . . . . . . . . 9

Plumbing Terminology . . . . . . . . . . . . . . . . . . . . . . . . . . . . . . . . . . . . . . . . . . . . . . . 24

Recommended Practice for Conversion to The International System of Units . . . . . 40

    Terminology and Abbreviations . . . . . . . . . . . . . . . . . . . . . . . . . . . . . . . . . . . 40

    Types of Conversion . . . . . . . . . . . . . . . . . . . . . . . . . . . . . . . . . . . . . . . . . . . 40

    SI Units and Symbols . . . . . . . . . . . . . . . . . . . . . . . . . . . . . . . . . . . . . . . . . . 41

        Base units . . . . . . . . . . . . . . . . . . . . . . . . . . . . . . . . . . . . . . . . . . . . . 41

        Supplementary units . . . . . . . . . . . . . . . . . . . . . . . . . . . . . . . . . . . . . 41

        Derived units . . . . . . . . . . . . . . . . . . . . . . . . . . . . . . . . . . . . . . . . . . 41

    Non-SI Units and Symbols for Use with the SI System . . . . . . . . . . . . . . . . . . 42

SI Unit Prefixes and Symbols . . . . . . . . . . . . . . . . . . . . . . . . . . . . . 42
SI Units Style and Use . . . . . . . . . . . . . . . . . . . . . . . . . . . . . . . . . 42
SI Unit Conversion Factors . . . . . . . . . . . . . . . . . . . . . . . . . . . . . . 43
    Acceleration, linear . . . . . . . . . . . . . . . . . . . . . . . . . . . . . . . 43
    Area . . . . . . . . . . . . . . . . . . . . . . . . . . . . . . . . . . . . . . . 43
    Bending moment (torque) . . . . . . . . . . . . . . . . . . . . . . . . . . . 43
    Bending moment (torque) per unit length . . . . . . . . . . . . . . . . . . 43
    Electricity and magnetism . . . . . . . . . . . . . . . . . . . . . . . . . . 43
    Energy (work) . . . . . . . . . . . . . . . . . . . . . . . . . . . . . . . . . 43
    Energy per unit area per unit time . . . . . . . . . . . . . . . . . . . . . . 43
    Force . . . . . . . . . . . . . . . . . . . . . . . . . . . . . . . . . . . . . . 43
    Force per unit length . . . . . . . . . . . . . . . . . . . . . . . . . . . . . 44
    Heat . . . . . . . . . . . . . . . . . . . . . . . . . . . . . . . . . . . . . . 44
    Length . . . . . . . . . . . . . . . . . . . . . . . . . . . . . . . . . . . . . 44
    Light (illuminance) . . . . . . . . . . . . . . . . . . . . . . . . . . . . . . 44
    Mass . . . . . . . . . . . . . . . . . . . . . . . . . . . . . . . . . . . . . . 44
    Mass per unit area . . . . . . . . . . . . . . . . . . . . . . . . . . . . . . 44
    Mass per unit length . . . . . . . . . . . . . . . . . . . . . . . . . . . . . 44
    Mass per unit time (flow) . . . . . . . . . . . . . . . . . . . . . . . . . . . 44
    Mass per unit volume (density) . . . . . . . . . . . . . . . . . . . . . . . 44
    Moment of inertia . . . . . . . . . . . . . . . . . . . . . . . . . . . . . . . 44
    Plane angle . . . . . . . . . . . . . . . . . . . . . . . . . . . . . . . . . . 44
    Power . . . . . . . . . . . . . . . . . . . . . . . . . . . . . . . . . . . . . 45
    Pressure (stress), force per unit area . . . . . . . . . . . . . . . . . . . . 45
    Temperature equivalent . . . . . . . . . . . . . . . . . . . . . . . . . . . 45
    Velocity (length per unit time) . . . . . . . . . . . . . . . . . . . . . . . . 45
    Volume . . . . . . . . . . . . . . . . . . . . . . . . . . . . . . . . . . . . 45
    Volume per unit time (flow) . . . . . . . . . . . . . . . . . . . . . . . . . . 45
References . . . . . . . . . . . . . . . . . . . . . . . . . . . . . . . . . . . . . . . . . 49
ILLUSTRATIONS
    Figure 1-1  Square . . . . . . . . . . . . . . . . . . . . . . . . . . . . . . . . 4
    Figure 1-2  Rectangle . . . . . . . . . . . . . . . . . . . . . . . . . . . . . . 4
    Figure 1-3  Rhombus . . . . . . . . . . . . . . . . . . . . . . . . . . . . . . 4
    Figure 1-4  Rhomboid . . . . . . . . . . . . . . . . . . . . . . . . . . . . . 4
    Figure 1-5  Trapezoid . . . . . . . . . . . . . . . . . . . . . . . . . . . . . 4
    Figure 1-6  Trapezium . . . . . . . . . . . . . . . . . . . . . . . . . . . . . 4
    Figure 1-7  Right-Angle Triangle . . . . . . . . . . . . . . . . . . . . . . . 4
    Figure 1-8  Isosceles Triangle . . . . . . . . . . . . . . . . . . . . . . . . . 5
    Figure 1-9  Ellipse . . . . . . . . . . . . . . . . . . . . . . . . . . . . . . . . 5
    Figure 1-10  Cylinder . . . . . . . . . . . . . . . . . . . . . . . . . . . . . 5
    Figure 1-11  Cube or Rectangular Solid . . . . . . . . . . . . . . . . . . . 5
    Figure 1-12  Pyramid . . . . . . . . . . . . . . . . . . . . . . . . . . . . . 5
    Figure 1-13  Cone . . . . . . . . . . . . . . . . . . . . . . . . . . . . . . . 5

# Table of Contents

    Figure 1-14  Circle .................................................. 5
    Figure 1-15  Triangle ................................................ 6

TABLES

    Table 1-1  Standard Plumbing and Piping Symbols ..................... 9
    Table 1-2  Symbols for Use in Architectural and Engineering Drawings and Insurance Diagrams ......................................... 15
    Table 1-3  Abbreviations for Text, Drawings, and Computer Programs ...... 17
    Table 1-4  Temperature Conversion Chart, °F – °C ..................... 46
    Table 1-5  Conversion to SI Units ..................................... 47

**CHAPTER 2  Standard Plumbing Materials and Equipment .............. 51**
    Table 2-1  Standards for Plumbing Materials and Equipment ............. 51
    Table 2-2  Organization Abbreviations, Addresses, and Phone Numbers ..... 60

**CHAPTER 3  Plumbing Specifications ................................ 63**
Introduction ...................................................... 63
Construction Contract Documents .................................... 63
    Project Manual .............................................. 64
CSI Format ....................................................... 65
    Document Categories ......................................... 65
    Masterformat Organization .................................... 66
    Page Formatting ............................................. 66
Specifications ..................................................... 66
    Writing the Specifications ..................................... 66
        Division 1 ............................................... 67
    Master Specifications ......................................... 67
    Spec Via Computer .......................................... 68
    Administration of Specifications ................................ 68
Plumbing Specifiers ................................................ 69
Appendix 3-A1 .................................................... 71
    Specifications ............................................... 71
        Division 1—General Requirements ......................... 71
        Division 2—Site Work ................................... 71
        Division 3—Concrete .................................... 71
        Division 4—Masonry .................................... 71
        Division 5—Metals ...................................... 71
        Division 6—Wood and Plastics ............................ 72
        Division 7—Thermal and Moisture Protection ................ 72
        Division 8—Doors and Windows .......................... 72
        Division 9—Finishes ..................................... 72
        Division 10—Specialties .................................. 72
        Division 11—Equipment ................................. 72
        Division 12—Furnishings ................................. 73

   Division 13—Special Construction . . . . . . . . . . . . . . . . . . . . . . . . . . . . . . . . . . . 73
   Division 14—Conveying Systems . . . . . . . . . . . . . . . . . . . . . . . . . . . . . . . . . . . 73
   Division 15—Mechanical . . . . . . . . . . . . . . . . . . . . . . . . . . . . . . . . . . . . . . . . . 73
   Division 16—Electrical . . . . . . . . . . . . . . . . . . . . . . . . . . . . . . . . . . . . . . . . . . 73
Appendix 3-A2 . . . . . . . . . . . . . . . . . . . . . . . . . . . . . . . . . . . . . . . . . . . . . . . . . . . . . . . 74
 Division 2—Site Work . . . . . . . . . . . . . . . . . . . . . . . . . . . . . . . . . . . . . . . . . . . . . 74
  02600 Utility Piping Materials . . . . . . . . . . . . . . . . . . . . . . . . . . . . . . . . . . . 74
  02660 Water Distribution . . . . . . . . . . . . . . . . . . . . . . . . . . . . . . . . . . . . . . 74
  02680 Fuel and Steam Distribution . . . . . . . . . . . . . . . . . . . . . . . . . . . . . . 74
  02700 Sewerage and Drainage . . . . . . . . . . . . . . . . . . . . . . . . . . . . . . . . . . 74
  02760 Restoration of Underground Pipe . . . . . . . . . . . . . . . . . . . . . . . . . . 74
  02800 Site Improvements . . . . . . . . . . . . . . . . . . . . . . . . . . . . . . . . . . . . . 74
 Division 15—Mechanical . . . . . . . . . . . . . . . . . . . . . . . . . . . . . . . . . . . . . . . . . . 74
  15050 Basic Mechanical Materials and Methods . . . . . . . . . . . . . . . . . . . . 74
  15250 Mechanical Insulation . . . . . . . . . . . . . . . . . . . . . . . . . . . . . . . . . . . 74
  15300 Fire Protection . . . . . . . . . . . . . . . . . . . . . . . . . . . . . . . . . . . . . . . . 74
  15400 Plumbing . . . . . . . . . . . . . . . . . . . . . . . . . . . . . . . . . . . . . . . . . . . . 74
Appendix 3-B
Section Shell Outline . . . . . . . . . . . . . . . . . . . . . . . . . . . . . . . . . . . . . . . . . . . . . . . . . 75
Section XXXXX . . . . . . . . . . . . . . . . . . . . . . . . . . . . . . . . . . . . . . . . . . . . . . . . . . . . . . 75
 Part 1—General . . . . . . . . . . . . . . . . . . . . . . . . . . . . . . . . . . . . . . . . . . . . . . . . . 75
 Part 2—Products . . . . . . . . . . . . . . . . . . . . . . . . . . . . . . . . . . . . . . . . . . . . . . . . 77
 Part 3—Execution . . . . . . . . . . . . . . . . . . . . . . . . . . . . . . . . . . . . . . . . . . . . . . . 78
References . . . . . . . . . . . . . . . . . . . . . . . . . . . . . . . . . . . . . . . . . . . . . . . . . . . . . . . . . 79
Bibliography . . . . . . . . . . . . . . . . . . . . . . . . . . . . . . . . . . . . . . . . . . . . . . . . . . . . . . . . 80
Sources of Additional Information . . . . . . . . . . . . . . . . . . . . . . . . . . . . . . . . . . . . . . 80

## CHAPTER 4 Plumbing Cost Estimation . . . . . . . . . . . . . . . . . . . . . . . . . . . . . 81
TABLES
 Table 4-1 Fittings and Their Joint Numbers . . . . . . . . . . . . . . . . . . . . . . . . . 82
 Table 4-2 Man-Hours for Installation of Pipes, Hangers, and Fixtures . . . . . . 83
 Table 4-3 Man-Hours for Installation of Plumbing Fixtures . . . . . . . . . . . . . 83
 Table 4-4 Man-Hour Tables (Miscellaneous) . . . . . . . . . . . . . . . . . . . . . . . . . 84
 Table 4-5 Labor Units for Estimating . . . . . . . . . . . . . . . . . . . . . . . . . . . . . . 84
 Table 4-6 Excavation Table . . . . . . . . . . . . . . . . . . . . . . . . . . . . . . . . . . . . . 85
 Table 4-7 Backfill Man-Hours Per Linear Foot . . . . . . . . . . . . . . . . . . . . . . . 86
Appendix — Worksheets
 Water Piping—Sample Estimating Form . . . . . . . . . . . . . . . . . . . . . . . . . . . . . 85
 Estimate Worksheet . . . . . . . . . . . . . . . . . . . . . . . . . . . . . . . . . . . . . . . . . . . . . 86
 Underground Piping Bill of Material . . . . . . . . . . . . . . . . . . . . . . . . . . . . . . . . 87

## CHAPTER 5 Job Preparation, Plumbing Drawing, and Field Checklists . . . . . 91
General . . . . . . . . . . . . . . . . . . . . . . . . . . . . . . . . . . . . . . . . . . . . . . . . . . . . . . . . . . . . 91

# Table of Contents

Suggested Items to be Checked and Information to be
Obtained at the Start of a Job .................................................. 91
Suggested Procedure for The Design of Plumbing Work
and The Preparation of Plumbing Drawings ................................ 92
Plumbing Drawing Checklist ........................................................ 95
Field Checklist ......................................................................... 96
    Underground Inspection ........................................................ 96
        Building drain, storm sewer ............................................. 96
        Water and gas ............................................................... 96
    Above-Ground Inspection ..................................................... 96
    Setting of Fixtures and Finish
    Inspection ......................................................................... 97

**CHAPTER 6 Plumbing for Physically Challenged Individuals ............ 99**
Introduction ............................................................................. 99
Background ............................................................................. 99
Legislation .............................................................................. 100
Design .................................................................................. 101
    Clear Floor or Ground Space for Wheelchairs ........................ 103
    Anthropometrics ................................................................. 105
        Forward reach. .............................................................. 103
        Side reach. ................................................................... 107
Plumbing Elements and Facilities ............................................... 107
    601 General ....................................................................... 107
        601.1 Scope .................................................................. 107
    602 Drinking Fountains and Water Coolers ............................ 107
        602.1 General ................................................................ 107
        602.2 Clear floor or ground space .................................... 107
        602.3 Operable parts ...................................................... 107
        602.4 Spout height ......................................................... 107
        602.5 Spout location ...................................................... 107
        602.6 Water flow ........................................................... 107
        602.7 Protruding objects ................................................. 107
    603 Toilet and Bathing Rooms .............................................. 108
        603.1 General ................................................................ 108
        603.2 Clearances ........................................................... 108
        603.3 Mirrors ................................................................. 108
        603.4 Coat hooks and shelves ......................................... 108
    604 Water Closets and Toilet Compartments .......................... 109
        604.1 General ................................................................ 109
        604.2 Location ............................................................... 109
        604.3 Clearance ............................................................. 109
        604.4 Height .................................................................. 110
        604.5 Grab bars ............................................................. 110

- 604.6 Flush controls ... 110
- 604.7 Dispensers ... 110
- 604.8 Toilet compartments ... 110

605 Urinals ... 113
- 605.1 General ... 113
- 605.2 Height ... 113
- 605.3 Clear floor or ground space ... 113
- 605.4 Flush controls ... 113

606 Lavatories and Sinks ... 113
- 606.1 General ... 113
- 606.2 Clear floor or ground space ... 15
- 606.3 Height and clearances ... 114
- 606.4 Faucets ... 114
- 606.5 Bowl depth ... 114
- 606.6 Exposed pipes and surfaces ... 114

607 Bathtubs ... 114
- 607.1 General ... 114
- 607.2 Clearance ... 114
- 607.3 Seat ... 114
- 607.4 Grab bars ... 114
- 607.5 Controls ... 114
- 607.6 Shower unit ... 114
- 607.7 Bathtub enclosures ... 115

608 Shower Compartments ... 115
- 608.1 General ... 115
- 608.2 Size and clearances ... 116
- 608.3 Grab bars ... 117
- 608.4 Seats ... 117
- 608.5 Controls ... 117
- 608.6 Shower unit ... 117
- 608.7 Thresholds ... 118
- 608.8 Shower enclosures ... 118

609 Grab Bars ... 119
- 609.1 General ... 119
- 609.2 Size ... 119
- 609.3 Spacing ... 119
- 609.4 Position of grab bars ... 119
- 609.5 Surface hazards. ... 119
- 609.6 Fittings ... 119
- 609.7 Installation ... 119
- 609.8 Structural strength ... 119

610 Seats ... 121

# Table of Contents

xiii

      610.1 General .................................................. 121
      610.2 Bathtub seats ........................................... 121
      610.3 Shower compartment seats ............................. 121
      610.4 Structural strength ..................................... 121
    611 Laundry Equipment ........................................... 122
      611.1 General .................................................. 122
      611.2 Clear floor or ground space ............................ 122
      611.3 Operable parts .......................................... 122
      611.4 Height ................................................... 122
References ........................................................... 122
ILLUSTRATIONS
    Figure 6-1   Dimensions of Adult-Sized Wheelchairs ............... 102
    Figure 6-2   Clear Floor Space for Wheelchairs ..................... 103
    Figure 6-3   Wheelchair Approaches ................................. 104
    Figure 6-4   Clear Floor Space in Alcoves ........................... 104
    Figure 6-5   Unobstructed Forward Reach Limit .................... 105
    Figure 6-6   Forward Reach Over an Obstruction ................... 105
    Figure 6-7   Unobstructed Side Reach Limit ........................ 106
    Figure 6-8   Obstructed Side Reach Limit ........................... 106
    Figure 6-9   Cantilevered Drinking Fountains and Water Coolers ... 106
    Figure 6-10  Horizontal Angle of Water Stream — Plan View ....... 107
    Figure 6-11  Leg Clearances ......................................... 108
    Figure 6-12  Ambulatory Accessible Stall ........................... 108
    Figure 6-13  Wheelchair Accessible Toilet Stalls — Door Swing Out ... 109
    Figure 6-14  Clear Floor Space at Water Closets .................... 109
    Figure 6-15  Water Closet — Side View .............................. 110
    Figure 6-16  Water Closet — Front View ............................ 110
    Figure 6-17  Wheelchair Accessible Toilet Stalls — Door Swing In ... 111
    Figure 6-18  Clear Floor Space at Lavatories and Sinks ............ 113
    Figure 6-19  Clear Floor Space at Bathtubs ......................... 115
    Figure 6-20  Bathtub Accessories .................................... 116
    Figure 6-21  Transfer Type Shower Stall ............................. 117
    Figure 6-22  Roll-in Type Shower Stall .............................. 117
    Figure 6-23  Grab Bars at Shower Stalls ............................. 118
    Figure 6-24  Size and Spacing of Grab Bars ......................... 119
    Figure 6-25  Shower Seat Design .................................... 121
TABLES
    Table 6-1   Graphic Conventions ..................................... 102

## CHAPTER 7   Energy Conservation in Plumbing Systems ........... 122

Introduction ......................................................... 122
Saving Energy in Plumbing Systems ................................. 122
    Reducing Service-Water Temperatures ........................... 122

  Reduced Water Flow Rates . . . . . . . . . . . . . . . . . . . . . . . . . . . . . . . . . . . . 123
  Economic Thermal Insulation Thickness . . . . . . . . . . . . . . . . . . . . . . . . . . . 125
  Hot Water System Improvement . . . . . . . . . . . . . . . . . . . . . . . . . . . . . . . . 127
    From the owner's viewpoint . . . . . . . . . . . . . . . . . . . . . . . . . . . . . 127
    From the plumbing engineer's viewpoint . . . . . . . . . . . . . . . . . . . . . 127
Standby Losses and Circulating Vs. Noncirculating Systems . . . . . . . . . . . . . . . 127
  Use of Waste Heat to Heat Water . . . . . . . . . . . . . . . . . . . . . . . . . . . . . . . 128
    Heat rejected from air conditioning and commercial
      refrigeration processes . . . . . . . . . . . . . . . . . . . . . . . . . . . . . . 128
    Heat reclaimed from steam condensate . . . . . . . . . . . . . . . . . . . . . . 128
    Cogeneration systems . . . . . . . . . . . . . . . . . . . . . . . . . . . . . . . . . . 128
    Heat pump and heat reclamation systems . . . . . . . . . . . . . . . . . . . . 128
Saving Utility Costs in Plumbing Systems . . . . . . . . . . . . . . . . . . . . . . . . . . . . 131
  Using Off-Peak Power . . . . . . . . . . . . . . . . . . . . . . . . . . . . . . . . . . . . . . . 131
  Using Minimum-Energy-Consuming Equipment . . . . . . . . . . . . . . . . . . . . . 131
  Water Reuse Systems . . . . . . . . . . . . . . . . . . . . . . . . . . . . . . . . . . . . . . . 131
Nondepletable and Alternate Energy Sources . . . . . . . . . . . . . . . . . . . . . . . . . 131
  Solar Energy . . . . . . . . . . . . . . . . . . . . . . . . . . . . . . . . . . . . . . . . . . . . . 132
  Geothermal Energy . . . . . . . . . . . . . . . . . . . . . . . . . . . . . . . . . . . . . . . . 132
  Solid Waste Disposal . . . . . . . . . . . . . . . . . . . . . . . . . . . . . . . . . . . . . . . 132
Performance Efficiency of Heating and Hot Water Storage Equipment . . . . . . . . 132
Energy Code Compliance . . . . . . . . . . . . . . . . . . . . . . . . . . . . . . . . . . . . . . . . 133
References . . . . . . . . . . . . . . . . . . . . . . . . . . . . . . . . . . . . . . . . . . . . . . . . . 133
Glossary . . . . . . . . . . . . . . . . . . . . . . . . . . . . . . . . . . . . . . . . . . . . . . . . . . . 133
ILLUSTRATIONS
  Figure 7-1 Energy Savings from Reduced Faucet Flow Rates . . . . . . . . . . . . 125
  Figure 7-2 Refrigeration Waste Heat Recovery . . . . . . . . . . . . . . . . . . . . . . 129
  Figure 7-3 Condenser Water Heat Recovery . . . . . . . . . . . . . . . . . . . . . . . . 129
  Figure 7-4 Condenser Water Heat Recovery with Storage Tank . . . . . . . . . . 130
  Figure 7-5 Wastewater Heat Recovery . . . . . . . . . . . . . . . . . . . . . . . . . . . . 130
TABLES
  Table 7-1 Energy Savings Chart for Steel Hot Water Pipes and Tanks . . . . . 126
  Table 7-2 Energy Savings Chart for Copper Hot Water Pipes and Tanks . . . 126
  Table 7-3 The Effect of Stopping Circulation . . . . . . . . . . . . . . . . . . . . . . . 127

**CHAPTER 8  Corrosion . . . . . . . . . . . . . . . . . . . . . . . . . . . . . . . . . . . . . . . . 135**
Introduction . . . . . . . . . . . . . . . . . . . . . . . . . . . . . . . . . . . . . . . . . . . . . . . . 135
Fundamental Corrosion Cell . . . . . . . . . . . . . . . . . . . . . . . . . . . . . . . . . . . . . 135
  Basic Relations . . . . . . . . . . . . . . . . . . . . . . . . . . . . . . . . . . . . . . . . . . . 135
  Electrochemical Equivalents . . . . . . . . . . . . . . . . . . . . . . . . . . . . . . . . . . 135
Common Forms of Corrosion . . . . . . . . . . . . . . . . . . . . . . . . . . . . . . . . . . . . . 137
The Galvanic Series . . . . . . . . . . . . . . . . . . . . . . . . . . . . . . . . . . . . . . . . . . . 140
Electromotive Force Series . . . . . . . . . . . . . . . . . . . . . . . . . . . . . . . . . . . . . . 140

# Table of Contents

Factors Affecting The Rate of Corrosion ................................. 140
    General ........................................................ 140
    Effect of the Metal Itself ...................................... 141
    Acidity ........................................................ 141
    Oxygen Content ................................................. 142
    Film Formation ................................................. 142
    Temperature .................................................... 142
    Velocity ....................................................... 142
    Homogeneity .................................................... 142
Corrosion Control ...................................................... 142
    Materials Selection ............................................ 142
    Design to Reduce Corrosion ..................................... 143
    Passivation .................................................... 143
    Coating ........................................................ 143
    Cathodic Protection ............................................ 144
        Galvanic anodes ............................................ 144
        Impressed current .......................................... 147
        Cathodic protection criteria ............................... 148
        Costs of cathodic protection ............................... 149
    Inhibitors (Water Treatment) ................................... 149
Glossary ............................................................... 149
References ............................................................. 152
ILLUSTRATIONS
    Figure 8-1   Basic Corrosion Cell ................................ 136
    Figure 8-2   Basic Cell Applied to an Underground Structure ...... 136
    Figure 8-3   Uniform Attack ...................................... 137
    Figure 8-4   Pitting Corrosion ................................... 137
    Figure 8-5   Galvanic Corrosion .................................. 138
    Figure 8-6   Concentration Cells ................................. 138
    Figure 8-7   Impingement Attack .................................. 138
    Figure 8-8   Stress Corrosion .................................... 138
    Figure 8-9   (A) Plug-Type Dezincification (B) Layer-Type Dezincification .... 138
    Figure 8-11  Corrosion by Differential Environmental Conditions ......... 139
    Figure 8-10  Stray Current Corrosion ............................ 139
    Figure 8-12  Cathodic Protection by the Sacrificial Anode Method ........ 145
    Figure 8-13  Typical Sacrificial Anode Installation .............. 146
    Figure 8-14  Cathodic Protection by the Impressed Current Method ........ 148
TABLES
    Table 8-1    Electrochemical Metal Losses of Some Common Metals ......... 137
    Table 8-2    Galvanic Series of Metals ........................... 140
    Table 8-3    Electromotive Force Series .......................... 141
    Table 8-4    Corrosion Rates for Common Metals ................... 141

**CHAPTER 9   Seismic Protection of Plumbing Equipment** .............. 155
Introduction ........................................................ 155
Causes and Effects of Earthquakes ................................... 156
    Plate Tectonics and Faults ...................................... 156
    Damage from Earthquakes ....................................... 158
Earthquake Measurement and Seismic Design ........................... 159
    Ground Shaking and Dynamic Response ............................ 159
    The Response Spectrum .......................................... 160
Learning from Past Earthquakes ...................................... 162
    Damage to Plumbing Equipment ................................... 162
    The 1964 Alaska Earthquake ..................................... 163
        Damage summary ............................................. 163
    The 1971 San Fernando Earthquake ............................... 163
        Damage summary ............................................. 163
Seismic Protection Techniques ....................................... 164
    General ........................................................ 164
    Equipment ...................................................... 164
        Fixed, floor-mounted equipment ............................. 165
        Fixed, suspended equipment ................................. 165
        Vibration-isolated, floor-mounted equipment ................ 165
        Vibration-isolated, suspended equipment .................... 167
    Piping Systems ................................................. 168
Codes ............................................................... 185
    Design Philosophy .............................................. 185
    Code Requirements .............................................. 185
        Sprinkler systems: NFPA 13 ................................. 186
Analysis Techniques ................................................. 186
    Determination of Seismic Forces ................................ 186
    Determination of Anchorage Forces .............................. 189
Computer Analysis of Piping Systems ................................. 189
Design Considerations ............................................... 191
    Loads in Structures ............................................ 191
Potential Problems .................................................. 191
    Additional Considerations ...................................... 193
Glossary ............................................................ 193
References .......................................................... 195
ILLUSTRATIONS
    Figure 9-1   Significant Earthquakes in the United States ............... 156
    Figure 9-2   (A) Seismic Zone Map of the United States;
               (B) Map of Seismic Zones and Effective, Peak-Velocity-Related
               Acceleration (Av) for Contiguous 48 States. .................... 157
    Figure 9-3   World Map Showing Relation Between the Major
               Tectonic Plates and Recent Earthquakes and Volcanoes. ......... 158

# Table of Contents

Figure 9-4    Elastic Rebound Theory of Earthquake Movement . . . . . . . . . . . 159
Figure 9-5    Earthquake Ground Accelerations in Epicentral Regions . . . . . . 160
Figure 9-6    Undamped Mechanical Systems . . . . . . . . . . . . . . . . . . . . . . . . 161
Figure 9-7    Response Spectrum . . . . . . . . . . . . . . . . . . . . . . . . . . . . . . . . . 162
Figure 9-8    Snubbing Devices . . . . . . . . . . . . . . . . . . . . . . . . . . . . . . . . . . 167
Figure 9-9    Isolators with Built-In Seismic Restraint . . . . . . . . . . . . . . . . . 167
Figure 9-10    Parameters to Be Considered for Pipe Bracing . . . . . . . . . . . . . 168
Figure 9-11    Pipe Bracing Systems . . . . . . . . . . . . . . . . . . . . . . . . . . . 172–179
Figure 9-12    Construction Details of Seismic Protection for Pipes . . . . . . . . . 180
Figure 9-13    Sway Bracing, 0.5 G Force . . . . . . . . . . . . . . . . . . . . . . . . . . . 182
Figure 9-14    A Seismic Bracing Method . . . . . . . . . . . . . . . . . . . . . . . . . . . 184
Figure 9-15    Acceptable Types of Sway Bracing . . . . . . . . . . . . . . . . . . . . . 190
Figure 9-16    Forces for Seismic Design . . . . . . . . . . . . . . . . . . . . . . . . . . . . 192
Figure 9-17    Potential Problems in Equipment Anchorage or Pipe Bracing . . 194

TABLES
Table 9-1    Piping Weights for Determining Horizontal Load . . . . . . . . . . . 186
Table 9-2    Assigned Load Table for Lateral and Longitudinal Sway Bracing . . 187
Table 9-3    Maximum Horizontal Loads for Sway Bracing . . . . . . . . . . . . . 187

**CHAPTER 10    Acoustics in Plumbing Systems . . . . . . . . . . . . . . . . . . . . . . 197**
Introduction . . . . . . . . . . . . . . . . . . . . . . . . . . . . . . . . . . . . . . . . . . . . . . . . . . 197
Acceptable Acoustical Levels in Buildings . . . . . . . . . . . . . . . . . . . . . . . . . . . . 197
Acoustical Performance of Building Materials . . . . . . . . . . . . . . . . . . . . . . . . 197
    Insulation Against Airborne Sound . . . . . . . . . . . . . . . . . . . . . . . . . . . . . . 197
Acoustical Ratings of Plumbing Fixtures and Appliances . . . . . . . . . . . . . . . . 198
      Valves . . . . . . . . . . . . . . . . . . . . . . . . . . . . . . . . . . . . . . . . . . . . . . . . . 198
      Water closets . . . . . . . . . . . . . . . . . . . . . . . . . . . . . . . . . . . . . . . . . . . 198
      Urinals . . . . . . . . . . . . . . . . . . . . . . . . . . . . . . . . . . . . . . . . . . . . . . . . 199
      Bathtubs . . . . . . . . . . . . . . . . . . . . . . . . . . . . . . . . . . . . . . . . . . . . . . . 199
      Showers . . . . . . . . . . . . . . . . . . . . . . . . . . . . . . . . . . . . . . . . . . . . . . . 199
      Dishwashers . . . . . . . . . . . . . . . . . . . . . . . . . . . . . . . . . . . . . . . . . . . 199
      Waste-disposal units . . . . . . . . . . . . . . . . . . . . . . . . . . . . . . . . . . . . . 199
      Washing machines . . . . . . . . . . . . . . . . . . . . . . . . . . . . . . . . . . . . . . 199
General Acoustical Design . . . . . . . . . . . . . . . . . . . . . . . . . . . . . . . . . . . . . . . 200
   Water Pipes . . . . . . . . . . . . . . . . . . . . . . . . . . . . . . . . . . . . . . . . . . . . . . . . 200
     Origin and spread of noise . . . . . . . . . . . . . . . . . . . . . . . . . . . . . . . . 200
     Reducing the noise at its origin . . . . . . . . . . . . . . . . . . . . . . . . . . . . 200
     Reducing the spread of noise . . . . . . . . . . . . . . . . . . . . . . . . . . . . . . 200
   Occupied Domestic Spaces . . . . . . . . . . . . . . . . . . . . . . . . . . . . . . . . . . . . 201
   Pumps . . . . . . . . . . . . . . . . . . . . . . . . . . . . . . . . . . . . . . . . . . . . . . . . . . . . 201
     Sources of noise . . . . . . . . . . . . . . . . . . . . . . . . . . . . . . . . . . . . . . . . 201
     Possible modifications . . . . . . . . . . . . . . . . . . . . . . . . . . . . . . . . . . . 202
     Plant noise . . . . . . . . . . . . . . . . . . . . . . . . . . . . . . . . . . . . . . . . . . . . 202

　　　　Estimating the noise level of a pump .............................. 203
　　Flow Velocity and Water Hammer .................................... 203
　　Design Procedures ................................................. 203
　　Noise and Vibration Control ....................................... 204
　　　　Equipment design .............................................. 205
Systsem Design ......................................................... 205
　　Equipment selection ............................................... 205
　　Pressure .......................................................... 206
　　Speed ............................................................. 206
　　Pipe sleeves ...................................................... 206
　　Water hammer ...................................................... 206
　　Pipe wrapping ..................................................... 206
　　System layout ..................................................... 207
　　Vibration isolation ............................................... 207
Glossary ............................................................... 215
ILLUSTRATIONS
　　Figure 10-1　Pipe-Sleeve Floor Penetration ........................ 207
　　Figure 10-2　Acoustical Treatment for Pipe-Sleeve Penetration at
　　　　Spaces with Inner Wall on Neoprene Isolators .................. 207
　　Figure 10-3　Acoustical Pipe-Penetration Seals .................... 207
　　Figure 10-4　Installation of an Air Lock in a Residential Plumbing System .. 207
　　Figure 10-5　Examples of Suction-Piping Installations ............. 208
　　Figure 10-6　Typical Vibration-Isolation Devices .................. 209
　　Figure 10-7　Typical Flexible Connectors .......................... 210
　　Figure 10-8　Bathtub and/or Shower Installation ................... 212
　　Figure 10-9　Suggested Mounting of Piping and Plumbing Fixture .... 212
　　Figure 10-10　Suggested Installation of Plumbing Fixtures ......... 213
　　Figure 10-11　Vibration Isolation of Flexible-Coupled,
　　　　Horizontally Split, Centrifugal Pumps ......................... 214
　　Figure 10-12　Common Errors Found in Installation of
　　　　Vibration-Isolated Pumps ...................................... 214
　　Figure 10-13　Vibration Isolation of a Sump Pump .................. 214
　　Figure 10-14　Typical Pipe Run Installations ...................... 215
　　Figure 10-15　Typical Flexible Pipe-Connector Installations ....... 216
TABLES
　　Table 10-1　Recommended Static Deflection for
　　　　Pump Vibration-Isolation Devices .............................. 211
　　Table 10-2　Typical Sound Levels .................................. 217

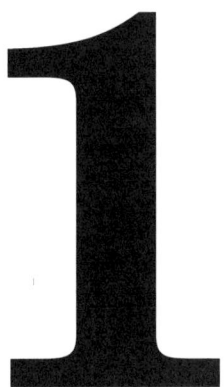

# Plumbing Formulae, Symbols, and Terminology

## FORMULAE COMMONLY USED IN PLUMBING ENGINEERING

For the convenience of ASPE members, the Society has gathered some of the basic formulae commonly referred to and utilized in plumbing engineering and design. It is *extremely important* to convert to values of the proper units whenever using these equations.

**Equation 1-1, the Manning Formula**  Used for determining the velocity (V) of uniform flow (defined as the flow that is achieved in open channels of constant shape and size and uniform slope) in sloping drains. Note that the slope of the water surface is equal to the slope of the channel and that the flows in such open channels do not depend on the pressure applied to the water but on the gravitational force induced by the slope of the drain and the height of the water in that drain.

*Equation 1-1*

$$V = \frac{1.486 \, R^{2/3} \, S^{1/2}}{n}$$

*where*

- V = Velocity of flow, ft/s (m/s)
- n = Coefficient representing roughness of pipe surface, degree of fouling, and pipe diameter
- R = Hydraulic radius, ft (m)
- S = Hydraulic slope of surface of flow, ft/ft (m/m)

The hydraulic radius (R) can be calculated using Equation 1-3. The roughness coefficient (n) and several values for the hydraulic radii are given in Baumeister and Marks's *Standard Handbook for Mechanical Engineers*.

**Equation 1-2, Rate of flow**  Used for determining the amount of water passing through a pipe. This quantity of water, for a given time, depends on the cross-sectional area of the pipe and the velocity of the water.

*Equation 1-2*

$$Q = AV$$

*where*

- Q = Flow rate of water, ft³/s (m³/s)
- A = Cross-sectional area of pipe, ft² (m²)
- V = Flow velocity of water, ft/s (m/s)

(a) Therefore, substituting Equation 1-2 in Equation 1-1, the Manning Formula can be represented as follows:

*Equation 1-2a*

$$Q = \frac{1.486 \, AR^{2/3} \, S^{1/2}}{n}$$

**Equation 1-3, Hydraulic radius (R)**  Usually referred to as the hydraulic mean depth of flow, the ratio of the cross-sectional area of flow to the wetted perimeter of pipe surface.

*Equation 1-3*

$$R = \text{Area of flow/Wetted perimeter}$$

For half-full (HF) and full-flow (FF) conditions, the hydraulic radii can be represented as:

### Equation 1-3a

$$R_{HF} = R_{FF} = \frac{D}{4}$$

where

- D = Diameter of pipe, ft (m)
- $R_{HF}$ = Hydraulic radius, half-full condition, ft (m)
- $R_{FF}$ = Hydraulic radius, full-flow condition, ft (m)

**Equation 1-4, Water flow in pipes** Two types of water flow exist: laminar and turbulent. Each type is characterized by the Reynolds number, a dimensionless quantity. The physical characteristics of the water, the velocity of the flow, and the internal diameter of the pipe are factors for consideration, and the Reynolds number is represented as:

### Equation 1-4

$$Re = \frac{VD}{\mu}$$

where

- Re = Reynolds number, dimensionless
- V = Velocity of flow, ft/s (m/s)
- D = Diameter of pipe, ft (m)
- $\mu$ = Absolute viscosity of fluid, lb-s/ft² (m²/s)

Values of viscosity are tabulated in the ASHRAE *Handbook of Fundamentals*. In laminar flow, the fluid particles move in layers in straight parallel paths, the viscosity of the fluid is dominant, and its upper limit is represented by Re = 2000. In turbulent flow, the fluid particles move in a haphazard fashion in all directions, the path of an individual fluid particle is not possible to trace, and Re is above 4000. Flows with Re between 2000 and 4000 are classified as critical flows. Re is necessary to calculate friction coefficients which, in turn, are used to determine pressure losses.

**Equation 1-5, Friction head loss** Whenever flow occurs, a continuous pressure loss exists along the piping in the direction of flow and this head loss is affected by the density of the fluid, its temperature, the pipe roughness, the length of the run, and the fluid velocity. The friction head loss is represented by Darcy's Friction Formula:

### Equation 1-5

$$h = \frac{fLV^2}{2gD}$$

where

- h = Friction head loss, ft (m)
- f = Friction coefficient, dimensionless
- L = Length of pipe, ft (m)
- V = Velocity of flow, ft/s (m/s)
- g = Gravitational acceleration, 32.2 ft/s/s (9.8 m/s²)
- D = Internal diameter of pipe, ft (m)

(a) The static head is the pressure (P) exerted at any point by the height of the substance above that point. To convert from feet (m) of head to pounds per square inch (kPa or kg/m2), the following relationship is used:

### Equation 1-5a

$$P = \frac{\gamma h}{144}$$

where

- P = Pressure, lb/in² (kPa)
- $\gamma$ = Density of substance, lb/ft³ (N/m³)
- h = Static head, ft (m)

(b) Therefore, Equation 1-5 may be represented as:

### Equation 1-5b

$$P = \frac{\gamma fLV^2}{288gD}$$

(c) To convert pressure in meters of head to pressure in kilopascals, use

### Equation 1-5c

$$kPa = 9.81 \text{ (m head)}$$

(d) To calculate the friction loss, the Hazen-Williams Formula is used:

### Equation 1-5d

$$h = 0.002082L\left(\frac{100}{C}\right)^{1.85}\left(\frac{q^{1.85}}{d^{4.8655}}\right)$$

where

- C = Friction factor for Hazen-Williams
- q = Flow rate, gpm (L/s)
- d = Actual inside diameter of pipe, in. (mm)

# Chapter 1 — Plumbing Formulae, Symbols, and Terminology

L = Length of pipe, ft (m)

f = Friction factor

Values for f and C are tabulated in Baumeister and Marks's *Handbook for Mechanical Engineers*.

**Equation 1-6, Potential energy (PE)** Defined as the energy of a body due to its elevation above a given level and expressed as:

### Equation 1-6

$$PE = Wh$$

where

PE = Potential energy, ft-lb (J)

W = Weight of body, lb (N)

h = Height above level, ft (m)

**Equation 1-7, Kinetic energy (KE)** Defined as the energy of a body due to its motion and expressed as:

### Equation 1-7

$$KE = \frac{mV^2}{2} = \frac{WV^2}{2g}$$

where

KE = Kinetic energy, ft-lb (J)

m = Mass of body, lb (kg)

V = Velocity, ft/s (m/s)

g = Gravitational acceleration, 32.2 ft/s/s (9.8 m/s$^2$)

W = Weight of body, lb (kg)

**Equation 1-8, Flow at outlet** Can be determined by using the following relationship:

### Equation 1-8

$$Q = 20 \, d^2 P^{1/2}$$

where

Q = Flow at outlet, gpm (L/s)

d = Inside diameter of outlet, in. (mm)

P = Flow pressure, lb/in$^2$ (kPa)

**Equation 1-9, Length of vent piping** Can be determined by combining Darcy's Friction Formula (Equation 1-5) and the flow equation and is expressed as:

### Equation 1-9

$$L = \frac{2226 d^5}{fQ^2}$$

where

L = Length of pipe, ft (m)

d = Diameter of pipe, in. (mm)

f = Friction coefficient, dimensionless

Q = Rate of flow, gpm (L/s)

### Equation 1-10, Stacks

(a) Terminal velocity

### Equation 1-10a

$$V_T = 3 \left(\frac{Q}{d}\right)^{2/5}$$

where

$V_T$ = Terminal velocity in stack, ft/s (m/s)

Q = Rate of flow, gpm (L/s)

d = Diameter of stack, in. (mm)

(b) Terminal length

### Equation 1-10b

$$L_T = 0.052 \, V_T^2$$

where

$L_T$ = Terminal length below point of flow entry, ft (m)

(c) Capacity

### Equation 1-10c

$$Q = 27.8 \, r^{5/3} \, d^{8/3}$$

where

Q = Maximum permissible flow rate in stack, gpm (L/s)

r = Ratio of cross-sectional area of the sheet of water to cross-sectional area of stack.

d = Diameter of stack, in. (mm)

**Equation 1-11, Flow rate in fixture drain** The flow rate in a fixture drain should equal the flow rate at the fixture outlet and is expressed as:

### Equation 1-11

$$Q = 13.17 \, d^2 h^{1/2}$$

*where*

- Q = Discharge flow rate, gpm (L/s)
- d = Diameter of outlet orifice, in. (mm)
- h = Mean vertical height of water surface above the point of outlet orifice, ft (m)

**Equation 1-12, Pipe expansion and contraction** All pipes that are subject to temperature changes expand and contract. Piping expands with an increase in temperature and contracts with a decrease in temperature. The rate of change in length due to temperature is referred to as the expansion coefficient. The changes in length can be calculated by using the following relation:

***Equation 1-12***

$$L_2 - L_1 = C_E L_1 (T_2 - T_1)$$

*where*

- $L_2$ = Final length of pipe, ft (m)
- $L_1$ = Initial length of pipe, ft (m)
- $C_E$ = Coefficient of expansion of material (A material's expansion coefficient may be obtained from the ASHRAE *Handbook of Fundamentals*.)
- $T_2$ = Final temperature, °F (°C)
- $T_1$ = Initial temperature, °F (°C)

**Equation 1-13, Various formulae for areas and volumes, in ft² (m²) and ft³ (m³), respectively.**

***Equation 1-13a, Square*** (See Figure 1-1.)

$$A = bh$$

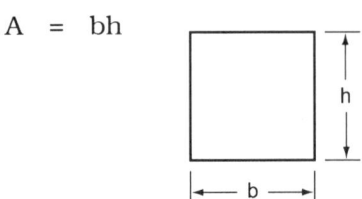

**Figure 1-1  Square**

***Equation 1-13b, Rectangle*** (See Figure 1-2.)

$$A = bh$$

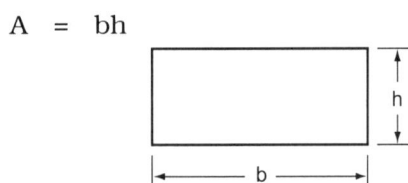

**Figure 1-2  Rectangle**

***Equation 1-13c, Rhombus*** (See Figure 1-3.)

$$A = bh$$

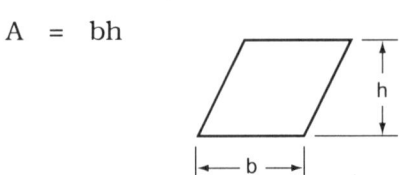

**Figure 1-3  Rhombus**

***Equation 1-13d, Rhomboid*** (See Figure 1-4.)

$$A = bh$$

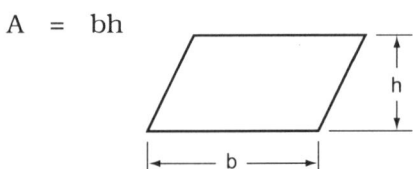

**Figure 1-4  Rhomboid**

***Equation 1-13e, Trapezoid*** (See Figure 1-5.)

$$A = \frac{h(a+b)}{2}$$

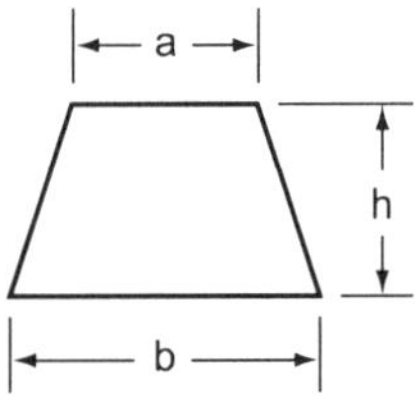

**Figure 1-5  Trapezoid**

***Equation 1-13f, Trapezium*** (See Figure 1-6.)

$$A = \frac{(H+h)a + bh + cH}{2}$$

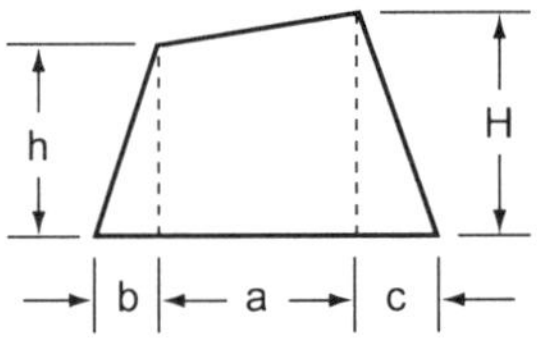

**Figure 1-6  Trapezium**

***Equation 1-13g, Right-angle triangle*** (See Figure 1-7.)

$$A = \frac{bh}{2}$$

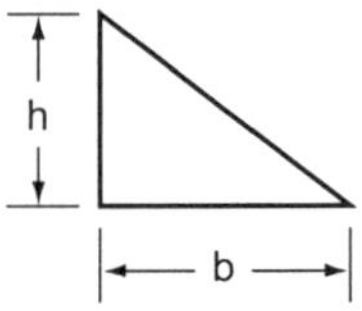

**Figure 1-7  Right-Angle Triangle**

# Chapter 1 — Plumbing Formulae, Symbols, and Terminology

**Equation 1-13h, Isosceles triangle** (See Figure 1-8.)

$$A = \frac{bh}{2}$$

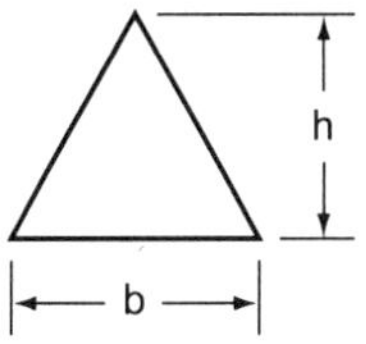

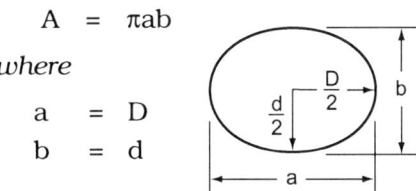

**Figure 1-8 Isosceles Triangle**

**Equation 1-13i, Ellipse** (See Figure 1-9.)

$$A = \pi ab$$

where

a = D
b = d

**Figure 1-9 Ellipse**

**Equation 1-13j, Cylinder** (See Figure 1-10.)

$$A = \pi Dh$$
$$V = \pi R^2 h$$

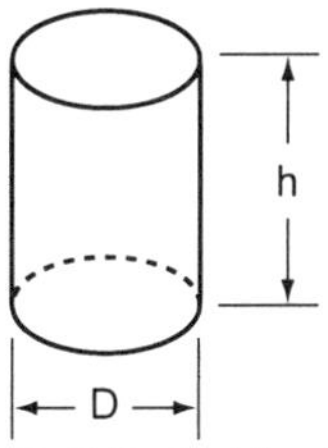

**Figure 1-10 Cylinder**

**Equation 1-13k, Cube or rectangular solid** (See Figure 1-11.)

$$V = whl$$

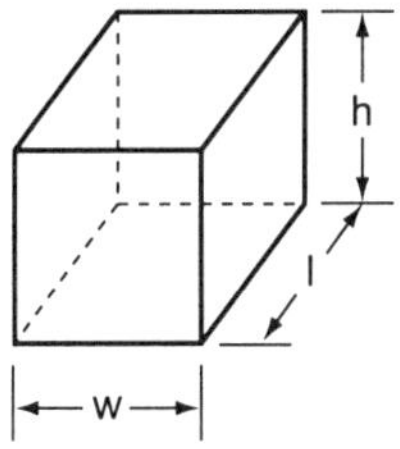

**Figure 1-11 Cube or Rectangular Solid**

**Equation 1-13l, Pyramid** (See Figure 1-12.)

$$V = \frac{abh}{3}$$

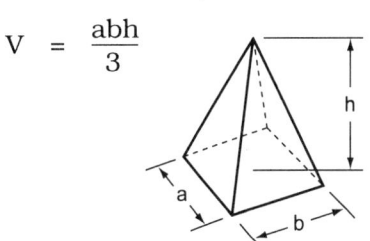

**Figure 1-12 Pyramid**

**Equation 1-13m, Cone** (See Figure 1-13.)

$$A = \frac{\pi Ds}{2}$$
$$V = \frac{\pi R^2 h}{3}$$

where

$D = b$
$R = \frac{b}{2}$

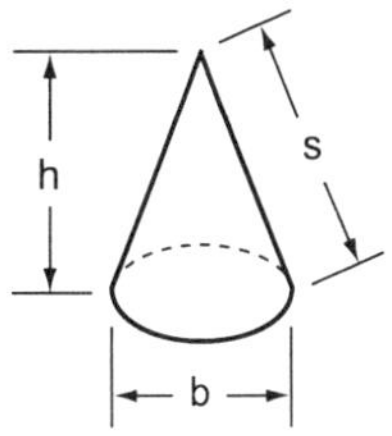

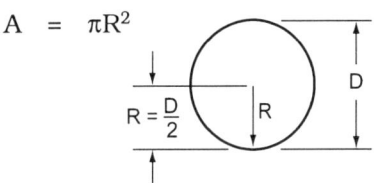

**Figure 1-13 Cone**

**Equation 1-13n, Circle** (See Figure 1-14.)

$$C = 2\pi R$$

**Equation 1-13o, Circle** (See Figure 1-14.)

$$A = \pi R^2$$

**Figure 1-14 Circle**

**Equation 1-13p, Triangle³** (See Figure 1-15.)

Known:    2 angles
Required: Third angle
Solution: $A = 180° - (B + C)$

**Equation 1-13q, Triangle³** (See Figure 1-15.)

Known:    3 sides
Required: Any angle
Solution: $\cos A = \dfrac{b^2 + c^2 - a^2}{2bc}$

**Equation 1-13r, Triangle³** (See Figure 1-15.)

Known:    2 sides and included angle
Required: Third side
Solution: $c = (a^2 + b^2 - 2ab \cos C)^{1/2}$

**Equation 1-13s, Triangle³** (See Figure 1-15.)

Known:    2 sides and included angle
Required: Third angle
Solution: $\tan A = \dfrac{a \sin C}{b - a \cos C}$

**Equation 1-13t, Triangle**[3] (See Figure 1-15.)

Known: 2 sides and excluded angle

Required: Third side

Solution: $c = b \cos A \pm (a^2 - b^2 \sin^2 A)^{1/2}$

**Equation 1-13u, Triangle**[3] (See Figure 1-15.)

Known: 1 side and adjacent angles

Required: Adjacent side

Solution: $c = \dfrac{a \sin C}{\sin A}$

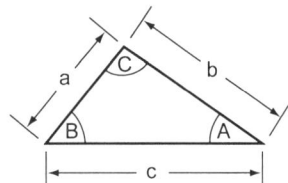

**Figure 1-15 Triangle**

**Equation 1-14, Flow rate in outlet** With Equation 1-11, we determined that the flow rate (Q) in the outlet should be equal to the flow rate in the fixture drain. The maximum discharge rate is expressed as:

*Equation 1-14*

$$Q_D = c_D Q_I$$

where

$Q_D$ = Actual discharge quantity, gpm (L/s)

$c_D$ = Discharge coefficient

$Q_I$ = Ideal discharge quantity, gpm (L/s)

The discharge coefficients ($c_D$) may be obtained from Baumeister and Marks's *Handbook for Mechanical Engineers*.

**Equation 1-15, Gravity circulation** This principle is used to keep the sanitary system free of foul odors and the growth of slime and fungi. The circulation is induced by the pressure difference between the outdoor air and the air in the vent piping. This pressure difference is due to the difference in temperature (T) and density (ρ) between the two and the height (h) of the air column in the vent piping. The gravity circulation is determined by using the following formula:

*Equation 1-15*

$$P = 0.1925 \, (\rho_O - \rho_I) \, h_s$$

where

P = Natural draft pressure, in. (mm)

$\rho_O$ = Density of outside air, lb/ft³ (N/m³)

$\rho_I$ = Density of air in pipe, lb/ft³ (N/m³)

$h_s$ = Height of air column in stack, ft (m)

The outside and inside air densities ($\rho_O$ and $\rho_I$) may be obtained from the ASHRAE *Handbook of Fundamentals*.

**Equation 1-16, Velocity head (h)** When the water in a piping system is at rest, it has potential energy (PE). When the water in a piping system is flowing, it has kinetic energy (KE). For the water to flow, some of the potential energy (PE) must be converted to kinetic energy (KE). The decrease in potential energy (static head) is referred to as the velocity head (h) and is expressed as:

*Equation 1-16*

$$h = \dfrac{V^2}{2g}$$

where

h = Height of the fall, ft (m)

V = Velocity at any moment, ft/s (m/s)

g = Gravitational acceleration, 32.2 ft/s/s (9.8 m/s²)

**Equation 1-17, Bernoulli's Equation** Since energy cannot be created or destroyed, Bernoulli developed a theorem to express this energy conservation. It is represented by the following equation:

*Equation 1-17*

$$Z + \dfrac{P}{\rho} + \dfrac{V^2}{2g} = C$$

where

Z = Height of point above datum, ft (m)

P = Pressure, lb/ft² (kPa)

ρ = Density, lb/ft³ (N/m³)

V = Velocity, ft/s (m/s)

g = Gravitational acceleration, 32.2 ft/s/s (9.8 m/s²)

C = Constant

---

[3]*Note*: Formulae may be used for all oblique triangles.

# Chapter 1 — Plumbing Formulae, Symbols, and Terminology

(a) For two points in the system, Equation 1-17 can be expressed as:

**Equation 1-17a**

$$Z_1 + P_{1/\rho} + \frac{V_1^2}{2g} = Z_2 + \frac{P_2}{\rho} + \frac{V_2^2}{2g}$$

Subscripts 1 and 2 represent points in the system.

**Equation 1-18, Friction head ($h_f$)** When water flows in a pipe, friction is produced by the rubbing of water particles against each other and against the walls of the pipe. This causes a pressure loss in the line of flow, called the friction head, which is expressed by using Bernoulli's equation:

**Equation 1-18**

$$h_f = \left(Z_1 + h_1 + \frac{V_1^2}{2g}\right) - \left(Z_2 + h_2 + \frac{V_2^2}{2g}\right)$$

where

- $h_f$ = Friction head, ft (m)
- $Z$ = Height of point, ft (m)
- $h$ = $P/\rho$ = static head or height of liquid column, ft (m)
- $V$ = Velocity at outlet, ft/s (m/s)
- $g$ = Gravitational acceleration, 32.2 ft/s/s (9.8 m/s²)

Subscripts 1 and 2 represent points in the system.

**Equation 1-19, Flow from outlets** This velocity can be expressed by the following:

**Equation 1-19**

$$V = C_D (2gh)^{1/2}$$

where

- $V$ = Velocity at outlet, ft/s (m/s)
- $C_D$ = Coefficient of discharge (usually 0.67)
- $g$ = Gravitational acceleration, 32.2 ft/s/s (9.8 m/s²)
- $h$ = Static head or height of liquid column, ft (m)

**Equation 1-20, Hydraulic shock** The magnitude of the pressure wave can be expressed by the following relationship:

**Equation 1-20**

$$P = \frac{\gamma a dV}{144g}$$

where

- $P$ = Pressure in excess of flow pressure, lb/in² (kPa)
- $\gamma$ = Specific weight of liquid, lb/ft³ (N/m³)
- $a$ = Velocity of propagation of elastic vibration in the pipe, ft/s (m/s)
- $dV$ = Change in flow velocity, ft/s (m/s)
- $g$ = Gravitational acceleration, 32.2 ft/s/s (9.8 m/s²)

(a) The velocity of propagation of elastic vibration in the pipe can be defined as:

**Equation 1-20a**

$$a = \frac{4660}{(1 + KB)^{1/2}}$$

where

- $a$ = Propagation velocity, ft/s (m/s)
- 4660 = Velocity of sound in water, ft/s (m/s)
- $K$ = Ratio of modulus of elasticity of fluid to modulus of elasticity of pipe
- $B$ = Ratio of pipe diameter to wall thickness

The values for specific weights ($\gamma$), K, and B are given or can be calculated from the ASHRAE *Handbook of Fundamentals*.

(b) The time interval required for the pressure wave to travel back and forth in the pipe can be expressed as:

**Equation 1-20b**

$$t = \frac{2L}{a}$$

where

- $t$ = Time interval, s
- $L$ = Length of pipe from point of closure to point of relief, ft (m)

**Equation 1-21, Pump affinity laws** Affinity laws describe the relationships among the capacity, head, brake horsepower, speed, and impeller diameter of a given pump.

The first law states the performance data of constant impeller diameter with change in speed.

### Equation 1-21a

$$\frac{Q_1}{Q_2} = \frac{N_1}{N_2} \quad \text{and} \quad \frac{H_1}{H_2} = \left(\frac{N_1}{N_2}\right)^2$$

and

$$\frac{BHP_1}{BHP_2} = \left(\frac{N_1}{N_2}\right)^3$$

or

$$\frac{N_1}{N_2} = \frac{Q_1}{Q_2} = \left(\frac{H_1}{H_2}\right)^{1/2} = \left(\frac{BHP_1}{BHP_2}\right)^{1/3}$$

where

Q = Capacity, gpm (m³/H)
N = Speed, rpm (r/s)
H = head, ft (m)
BHP = Brake horspower, W

The second law assumes the performance data of constant speed with change in diameter of the impeller.

### Equation 1-21b

$$\frac{Q_1}{Q_2} = \frac{D_1}{D_2} \quad \text{and} \quad \frac{H_1}{H_2} = \frac{D_1^2}{D_2^2} \quad \text{and} \quad \frac{BHP_1}{BHP_2} = \frac{D_1^3}{D_2^3}$$

or

$$\frac{D_1}{D_2} = \frac{Q_1}{Q_2} = \left(\frac{H_1}{H_2}\right)^{1/2} = \left(\frac{BHP_1}{BHP_2}\right)^{1/3}$$

where

D = Impeller diameter, in. (m)

**Equation 1-22, Pump efficiency** The efficiency of a pump is represented by the following equation:

### Equation 1-22

$$Ep = \frac{WHP}{BHP}$$

where

Ep = Pump efficiency as a decimal equivalent
WHP = Water horsepower derived from:

$$WHP = ft\ Hd \times \frac{gal}{min} \times \frac{8.33\ lb}{gal} \times \frac{HP}{33{,}000\ ft\text{-}lb/min}$$

BHP = Brake horsepower input to pump

From Equation 1-22, the brake horsepower can be represented as:

### Equation 1-22a

$$BHP = \frac{WHP}{Ep} \text{ or } \frac{ft\ Hd \times gpm}{3960 \times Ep}$$

**Equation 1-23, Rational method of storm design** Calculates the peak storm-water runoff.

### Equation 1-23

Q = CIA

where

Q = Runoff, ft³/s (m³/s)
C = Runoff coefficient (surface roughness in drained area)
I = Rainfall intensity, in/h (mm/h)
A = Drainage area, acres (m²)

**Equation 1-24, Spitzglass Formula** Used to size gas piping in systems operating at a pressure of less than 1 psi.

### Equation 1-24

$$Q = 3550 \left(\frac{d^5}{1 + 3.6/d + 0.03d}\right)^{1/2} \left(\frac{h}{SL}\right)^{1/2}$$

where

Q = Flow rate, f³/h (m³/h)
d = Diameter of pipe, in. (mm)
h = Pressure drop over length, in. wc
S = Specific gravity
L = Length of pipe, ft (m)

**Equation 1-25, Weymouth Formula** Used to size gas piping in systems operating at a pressure in excess of 1 psi.

### Equation 1-25

$$Q = 28.05 \left[\frac{(P_1^2 - P_2^2)\ d^{16/3}}{SL}\right]^{1/2}$$

where

Q = Flow rate, ft³/h (m³/h)
P₁ = Initial gas pressure, psi
P₂ = Final gas pressure, psi
d = Diameter of pipe, in. (mm)
S = Specific gravity
L = Length of pipe, mi (km)

# Chapter 1 — Plumbing Formulae, Symbols, and Terminology

## SYMBOLS

The standardized plumbing and piping-related symbols in Tables 1-1 and 1-2 and the abbreviations in Table 1-3 have been tabulated by the American Society of Plumbing Engineers for use in the design and preparation of drawings. Users of these symbols are cautioned that some governmental agencies, industry groups, and other clients may have a list of symbols that are required for their projects. All symbols should be applied with a consideration for drafting and clarity if drawings are to be reduced.

### Table 1-1  Standard Plumbing and Piping Symbols

| Symbol | Description | Abbreviation |
|---|---|---|
| ——— SD ——— | Storm drain, rainwater drain | SD, ST |
| ——— SSD ——— | Subsoil drain, footing drain | SSD |
| ——— SS ——— | Soil, waste, or sanitary sewer | S, W, SAN, SS |
| — — — — — — — — | Vent | V |
| ——— AW ——— | Acid waste | AW |
| — — AV — — — | Acid vent | AV |
| ——— D ——— | Indirect drain | D |
| ——— PD ——— | Pump discharge line | PD |
| ——— - - ——— | Cold water | CW |
| ——— - - - ——— | Hot water supply (140°F)[a] | HW |
| ——— - - - - ——— | Hot water recirculating (140°F)[a] | HWR |
| ——— TW ——— | Tempered hot water (temp. °F)[b] | TEMP. HW, TW |
| ——— TWR ——— | Tempered hot water recirculating (temp. °F)[b] | TEMP. HWR, TWR |
| ——— DWS ——— | (Chilled) drinking water supply | DWS |
| ——— DWR ——— | (Chilled) drinking water recirculating | DWR |
| ——— SW ——— | Soft water | SW |
| ——— CD ——— | Condensate drain | CD |
| ——— DI ——— | Distilled water | DI |
| ——— DE ——— | Deionized water | DE |

*(Continued)*

*(Table 1-1 continued)*

| Symbol | Description | Abbreviation |
|---|---|---|
| ——— CWS ——— | Chilled water supply | CWS |
| ——— CWR ——— | Chilled water return | CWR |
| ——— LS ——— | Lawn sprinkler supply | LS |
| ——— F ——— | Fire protection water supply | F |
| ——— G ——— | Gas–low-pressure | G |
| ——— MG ——— | Gas–medium-pressure | MG |
| ——— HG ——— | Gas–high-pressure | HG |
| – – – GV – – – | Gas vent | GV |
| ——— FOS ——— | Fuel oil supply | FOS |
| ——— FOR ——— | Fuel oil return | FOR |
| – – – FOV – – – | Fuel oil vent | FOV |
| ——— LO ——— | Lubricating oil | LO |
| – – – LOV – – – | Lubricating oil vent | LOV |
| ——— WO ——— | Waste oil | WO |
| – – – WOV – – – | Waste oil vent | WOV |
| ——— OX ——— | Oxygen | OX |
| ——— LOX ——— | Liquid oxygen | LOX |
| ——— A ——— | Compressed air[c] | A |
| ——— X#A ——— | Compressed air–X#[c] | X#A |
| ——— MA ——— | Medical compressed air | MA |
| ——— LA ——— | Laboratory compressed air | LA |
| ——— HHWS ——— | (Heating) hot water supply | HHWS |
| ——— HHWR ——— | (Heating) hot water return | HHWR |
| ——— V ——— | Vacuum | VAC |

*(Continued)*

# Chapter 1 — Plumbing Formulae, Symbols, and Terminology

*(Table 1-1 continued)*

| Symbol | Description | Abbreviation |
|---|---|---|
| —— MV —— | Medical vacuum | MV |
| —— SV —— | Surgical vacuum | SV |
| —— LV —— | Laboratory vacuum | LV |
| —— N —— | Nitrogen | N |
| —— $N_2O$ —— | Nitrous oxide | $N_2O$ |
| —— $CO_2$ —— | Carbon dioxide | $CO_2$ |
| —— WVC —— | Wet vacuum cleaning | WVC |
| —— DVC —— | Dry vacuum cleaning | DVC |
| —— LPS —— | Low-pressure steam supply | LPS |
| – – – LPC – – – | Low-pressure condensate | LPC |
| —— MPS —— | Medium-pressure steam supply | MPS |
| – – – MPC – – – | Medium-pressure condensate | MPC |
| —— HPS —— | High-pressure steam supply | HPS |
| – – – HPC – – – | High-pressure condensate | HPC |
| – – – ATV – – – | Atmospheric vent (steam or hot vapor) | ATV |
| ⋈ | Gate valve | GV |
| ⋈ | Globe valve | GLV |
| ⋈ | Angle valve | AV |
| ⋈ | Ball valve | BV |
| ⋈ | Butterfly valve | BFV |
| ⋈ | Gas cock, gas stop | |
| ⋈ | Balancing valve (specify type) | BLV |
| ⋈ | Check valve | CV |
| ⋈ | Plug valve | PV |

*(Continued)*

*(Table 1-1 continued)*

| Symbol | Description | Abbreviation |
|---|---|---|
| | Solenoid valve | |
| | Motor-operated valve (specify type) | |
| | Pressure-reducing valve | PRV |
| | Pressure-relief valve | RV |
| | Temperature-pressure-relief valve | TPV |
| | Reduced zone backflow preventer | RZBP |
| | Double-check backflow preventer | DCBP |
| | Hose bibb | HB |
| | Recessed-box hose bibb or wall hydrant | WH |
| | Valve in yard box (valve type symbol as required for valve use) | YB |
| | Union (screwed) | |
| | Union (flanged) | |
| | Strainer (specify type) | |
| | Pipe anchor | PA |
| | Pipe guide | |
| | Expansion joint | EJ |
| | Flexible connector | FC |
| | Tee | |
| | Concentric reducer | |
| | Eccentric reducer | |

*(Continued)*

# Chapter 1 — Plumbing Formulae, Symbols, and Terminology

*(Table 1-1 continued)*

| Symbol | Description | Abbreviation |
|---|---|---|
| | Aquastat | |
| | Flow switch | FS |
| | Pressure switch | PS |
| | Water hammer arrester | WHA |
| | Pressure gauge with gauge cock | PG |
| | Thermometer (specify type) | |
| | Automatic air vent | AAV |
| | Valve in riser (type as specified or noted) | |
| | Riser down (elbow) | |
| | Riser up (elbow) | |
| | Air chamber | AC |
| | Rise or drop | |
| | Branch–top connection | |
| | Branch–bottom connection | |
| | Branch–side connection | |
| | Cap on end of pipe | |
| | Flow indicator for stationary meter (orifice) | |
| | Flow indicator for portable meter (specify flow rate) | |
| | Upright fire sprinkler head | |
| | Fire hose rack | FHR |

*(Continued)*

*(Table 1-1 continued)*

| Symbol | Description | Abbreviation |
|---|---|---|
|  | Fire hose cabinet (surface-mounted) | FHC |
|  | Fire hose cabinet (recessed) | FHC |
|  | Cleanout plug | CO |
|  | Floor cleanout | FCO |
|  | Wall cleanout | WCO |
|  | Yard cleanout or cleanout to grade | CO |
|  | Drain (all types) (specify) | D |
|  | Pitch down or up–in direction of arrow | |
|  | Flow–in direction of arrow | |
|  | Point of connect | POC |
|  | Outlet (specify type) | |
|  | Steam trap (all types) | |
|  | Floor drain with p-trap | FD |

[a]Hot water (140°F) and hot water return (140°F). Use for normal hot water distribution system, usually but not necessarily (140°F). Change temperature designation if required.

[b]Hot water (temp. °F) and hot water return (temp. °F). Use for any domestic hot water system (e.g., tempered or sanitizing) required in addition to the normal system (see note "a" above). Insert system supply temperature where "temp." is indicated.

[c]Compressed air and compressed air X#. Use pressure designations (X#) when compressed air is to be distributed at more than one pressure.

# Chapter 1 — Plumbing Formulae, Symbols, and Terminology

## Table 1-2  Standard Fire Protection Piping Symbols

| Referent (Synonym) | Symbol | Comments |
|---|---|---|

**Water supply and distribution symbols**

*Mains, pipe*

    Riser

*Hydrants*

| | | |
|---|---|---|
| Public hydrant, two hose outlets | | Indicate size,[a] type of thread or connection. |
| Public hydrant, two hose outlets, and pumper connection | | Indicate size,[a] type of thread or connection. |
| Wall hydrant, two hose outlets | | Inditate size,[a] type of thread or connection. |

*Fire department connections*

| | | |
|---|---|---|
| Siamese fire department connection | | Specify type, size, and angle. |
| Free-standing siamese fire department connection | | Sidewalk or pit type, specify size. |

*Fire pumps*

| | | |
|---|---|---|
| Fire pump | | Free-standing. Specify number and sizes of outlets. |
| Test header | | Wall |

**Symbols for control panels**

| | | |
|---|---|---|
| Control panel | | Basic shape. |
| (a) | FCP | Fire alarm control panel |

**Symbols for fire extinguishing systems**

*Symbols for various types of extinguishing systems[b]*

    *Supplementary symbols*

| | | |
|---|---|---|
| Fully sprinklered space | (AS) | |
| Partially sprinklered space | (AS) | |
| Nonsprinklered space | ⟨NS⟩ | |

*(Continued)*

*(Table 1-2 continued)*

| Referent (Synonym) | Symbol | Comments |
|---|---|---|
| *Symbols for fire sprinkler heads* | | |
| Upright sprinkler[c] | | |
| Pendent sprinkler[c, d] | | |
| Upright sprinkler, nippled up[c] | | |
| Pendent sprinkler, on drop nipple[c, d] | | |
| Sidewall sprinkler[c] | | |
| *Symbols for piping, valves, control devices, and hangers.[e]* | | |
| Pipe hanger | | This symbol is a diagonal stroke imposed on the pipe that it supports. |
| Alarm check valve | | Specify size, direction of flow. |
| Dry pipe valve | | Specify size. |
| Deluge valve | | Specify size and type. |
| Preaction valve | | Specify size and type. |
| **Symbols for portable fire extinguishers** | | |
| Portable fire extinguisher | | Basic shape. |
| **Symbols for firefighting equipment** | | |
| Hose station, dry standpipe | | |
| Hose station, changed standpipe | | |

*Source*: National Fire Protection Association (NFPA), Standard 170.

[a] Symbol element can be utilized in any combination to fit the type of hydrant.

[b] These symbols are intended for use in identifying the type of system installed to protect an area within a building.

[c] Temperature rating of sprinkler and other characteristics can be shown via legends where a limited number of an individual type of sprinkler is called for by the design.

[d] Can notate "DP" on drawing and/or in specifications where dry pendent sprinklers are employed.

[e] See also NFPA Standard 170, Section 5-4, for related symbols.

# Chapter 1 — Plumbing Formulae, Symbols, and Terminology

### Table 1-3  Abbreviations for Text, Drawings, and Computer Programs

| Term | Text | Drawings | Program |
|---|---|---|---|
| Above finished floor | – | AFF | – |
| Absolute | abs | ABS | ABS |
| Accumulat(-e, -or) | acc | ACCUM | ACCUM |
| Air condition(-ing, -ed) | – | AIR COND | – |
| Air-conditioning unit(s) | – | ACU | ACU |
| Air-handling unit | – | AHU | AHU |
| Air horsepower | ahp | AHP | AHP |
| Alteration | altrn | ALTRN | – |
| Alternating current | ac | AC | AC |
| Altitude | alt | ALT | ALT |
| Ambient | amb | AMB | AMB |
| American National Standards Institute[a] | ANSI | ANSI | – |
| American wire gage | AWG | AWG | – |
| Ampere (amp, amps) | amp | AMP | AMP, AMPS |
| Angle | – | – | ANG |
| Angle of incidence | – | – | ANGI |
| Apparatus dew point | adp | ADP | ADP |
| Approximate | approx. | APPROX | – |
| Area | – | – | A |
| Atmosphere | atm | ATM | – |
| Average | avg | AVG | AVG |
| Azimuth | az | AZ | AZ |
| Azimuth, solar | – | – | SAZ |
| Azimuth, wall | – | – | WAZ |
| Baromet(-er, -ric) | baro | BARO | – |
| Bill of material | b/m | BOM | – |
| Boiling point | bp | BP | BP |
| Brake horsepower | bhp | BHP | BHP |
| Brown & Sharpe wire gage | B&S | B&S | – |
| British thermal unit | Btu | BTU | BTU |
| Celsius | °C | °C | °C |
| Center to center | c to c | C TO C | – |
| Circuit | ckt | CKT | CKT |
| Clockwise | cw | CW | – |
| Coefficient | coeff. | COEF | COEF |
| Coefficient, valve flow | $C_v$ | $C_v$ | CV |
| Coil | – | – | COIL |
| Compressor | cprsr | CMPR | CMPR |
| Condens(-er, -ing, -ation) | cond | COND | COND |

*(Continued)*

*(Table 1-3 continued)*

| Term | Text | Drawings | Program |
|---|---|---|---|
| Conductance | – | – | C |
| Conductivity | cndct | CNDCT | K |
| Conductors, number of (3) | 3/c | 3/c | – |
| Contact factor | – | – | CF |
| Cooling load | clg load | CLG LOAD | CLOAD |
| Counterclockwise | ccw | CCW | – |
| Cubic feet | ft$^3$ | CU FT | CUFT, CFT |
| Cubic inch | in$^3$ | CU IN | CUIN, CIN |
| Cubic feet per minute | cfm | CFM | CFM |
| cfm, standard conditions | scfm | SCFM | SCFM |
| Cubic ft per sec, standard | scfs | SCFS | SCFS |
| Decibel | dB | DB | DB |
| Degree | deg. or ° | DEG or ° | DEG |
| Density | dens | DENS | RHO |
| Depth or deep | dp | DP | DPTH |
| Dew-point temperature | dpt | DPT | DPT |
| Diameter | dia. | DIA | DIA |
| Diameter, inside | ID | ID | ID |
| Diameter, outside | OD | OD | OD |
| Difference or delta | diff., Δ | DIFF | D, DELTA |
| Diffuse radiation | – | – | DFRAD |
| Direct current | dc | DC | DC |
| Direct radiation | dir radn | DIR RADN | DIRAD |
| Dry | – | – | DRY |
| Dry-bulb temperature | dbt | DBT | DB, DBT |
| Effectiveness | – | – | EFT |
| Effective temperature[b] | ET* | ET* | ET |
| Efficiency | eff | EFF | EFF |
| Efficiency, fin | – | – | FEFF |
| Efficiency, surface | – | – | SEFF |
| Electromotive force | emf | EMF | – |
| Elevation | elev. | EL | ELEV |
| Entering | entr | ENT | ENT |
| Entering air temperature | EAT | EAT | EAT |
| Entering water temperature | EWT | EWT | EWT |
| Enthalpy | – | – | H |
| Entropy | – | – | S |
| Equivalent direct radiation | edr | EDR | – |
| Equivalent feet | eqiv ft | EQIV FT | EQFT |
| Equivalent inches | eqiv in | EQIV IN | EQIN |
| Evaporat(-e, -ing, -ed, -or) | evap | EVAP | EVAP |

*(Continued)*

# Chapter 1 — Plumbing Formulae, Symbols, and Terminology

*(Table 1-3 continued)*

| Term | Text | Drawings | Program |
|---|---|---|---|
| Expansion | exp | EXP | XPAN |
| Face area | fa | FA | FA |
| Face to face | f to f | F to F | – |
| Face velocity | fvel | FVEL | FV |
| Factor, correction | – | – | CFAC, CFACT |
| Factor, friction | – | – | FFACT, FF |
| Fahrenheit | °F | °F | F |
| Fan | – | – | FAN |
| Feet per minute | fpm | FPM | FPM |
| Feet per second | fps | FPS | FPS |
| Film coefficient, inside[c] | – | – | FI, HI |
| Film coefficient, outside[c] | – | – | FO, HO |
| Flow rate, air | – | – | QAR, QAIR |
| Flow rate, fluid | – | – | QFL |
| Flow rate, gas | – | – | QGA, QGAS |
| Foot or feet | ft | FT | FT |
| Foot-pound | ft-lb | FT LB | – |
| Freezing point | fp | FP | FP |
| Frequency | Hz | HZ | – |
| Gage or gauge | ga | GA | GA, GAGE |
| Gallons | gal | GAL | GAL |
| Gallons per hour | gph | GPH | GPH |
| gph, standard | std gph | SGPH | SGPH |
| Gallons per day | gpd | GPD | GPD |
| Grains | gr | GR | GR |
| Gravitational constant | $g$ | G | G |
| Greatest temperature difference | GTD | GTD | GTD |
| Head | hd | HD | HD |
| Heat | – | – | HT |
| Heater | – | – | HTR |
| Heat gain | HG | HG | HG, HEATG |
| Heat gain, latent | LHG | LHG | HGL |
| Heat gain, sensible | SHG | SHG | HGS |
| Heat loss | – | – | HL, HEATL |
| Heat transfer | – | – | Q |
| Heat transfer coefficient | $U$ | U | U |
| Height | hgt | HGT | HGT, HT |
| High-pressure steam | hps | HPS | HPS |
| High-temperature hot water | hthw | HTHW | HTHW |
| Horsepower | hp | HP | HP |
| Hour(s) | h | HR | HR |

*(Continued)*

*(Table 1-3 continued)*

| Term | Text | Drawings | Program |
|---|---|---|---|
| Humidity ratio | *W* | W | W |
| Humidity, relative | rh | RH | RH |
| Incident angle | – | – | INANG |
| Indicated horsepower | ihp | IHP | – |
| International Pipe Std. | IPS | IPS | – |
| Iron pipe size | ips | IPS | – |
| Kelvin | K | K | K |
| Kilowatt | kW | kW | KW |
| Kilowatt hour | kWh | KWH | KWH |
| Latent heat | LH | LH | LH, LHEAT |
| Least mean temp. difference[d] | LMTD | LMTD | LMTD |
| Least temperature difference[d] | LTD | LTD | LTD |
| Leaving air temperature | lat | LAT | LAT |
| Leaving water temperature | lwt | LWT | LWT |
| Length | lg | LG | LG, L |
| Linear feet | lin ft | LF | LF |
| Liquid | liq | LIQ | LIQ |
| Logarithm (natural) | ln | LN | LN |
| Logarithm to base 10 | log | LOG | LOG |
| Low-pressure steam | lps | LPS | LPS |
| Low-temperature hot water | lthw | LTHW | LTHW |
| Mach number | Mach | MACH | – |
| Mass flow rate | mfr | MFR | MFR |
| Maximum | max. | MAX | MAX |
| Mean effective temperature | MET | MET | MET |
| Mean temp. difference | MTD | MTD | MTD |
| Medium-pressure steam | mps | MPS | MPS |
| Medium-temperature hot water | mthw | MTHW | MTHW |
| Mercury | Hg | HG | HG |
| Miles per hour | mph | MPH | MPH |
| Minimum | min. | MIN | MIN |
| Noise criteria | NC | NC | – |
| Normally closed | n c | N C | – |
| Normally open | n o | N O | – |
| Not applicable | na | N/A | – |
| Not in contract | n i c | N I C | – |
| Not to scale | – | NTS | – |
| Number | no. | NO | N, NO |
| Number of circuits | – | – | NC |
| Number of tubes | – | – | NT |
| Ounce | oz | OZ | OZ |

*(Continued)*

# Chapter 1 — Plumbing Formulae, Symbols, and Terminology

*(Table 1-3 continued)*

| Term | Text | Drawings | Program |
|---|---|---|---|
| Outside air | oa | OA | OA |
| Parts per million | ppm | PPM | PPM |
| Percent | % | % | PCT |
| Phase (electrical) | ph | PH | – |
| Pipe | – | – | PIPE |
| Pounds | lb | LBS | LBS |
| Pounds per square foot | psf | PSF | PSF |
| psf absolute | psfa | PSFA | PSFA |
| psf gage | psfg | PSFG | PSFG |
| Pounds per square inch | psi | PSI | PSI |
| psi absolute | psia | PSIA | PSIA |
| psi gage | psig | PSIG | PSIG |
| Pressure | – | PRESS | PRES, P |
| Pressure, barometric | baro pr | BARO PR | BP |
| Pressure, critical | – | – | CRIP |
| Pressure drop or difference | PD | PD | PD, DELTP |
| Pressure, dynamic (velocity) | vp | VP | VP |
| Pressure, static | sp | SP | SP |
| Pressure, vapor | vap pr | VAP PR | VAP |
| Primary | pri | PRI | PRIM |
| Quart | qt | QT | QT |
| Radian | – | – | RAD |
| Radiat(-e, -or) | – | RAD | – |
| Radiation | – | RADN | RAD |
| Radius | – | – | R |
| Rankine | °R | °R | R |
| Receiver | rcvr | RCVR | REC |
| Recirculate | recirc. | RECIRC | RCIR, RECIR |
| Refrigerant (12, 22, etc.) | R-12, R-22 | R12, R22 | R12, R22 |
| Relative humidity | rh | RH | RH |
| Resist(-ance, -ivity, -or) | res | RES | RES, OHMS |
| Return air | ra | RA | RA |
| Revolutions | rev | REV | REV |
| Revolutions per minute | rpm | RPM | RPM |
| Revolutions per second | rps | RPS | RPS |
| Roughness | rgh | RGH | RGH, E |
| Safety factor | sf | SF | SF |
| Saturation | sat. | SAT | SAT |
| Saybolt seconds Furol | ssf | SSF | SSF |
| Saybolt seconds Universal | ssu | SSU | SSU |
| Sea level | sl | SL | SE |

*(Continued)*

*(Table 1-3 continued)*

| Term | Text | Drawings | Program |
|---|---|---|---|
| Second | s | s | SEC |
| Sensible heat | SH | SH | SH |
| Sensible heat gain | SHG | SHG | SHG |
| Sensible heat ratio | SHR | SHR | SHR |
| Shading coefficient | – | – | SC |
| Shaft horsepower | sft hp | SFT HP | SHP |
| Solar | – | – | SOL |
| Specification | spec | SPEC | – |
| Specific gravity | SG | SG | – |
| Specific heat | sp ht | SP HT | C |
| sp ht at constant pressure | $c_p$ | $c_p$ | CP |
| sp ht at constant volume | $c_v$ | $c_v$ | CV |
| Specific volume | sp vol | SP VOL | V, CVOL |
| Square | sq. | SQ | SQ |
| Standard | std | STD | STD |
| Standard time meridian | – | – | STM |
| Static pressure | SP | SP | SP |
| Suction | suct. | SUCT | SUCT, SUC |
| Summ(-er, -ary, -ation) | – | – | SUM |
| Supply | sply | SPLY | SUP, SPLY |
| Supply air | sa | SA | SA |
| Surface | – | – | SUR, S |
| Surface, dry | – | – | SURD |
| Surface, wet | – | – | SURW |
| System | – | – | SYS |
| Tabulat(-e, -ion) | tab | TAB | TAB |
| Tee | – | – | TEE |
| Temperature | temp. | TEMP | T, TEMP |
| Temperature difference | TD, $\Delta t$ | TD | TD, TDIF |
| Temperature entering | TE | TE | TE, TENT |
| Temperature leaving | TL | TL | TL, TLEA |
| Thermal conductivity | $k$ | K | K |
| Thermal expansion coeff. | – | – | TXPC |
| Thermal resistance | $R$ | R | RES, R |
| Thermocouple | tc | TC | TC, TCPL |
| Thermostat | T STAT | T STAT | T STAT |
| Thick(-ness) | thkns | THKNS | THK |
| Thousand circular mils | Mcm | MCM | MCM |
| Thousand cubic feet | Mcf | MCF | MCF |
| Thousand foot-pounds | kip ft | KIP FT | KIPFT |
| Thousand pounds | kip | KIP | KIP |

*(Continued)*

# Chapter 1 — Plumbing Formulae, Symbols, and Terminology

*(Table 1-3 continued)*

| Term | Text | Drawings | Program |
|---|---|---|---|
| Time | – | T | T |
| Ton | – | – | TON |
| Tons of refrigeration | tons | TONS | TONS |
| Total | – | – | TOT |
| Total heat | tot ht | TOT HT | – |
| Transmissivity | – | – | TAU |
| U-factor | – | – | U |
| Unit | – | – | UNIT |
| Vacuum | vac | VAC | VAC |
| Valve | v | V | VLV |
| Vapor proof | vap prf | VAP PRF | – |
| Variable | var | VAR | VAR |
| Variable air volume | VAV | VAV | VAV |
| Velocity | vel. | VEL | VEL, V |
| Velocity, wind | w vel. | W VEL | W VEL |
| Ventilation, vent | vent | VENT | VENT |
| Vertical | vert. | VERT | VERT |
| Viscosity | visc | VISC | MU, VISC |
| Volt | V | V | E, VOLTS |
| Volt ampere | VA | VA | VA |
| Volume | vol. | VOL | VOL |
| Volumetric flow rate | – | – | VFR |
| Wall | – | – | W, WAL |
| Water | – | – | WTR |
| Watt | W | W | WAT, W |
| Watt-hour | Wh | WH | WHR |
| Weight | wt | WT | WT |
| Wet bulb | wb | WB | WB |
| Wet-bulb temperature | wbt | WBT | WBT |
| Width | – | – | WI |
| Wind | – | – | WD |
| Wind direction | wdir | WDIR | WDIR |
| Wind pressure | wpr | WPR | WP, WPRES |
| Yard | yd | YD | YD |
| Year | yr | YR | YR |
| Zone | z | Z | Z, ZN |

*Source*: ASHRAE, 1997, *Handbook of fundamentals.*

[a] Abbreviations of most proper names use capital letters in both text and drawings.

[b] The asterisk (*) is used with "ET," effective temperature.

[c] These are surface heat transfer coefficients.

[d] The letter "L" is also used for "logarithm of" these temperature differences in computer programming.

# PLUMBING TERMINOLOGY[4]

The following list of definitions and abbreviations that are frequently used in the plumbing industry has been compiled by the American Society of Plumbing Engineers for use by those working in this and related fields.

**ABS** Abbreviation for "acrylonitrile-butadiene-styrene."

***Absolute pressure*** The total pressure measured from absolute vacuum. It equals the sum of gauge pressure and atmospheric pressure corresponding to the barometer and is expressed in pounds per square inch (kiloPascals).

***Absolute temperature*** Temperature measured from absolute zero. A point of temperature theoretically equal to -459.73°F (-273.18°C). The hypothetical point at which a substance would have no molecular motion and no heat.

***Absolute zero*** Zero point on the absolute temperature scale. A point at which there is a total absence of heat, equivalent to -459.72°F (-273.18°C).

***Absorption*** Immersion in a fluid for a definite period of time, usually expressed as a percent of the weight of the dry pipe.

***Access door*** Hinged panel mounted in a frame with a lock, normally in a wall or ceiling, to provide access to concealed valves or equipment that require frequent attention.

***Accessible*** 1. a) (When applied to a fixture, connection, appliance, or piece of equipment) Having access thereto, though access may necessitate the removal of an access panel, door, or similar obstruction; b) (readily accessible) having direct access to without the necessity of removing or moving any panel, door, or similar obstruction. 2. (re: the physically challenged) Term used to describe a site, building, facility, or portion thereof, or a plumbing fixture that can be approached, entered, and/or used by physically challenged individuals.

***Accumulator*** A container in which fluid or gas is stored under pressure as a source of power.

***Acid vent*** A pipe venting an acid waste system.

***Acid waste*** A pipe that conveys liquid waste matter containing a pH of less than 7.0.

***Acme thread*** A screw thread, the thread section of which is between the square and V threads, used extensively for feed screws. The included angle of space is 29°, compared to 60° of the National Coarse of US Thread.

***Acrylonitrile-butadiene-styrene*** A thermoplastic compound from which fittings, pipe, and tubing are made.

***Active sludge*** Sewage sediment, rich in destructive bacteria, that can be used to break down fresh sewage more quickly.

***Adapter fitting*** 1. Any of various fittings designed to mate, or fit to each other, two pipes or fittings that are different in design, when the connection would otherwise be impossible. 2. A fitting that serves to connect two different tubes or pipes to each other, such as copper tube to iron pipe.

***Administrative authority*** The individual official, board, department, or agency established and authorized by a state, county, city, or other political subdivision created by law to administer and enforce the provisions of the plumbing code. *Also known as* AUTHORITY HAVING JURISDICTION.

***Aeration*** An artificial method of bringing water and air into direct contact with each other. One purpose is to release certain dissolved gases that often cause water to have obnoxious odors or disagreeable tastes. Also used to furnish oxygen to waters that are oxygen deficient. The process may be accomplished by spraying the liquid in the air, bubbling air through the liquid, or agitating the liquid to promote surface absorption of the air.

***Aerobic*** (re: bacteria) Living or active only in the presence of free oxygen.

**AGA** Abbreviation for "American Gas Association."

***Air break*** A physical separation in which a drain from a fixture, appliance, or device indirectly discharges into a fixture, receptacle, or interceptor at a point below the flood level rim of the receptacle to prevent backflow or back-siphonage. *Also known as* AIR GAP.

***Air chamber*** A continuation of the water piping beyond the branch to fixtures that is finished with a cap designed to eliminate shock or vibration (water hammer) of the piping when the faucet is closed suddenly.

---

[4]*Note:* The symbol "FP" is used to indicate definition or usage in the field of fire protection.

# Chapter 1 — Plumbing Formulae, Symbols, and Terminology

***Air, compressed*** Air at any pressure greater than atmospheric pressure.

***Air, free*** Air that is not contained and that is subject only to atmospheric conditions.

***Air gap*** The unobstructed vertical distance, through the free atmosphere, between the lowest opening from a pipe or faucet conveying water or waste to a tank, plumbing fixture receptor, or other device and the flood level rim of the receptacle. (Usually required to be a minimum of twice the diameter of the inlet.)

***Air, standard*** Air having a temperature of 70°F (21.1°C) at standard density of 0.0075 lb/ft (0.11 kg/m) and under pressure of 14.70 psia (101.4 kPa). The gas industry usually considers 60°F (15.6°C) the temperature of standard air.

***Air test*** A test using compressed air or nitrogen applied to a plumbing system upon its completion but before the building is sheetrocked.

***Alarm*** (FP) 1. Any audible or visible signal indicating existence of a fire or emergency requiring evacuation of occupants and response and emergency action on the part of the firefighting service. 2. The alarm device(s) by which fire and emergency signals are received.

***Alarm check valve*** (FP) A check valve, equipped with a signaling device, that will annunciate a remote alarm when a sprinkler head(s) is discharging.

***Alloy*** A substance composed of two or more metals or a metal and nonmetal intimately united, usually fused together and dissolving in each other when molten.

***Alloy pipe*** A steel pipe with one or more elements, other than carbon, that give it greater resistance to corrosion and more strength than carbon steel pipe.

***Ambient temperature*** The prevailing temperature in the immediate vicinity of or the temperature of the medium surrounding an object.

***American standard pipe thread*** A type of screw thread commonly used on pipe and fittings.

***Anaerobic*** (re: bacteria) Living or active in the absence of free oxygen.

***Anchor*** A device used to fasten or secure pipes to the building or structure.

***Angle of bend*** In a pipe, the angle between radial lines from the beginning and end of the bend to the center.

***Angle stop*** Common term for right-angle valves used to control water supplies to plumbing fixtures.

***Angle valve*** A device, usually of the globe type, in which the inlet and outlet are at right angles.

***ANSI*** Abbreviation for "American National Standards Institute."

***Approved*** Accepted or acceptable under an applicable specification or standard stated or cited for the proposed use under the procedures and authority of the administrative authority.

***Approved testing agency*** An organization established for purposes of testing to approved standards and acceptable to the administrative authority.

***Area drain*** A receptacle designed to collect surface or rainwater from a determined or calculated open area.

***Arterial vent*** A vent serving the building drain and the public sewer.

***ASHRAE*** Abbreviation for "American Society of Heating, Refrigerating and Air Conditioning Engineers."

***ASME*** Abbreviation for "American Society of Mechanical Engineers."

***ASPE*** Abbreviation for "American Society of Plumbing Engineers."

***ASPERF*** Abbreviation for "American Society of Plumbing Engineers Research Foundation."

***Aspirator*** A fitting or device supplied with water or other fluid under positive pressure that passes through an integral orifice or "constriction," causing a vacuum.

***ASSE*** Abbreviation for "American Society of Sanitary Engineering" or "American Society of Safety Engineers."

***ASTM*** Abbreviation for "American Society for Testing and Materials."

***Atmospheric vacuum breaker*** A mechanical device consisting of a check valve that opens to the atmosphere when the pressure in the piping drops to atmospheric.

***Authority having jurisdiction*** (FP) The organization, office, or individual responsible for approving equipment, materials, installation, or procedure.

***AWWA*** Abbreviation for "American Water Works Association."

***Backfill*** Material used to cover piping laid in an earthen trench.

***Backflow*** The flow of water or other liquids, mixtures, or substances from any source(s) other than the one(s) intended into the distributing pipes of a potable supply of water. *See* BACK-SIPHONAGE.

***Backflow connection*** A connection in any arrangement whereby backflow can occur.

***Backflow preventer*** A device or means to prevent backflow into the potable water system.

***Backing ring*** A metal strip used to prevent melted metal, from the welding process, from entering a pipe in the process of making a butt-welded joint.

***Back-siphonage*** The flowing back of used, contaminated, or polluted water from a plumbing fixture or vessel into the potable water supply pipe due to a negative pressure in the pipe. *See* BACKFLOW.

***Backup*** A condition where the waste water may flow back into another fixture or compartment but not backflow into the potable water system.

***Backwater valve*** A device that permits drainage in one direction but has a check valve that closes against back pressure. Sometimes used conjunctively with gate valves designed for sewage.

***Baffle plate*** A tray or partition placed in process equipment or tanks to direct or change the direction of flow.

***Ball check valve*** A device used to stop the flow of media in one direction while allowing flow in an opposite direction. The closure member used is spherical or ball-shaped.

***Ball valve*** A spherical gate valve providing very tight shut-off; a quick-closing (quarter-turn) valve.

***Barrier free*** *See* ACCESSIBLE, def. 2.

***Base*** The lowest portion or lowest point of a stack of vertical pipe.

***Battery of fixtures*** Any group of two or more similar, adjacent fixtures that discharge into a common horizontal waste or soil branch.

***Bell*** That portion of a pipe that, for a short distance, is sufficiently enlarged to receive the end of another pipe of the same diameter for the purpose of making a joint.

***Bell-and-spigot joint*** A commonly used joint in cast-iron soil pipe. Each piece is made with an enlarged diameter or bell at one end into which the plain or spigot end of another piece is inserted. The joint is then made tight by cement, oakum, lead, or rubber caulked into the bell around the spigot. *See also* HUB-AND-SPIGOT.

***Black pipe*** Steel pipe that has not been galvanized.

***Blank flange*** A solid plate flange used to seal off flow in a pipe.

***Boiler blow-off*** An outlet on a boiler to permit emptying or discharge of sediment.

***Boiler blow-off tank*** A vessel designed to receive the discharge from a boiler blow-off outlet and cool the discharge to a temperature that permits its safe discharge to the drainage system.

***Bonnet*** That part of a valve that connects the valve actuator to the valve body; in some valves, it may also contain the stem packing.

***Branch*** Any part of a piping system other than a main, riser, or stack.

***Branch interval*** A length of soil or waste stack corresponding, in general, to a story height, but in no case less than 8 ft (2.4 m), within which the horizontal branches from one floor or story of a building are connected to the stack.

***Branch tee*** A tee having one side branch.

***Branch vent*** A vent connecting one or more individual vents with a vent stack or stack vent.

***Brazing ends*** The ends of a valve or fitting that are prepared for silver brazing.

***Bronze trim or bronze-mounted*** An indication that certain internal (water contact) parts of the valves known as trim materials (stem, disc, seat rings, etc.) are made of copper alloy.

***Btu*** Abbreviation for "British thermal unit." The amount of heat required to raise the temperature of 1 pound (0.45 kg) of water 1 degree Fahrenheit (0.565°C).

***Btu/h*** Abbreviation for "British thermal units per hour."

***Bubble tight*** The condition of a valve seat that prohibits the leakage of visible bubbles when the valve is closed.

***Building (house)*** A structure built, erected, and framed of component structural parts designed for the housing, shelter, enclosure, or support of persons, animals, or property of any kind.

# Chapter 1 — Plumbing Formulae, Symbols, and Terminology

***Building (house) drain*** That part of the lowest piping of a drainage system that receives the discharge from soil, waste, and other drainage pipes inside the walls of the building (house) and conveys it to the building (house) sewer, which begins outside the building (house) walls.

***Building (house) drain, combined*** A building (house) drain that conveys both sewage and storm water or other drainage.

***Building (house) drain, sanitary*** A building (house) drain that conveys only sewage.

***Building (house) drain, storm*** A building (house) drain that conveys storm water or other drainage but no sewage.

***Building (house) sewer*** That part of the horizontal piping of a drainage system that extends from the end of the building (house) drain and receives the discharge from the building (house) drain and conveys it to a public sewer, private sewer, individual sewage-disposal system or other approved point of disposal.

***Building (house) subdrain*** That portion of a drainage system below the building (house) sewer that cannot drain by gravity in the building (house) sewer.

***Building (house) trap*** A device, fitting, or assembly of fittings installed in the building (house) drain to prevent circulation of air between the drainage of the building (house) and the building (house) sewer. It is usually installed as a running trap.

***Bull head tee*** A tee in which the branch is larger than the run.

***Burst pressure*** That pressure that can slowly be applied to a valve at room temperature for 30 seconds without causing rupture.

***Bushing*** A pipe fitting for connecting a pipe with a female fitting of a larger size. It is a hollow plug with internal and external threads. Used in lieu of a reducer/increaser.

***Butterfly valve*** A device deriving its name from the wing-like action of the disc, which operates at right angles to the flow. The disc impinges against the resilient liner with low-operating torque.

***Butt weld joint*** A welded pipe joint made with the ends of the two pipes butting each other, the weld being around the periphery.

***Butt weld pipe*** Pipe welded along a seam butted edge to edge and not scarfed or lapped.

***Bypass*** An auxiliary loop in a pipeline intended for diverting flow around a valve or other piece of equipment.

***Bypass valve*** A device used to divert the flow past the part of the system through which it normally passes.

***Capacity*** 1. The maximum or minimum flow obtainable under given conditions of media, temperature, pressure, velocity, etc. 2. The volume of media that may be stored in a container or receptacle.

***Capillary*** The action by which the surface of a liquid, where it is in contact with a solid, is elevated or depressed, depending on the relative attraction of the molecules of the liquid for each other and for those of the solid.

***Cathodic protection*** 1. The control of the electrolytic corrosion of an underground or underwater metallic structure by the application of an electric current in such a way that the structure is made to act as the cathode instead of the anode of an electrolytic cell. 2. The use of materials and liquid to cause electricity to flow to avoid corrosion.

***Caulking*** The method of rendering a joint tight against water or gas by means of applying plastic substances such as lead and oakum; a method of sealing between fixtures and adjacent surfaces.

***Cavitation*** A localized gaseous condition (usually involving air) that is found within a liquid stream.

***CDA*** Abbreviation for "Copper Development Association."

***Cement joint*** The union of two fittings by the insertion of material. Sometimes this joint is accomplished mechanically, sometimes chemically.

***Cesspool*** A lined excavation in the ground that receives the discharge of a drainage system, or part thereof, and is so designed to retain the organic matter and solids discharged therein but permit the liquids to seep through the bottom and sides.

***Chainwheel-operated valve*** A device that is operated by a chain-driven wheel that opens and closes the valve seats. Usually required for larger valves.

***Channel*** That trough through which any media may flow.

***Chase*** A recess in a wall or a space between two walls in which pipes can be run.

***Check valve*** A device designed to allow a fluid to pass through in one direction only.

***Chemical waste system*** Piping that conveys corrosive or harmful industrial, chemical, or processed wastes to the drainage system.

***Circuit*** The directed route taken by a flow of media from one point to another.

***Circuit vent*** A branch vent that serves two or more traps and extends from in front of the last fixture connection of a horizontal branch to the vent stack.

***CISPI*** Abbreviation for "Cast Iron Soil Pipe Institute."

***Clamp gate valve*** A gate valve whose body and bonnet are held together by a U-bolt clamp.

***Cleanout*** A plug or cover joined to an opening in a pipe that can be removed for the purpose of cleaning or examining the interior of the pipe.

***Clear-water waste*** Cooling water and condensate drainage from refrigeration and air-conditioning equipment; cooled condensate from steam-heating systems; cooled boiler blowdown water; waste-water drainage from equipment rooms and other areas where water is used without an appreciable addition of oil, gasoline, solvent, acid, etc.; and treated effluent in which impurities have been reduced below a minimum concentration considered harmful.

***Close nipple*** A nipple with a length twice the length of a standard pipe thread.

***Cock*** An original form of valve having a hole in a tapered plug that is rotated to provide passageway for fluid.

***Code*** Those regulations, subsequent amendments thereto, and any emergency rule or regulation that the department having jurisdiction may lawfully adopt.

***Coefficient of expansion*** The increase in unit length, area of volume for a 1-degree rise in temperature.

***Coliform group of bacteria*** All organisms considered in the coli aerogenes group as set forth by the American Water Works Association.

***Combination fixture*** A fixture that combines one sink and tray or a two- or three-compartment sink and/or tray in one unit.

***Combined waste and vent system*** A specially designed system of waste piping, embodying the horizontal wet venting of one or more sinks, floor sinks, or floor drains by means of a common waste and vent pipe, adequately sized to provide free movement of air above the flow line of the drain.

***Combustion efficiency*** The rated efficiency of a water heater or boiler determined by the equipment's ability to completely burn fuel, leaving no products of combustion in the flue gas.

***Common vent*** A vent that connects at the junction of two fixture drains and serves as a vent for both fixtures. *Also known as a* DUAL VENT.

***Companion flange*** A pipe flange to connect with another flange or with a flanged valve or fitting. It is attached to the pipe by threads, welding, or another method and differs from a flange that is an integral part of a pipe or fitting.

***Compression joint*** A multi-piece joint with cup-shaped, threaded nuts that, when tightened, compress tapered sleeves so that they form a tight joint on the periphery of the tubing they connect.

***Compressor*** A mechanical device for increasing the pressure of air or gas.

***Condensate*** Water that has liquefied (cooled) from steam.

***Conductor*** The piping from the roof to the building storm drain, combined building sewer or other approved means of disposal; it is located inside the building.

***Conduit*** A pipe or channel for conveying media.

***Confluent vent*** A vent serving more than one fixture vent or stack vent.

***Contaminator*** A medium or condition that spoils the nature or quality of another medium.

***Continuous vent*** A vent that is a continuation of the drain to which it connects.

***Continuous waste*** A drain from two or three fixtures connected to a single trap.

***Control*** A device used to regulate the function of a component or system.

***Controller*** (FP) The cabinet containing motor starter(s), circuit breaker(s), disconnect switch(es), and other control devices for the control of electric motors and internal-combustion-engine-driven fire pumps.

***Corporation cock*** A stopcock screwed into the street water main to supply the house service connection.

***Coupling*** A pipe fitting with female threads used only to connect two pipes in a straight line.

***CPVC*** Abbreviation for "chlorinated polyvinylchloride."

# Chapter 1 — Plumbing Formulae, Symbols, and Terminology

***Critical level*** The point on a backflow prevention device or vacuum breaker that determines the minimum elevation above the flood level rim of the fixture or receptacle served at which the device may be installed; the point conforms to approved standards and is established by the recognized (approved) testing laboratory (usually stamped or marked CL or C/L on the device by the manufacturer). When a backflow prevention device does not bear critical level marking, the bottom of the vacuum breaker or combination valve or the bottom of any such approved device shall constitute the critical level.

***Cross*** A pipe fitting with four branches in pairs, each pair on one axis, and the axis at right angles.

***Cross connection*** Any physical connection or arrangement between two otherwise separated piping systemsñone of which contains potable water and the other of which contains water or another substance of unknown or questionable safetyñwhereby flow may occur from one system to the other, the direction of flow depending on the pressure differential between the two systems. See BACKFLOW and BACK-SIPHONAGE.

***Crossover*** A pipe fitting with a double offset, or shaped like the letter "U" with the ends turned out, used to pass the flow of one pipe past another when the pipes are in the same plane.

***Cross valve*** A valve fitted on a transverse pipe so as to open communication between two parallel pipes.

***Crown*** That part of a trap in which the direction of flow is changed from upward to horizontal.

***Crown vent*** A vent pipe connected at the topmost point in the crown of a trap.

***CS*** Abbreviation for "Commercial Standards."

***Curb box*** A device at the curb that contains a valve that is used to shut off a supply line, usually of gas or water.

***Dampen*** 1. To check or reduce. 2. To deaden vibration.

***Dead end*** A branch leading from a soil, waste, or vent pipe; building (house) drain; or building (house) sewer that is terminated at a developed distance of 2 ft (0.6 m) or more by means of a plug or other closed fitting.

***Department having jurisdiction*** The administrative authority—and any other law enforcement agency affected by any provision of a code, whether such agency is specifically named or not.

***Detector, smoke*** (FP) Listed device for sensing visible or invisible products of combustion.

***Developed length*** The length along the center line of the pipe and fittings.

***Dewpoint*** The temperature of a gas or liquid at which condensation or evaporation occurs.

***Diameter*** Unless specifically stated otherwise, the nominal diameter as designated commercially.

***Diaphragm*** A flexible disc that is used to separate the control medium from the controlled medium and that actuates the valve stem.

***Diaphragm control valve*** A control valve having a spring-diaphragm actuator.

***Dielectric fitting*** A fitting having insulating parts or material that prohibits the flow of electric current. Used to separate dissimilar metals.

***Differential*** The variance between two target values, one of which is the high value of conditions, the other of which is the low value of conditions.

***Digestion*** That portion of the sewage treatment process where biochemical decomposition of organic matter takes place, resulting in the formation of simple organic and mineral substances.

***Disc*** That part of a valve that actually closes off the flow.

***Dishwasher*** An appliance for washing dishes, glassware, flatware, and some utensils.

***Displacement*** The volume or weight of a fluid, such as water, displaced by a floating body.

***Disposer*** A motor-driven appliance for reducing food and other waste by grinding so that it can flow through the drainage system.

***Domestic sewage*** The liquid and waterborne wastes derived from ordinary living processes that are free of industrial wastes and of such a character as to permit satisfactory disposal, without special treatment, into the public sewer or by means of a private sewage disposal system.

***Dosing tank*** A watertight tank in a septic system placed between the septic tank and the distribution box and equipped with a pump or automatic siphon designed to discharge sewage intermittently to a disposal field. This is done so that rest periods may be provided between discharges.

***Double disc*** A two-piece disc used in the gate valve. The wedges between the disc faces, upon contact with the seating faces in the valve, force them against the body seats to shut off the flow.

**Double offset** Two changes of direction installed in succession, or series, in continuous pipe.

**Double-ported valve** A valve having two ports to overcome line-pressure imbalance.

**Double-sweep tee** A tee made with easy (lon-radius) curves between body and branch.

**Double wedge** A device used in gate valves that is similar to a double disc in that the last downward turn of the stem spreads the split wedges, and each seals independently.

**Down** Term referring to piping running through the floor to a lower level.

**Downspout** The rainleader from the roof to the building storm drain, combined building sewer, or other means of disposal; it is located outside of the building.

**Downstream** Term referring to a location in the direction of flow after passing a referenced point.

**Drain** Any pipe that carries waste water or waterborne wastes in a building drainage system.

**Drain field** The area of a piping system arranged in troughs for the purpose of disposing unwanted liquid waste.

**Drainage fitting** A type of fitting used for draining fluid from pipes. The fitting makes possible a smooth and continuous interior surface for the piping system.

**Drainage system** The drainage piping within public or private premises (usually to 5 ft outside building walls) that conveys sewage, rainwater, or other liquid wastes to an approved point of disposal but does not include the mains of a public sewer system or a private or public sewage-treatment or disposal plant.

**Drift** The sustained deviation in a corresponding controller resulting from the predetermined relation between values and the controlled variable and positions of the final control element. *Also known as* WANDER.

**Droop** The amount by which the controlled variable pressure, temperature, liquid level, or differential pressure deviates from the set value at minimum controllable flow to the rated capacity.

**Drop** Term referring to piping running to a lower elevation within the same floor level.

**Drop elbow** A small elbow having wings cast on each side, the wings having countersunk holes so that they may be fastened by wood screws to a ceiling, wall, or framing timbers.

**Drop tee** A tee having wings of the same type as the drop elbow.

**Dross** 1. The solid scum that forms on the surface of a metal, as lead or antimony, when it is molten or melting, largely as a result of oxidation but sometimes because of the rising of dirt and impurities to the surface. 2. Waste or foreign matter mixed with a substance or left as a residue after that substance has been used or processed.

**Dry-bulb temperature** The temperature of air as measured by an ordinary thermometer.

**Dry-pipe valve** (FP) A valve used with a dry-pipe sprinkler system where water is on one side of the valve and air is on the other side. When a sprinkler head's fusible link melts, releasing air from the system, this valve opens, allowing water to flow to the sprinkler head.

**Dry-weather flow** Sewage collected during the summer that contains little or no ground water by infiltration and no storm water at the time of collection.

**Dry well** *See* LEACHING WELL.

**Dual vent** *See* COMMON VENT.

**Durham system** A term used to describe soil or waste systems where all piping is of threaded pipe, tubing or, other such material of rigid construction, and where recessed draining fittings corresponding to the type of piping are used.

**Durion** A high-silicon alloy that is resistant to practically all corrosive wastes. The silicon content is approximately 14.5% and the acid resistance is in the entire thickness of the metal.

**Dwelling** A one-family unit with or without accessory buildings.

**DWV** Abbreviation for "drainage, waste, and vent." A name for copper or plastic tubing used for drain, waste, or venting pipe.

**Eccentric fittings** Fittings where the openings are offset, allowing liquid to flow freely.

**Effective opening** The minimum cross-sectional area at the point of water-supply discharge, measured or expressed in terms of the diameter of a circle or, if the opening is not circular, the diameter of a circle of equivalent cross-sectional area. (This is applicable to an AIR GAP.)

**Effluent** Sewage, treated or partially treated, flowing out of sewage-treatment equipment.

**Elastic limit** The greatest stress that a material can withstand without permanent deformation after the release of the stress.

**Elbow (Ell)** A fitting that makes an angle between adjacent pipes. The angle is 90°, unless another angle is specified.

**Electrolysis** The process of producing chemical changes by passage of an electric current through an electrolyte (as in a cell), the ions present carrying the current by migrating to the electrodes where they may form new substances (as in the deposition of metals or the liberation of gases).

**Elutriation** A process of sludge conditioning in which certain constituents are removed by successive decontaminations with fresh water or plant effluent, thereby reducing the demand for conditioning chemicals.

**End connection** A reference to the method of connecting the parts of a piping system, e.g., threaded, flanged, butt-weld, socket-weld.

**Engineered plumbing system** Plumbing system designed by use of scientific engineering design criteria other than design criteria normally given in plumbing codes.

**Erosion** The gradual destruction of metal or other material by the abrasive action of liquids, gases, solids, or mixtures of these materials.

**Evapotranspiration** Loss of water from the soil by both evaporation and transpiration from the plants growing thereon.

**Existing work** A plumbing system, or any part thereof, that was installed prior to the effective date of an applicable code.

**Expansion joint** A joint whose primary purpose is to absorb longitudinal thermal expansion in the pipe line due to heat.

**Expansion loop** A large radius bend in a pipe line to absorb longitudinal thermal expansion in the line due to heat.

**Extra heavy** Description of piping material, usually cast-iron, indicating piping that is thicker than standard pipe.

**Face-to-face dimensions** The dimensions from the face of the inlet port to the face of the outlet port of a valve or fitting.

**Female thread** Internal thread in pipe fittings, valves, etc., for making screwed connections.

**Filter** A device through which fluid is passed to separate contaminants from it.

**Filter element or media** A porous device that performs the process of filtration, or filtering.

**Fire alarm system** (FP) A functionally related group of devices that, when automatically or manually activated, will sound audio or visual warning devices on or off the protected premises, signaling a fire.

**Fire department connection** (FP) A piping connection for fire department use to supplement in supplying water for standpipes and sprinkler systems. *See* STANDPIPE SYSTEM.

**Fire hazard** (FP) Any thing or act that increases, or will cause an increase of, the hazard or menace of fire to a degree greater than what is customarily recognized as normal by persons in the public service regularly engaged in preventing, suppressing, or extinguishing fire; or that will obstruct, delay, hinder, or interfere with the operations of the fire department or the egress of occupants in the event of fire.

**Fire hydrant valve** (FP) A valve that, when closed, drains at an underground level to prevent freezing.

**Fire line** (FP) A system of pipes and equipment used exclusively to supply water for extinguishing fires.

**Fire pump types**

   **Can pump** (FP) A vertical-shaft, turbine-type pump in a can (suction vessel) for installation in a pipeline to raise water pressure.

   **Centrifugal pump** (FP) A pump in which the pressure is developed principally by the action of centrifugal force.

   **End-suction pump** (FP) A single-suction pump having its suction nozzle on the opposite side of the casing from the stuffing box and having the face of the suction nozzle perpendicular to the longitudinal axis of the shaft.

   **Excess pressure pump** (FP) UL-listed and/or FM-approved, low-flow, high-head pump for sprinkler systems not being supplied from a fire pump. The pump pressurizes the sprinkler system so that the loss of water supply pressure will not cause a false alarm.

   **Fire pump** (FP) UL-listed and/or FM-approved pump with driver, controls, and accessories used for fire protection service. Fire pumps are of the centrifugal or turbine type and usually have an electric-motor or diesel-engine driver.

**Horizontal pump** (FP) A pump with the shaft normally in a horizontal position.

**Horizontal split-case pump** (FP) A centrifugal pump characterized by a housing that is split parallel to the shaft.

**In-line pump** (FP) A centrifugal pump in which the drive unit is supported by the pump, having its suction and discharge flanges on approximately the same center line.

**Pressure maintenance (jockey) pump** (FP) Pump with controls and accessories used to maintain pressure in a fire protection system without the operation of the fire pump. Does not have to be a listed pump.

**Vertical shaft turbine pump** (FP) A centrifugal pump with one or more impellers discharging into one or more bowls and a vertical educator or column pipe used to connect the bowl(s) to the discharge head on which the pump driver is mounted.

**Fitting** The connector or closure for fluid lines and passages.

**Fitting, compression** A fitting designed to join pipe or tubing by means of pressure or friction.

**Fitting, flange** A fitting that utilizes a radially extending collar for sealing and connection.

**Fitting, welded** A fitting attached by welding.

**Fixture branch** A pipe connecting several fixtures.

**Fixture carrier** A metal unit designed to support an off-the-floor plumbing fixture.

**Fixture carrier fittings** Special fittings for wall-mounted fixture carriers. Fittings have a sanitary drainage waterway with a minimum angle of 30 – 45° so that there are no fouling areas.

**Fixture drain** The drain from the trap of a fixture to the junction of that drain with any other drain pipe.

**Fixture, plumbing** See PLUMBING FIXTURE.

**Fixture supply** A water supply pipe connecting the fixture to the fixture branch or directly to a main water supply pipe.

**Fixture unit, drainage (dfu)** A measure of probable discharge into the drainage system by various types of plumbing fixtures. The drainage fixture unit value for a particular fixture depends on its volume rate of drainage discharge, on the time duration of a single drainage operation and on the average time between successive operations. Laboratory tests have shown that the rate of discharge of an ordinary lavatory with a nominal 1.2- in. (31.8-mm) outlet, trap, and waste is about 7.5 gpm (0.5 L/s). This figure is so near to 1 ft$^3$/min (0.5 L/s) that "1 ft$^3$/min" (0.5 L/s) has become the accepted flow rate of one fixture unit.

**Fixture unit, supply (sfu)** A measure of the probable hydraulic demand on the water supply by various types of plumbing fixtures. The supply fixture unit value for a particular fixture depends on its volume rate of supply, on the time duration of a single supply operation, and on the average time between successive operations.

**Flange** In pipe work, a ring-shaped plate on the end of a pipe at right angles to the end of the pipe and provided with holes for bolts to allow fastening the pipe to a similarly equipped adjoining pipe. The resulting joint is a flanged joint.

**Flange bonnet** A valve bonnet having a flange through which bolts connect it to a matching flange on the valve body.

**Flange ends** A valve or fitting having flanges for joining to other piping elements. Flange ends can be plain-faced, raised-face, large male-and-female, large tongue-and-groove, small tongue-and-groove or ring-joint.

**Flange faces** Pipe flanges that have the entire surface of the flange faced straight across and that use either a full-face or ring gasket.

**Flap valve** A non-return valve in the form of a hinged disc or flap, sometimes having leather or rubber faces.

**Flash point** The temperature at which a fluid first gives off sufficient flammable vapor to ignite when approached with a flame or spark.

**Float valve** A valve that is operated by means of a bulb or ball floating on the surface of a liquid within a tank. The rising and falling action operates a lever, which opens and closes the valve.

**Flooded** The condition when liquid rises to the flood level rim of a fixture.

**Flood level rim** The top edge of a receptacle or fixture from which water overflows.

**Flow pressure** The pressure in the water supply pipe near the water outlet while the faucet or water outlet is fully open and flowing.

**Flue** An enclosed passage, primarily vertical, for removal of gaseous products of combustion to the outer air.

# Chapter 1 — Plumbing Formulae, Symbols, and Terminology

***Flush valve*** A device located at the bottom of a tank for the purpose of flushing water closets and similar fixtures.

***Flushing-type floor drain*** A floor drain that is equipped with an integral water supply, enabling flushing of the drain receptor and trap.

***Flushometer valve*** A device that discharges a predetermined quantity of water to fixtures for flushing purposes and is actuated by direct water pressure.

***Footing*** The part of a foundation wall or column resting on the bearing soil, rock or piling that transmits the superimposed load to the bearing material.

***Foot valve*** A check valve installed at the base of a pump-suction pipe. Its purpose is to maintain pump prime by preventing pumped liquid from draining away from the pump.

***French drain*** A drain consisting of an underground passage made by filling a trench with loose stones and covering with earth. *Also known as* RUBBLE DRAIN.

***Fresh-air inlet*** A vent line connected with the building drain just inside the house trap and extending to the outer air. It provides fresh air at the lowest point of the plumbing system and with the vent stacks provides a ventilated system. A fresh-air inlet is not required where a septic-tank system of sewage disposal is employed.

***Frostproof closet*** A hopper that has no water in the bowl and has the trap and the control valve for its water supply installed below the frost line.

***FS*** Abbreviation for "federal specifications."

***Galvanic action*** When two dissimilar metals are immersed in the same electrolytic solution and connected electrically, there is an interchange of atoms carrying an electric charge between them. The anode metal with the higher electrode potential corrodes, the cathode is protected. Thus magnesium will protect iron. Iron will protect copper. *See also* ELECTROLYSIS.

***Galvanizing*** A process where the surface of iron or steel piping or plate is covered with a layer of zinc.

***Generally accepted standard*** A document referred to in a code that covers a particular subject and is accepted by the administrative authority.

***Grade*** The slope or fall of a line of pipe in reference to a horizontal plane. In drainage, it is expressed as the fall in a fraction of an inch or percentage slope per foot (mm/m) length of pipe.

***Grease interceptor*** An automatic or manual device used to separate and retain grease, with a capacity greater than 50 gal (227.3 L), and generally located outside a building.

***Grease trap*** An automatic or manual device used to separate and retain grease, with a capacity of 50 gal (227.3 L) or less, and generally located inside a building.

***Grinder pump*** A special class of solids-handling pump that grinds sewage solids to a fine slurry, rather than passing through entire spherical solids.

***Halon 1301*** (FP) Halon 1301 (bromtrifluoromethane CBrF3) is a colorless, odorless, electrically non-conductive gas that is an effective medium for extinguishing fires.

***Halon system types*** (FP) There are two types of systems recognized in this standard: "total flooding systems" and "local application systems."

   ***Total flooding system*** Consists of a supply of Halon 1301 arranged to discharged into, and fill to the proper concentration, an enclosed space or enclosure around the hazard.

   ***Local application system*** Consists of a supply of Halon 1301 arranged to discharge directly on the burning material.

***Hangers*** *See* SUPPORTS.

***Hub-and-spigot*** Piping made with an enlarged diameter or hub at one end and being plain or having a spigot at the other end. The joint is made tight by oakum and lead or by use of a neoprene gasket caulked or inserted in the hub around the spigot.

***Hubless*** Soil piping with plain ends. The joint is made tight with a stainless steel or cast-iron clamp and neoprene gasket assembly.

***Indirect waste pipe*** A pipe that does not connect directly with the drainage system but conveys liquid waste by discharging into a plumbing fixture or receptacle that is directly connected to the drainage system.

***Individual vent*** A pipe that is installed to vent a fixture trap and that connects with the vent system above the fixture served or terminates in the open air.

***Induced siphonage*** Loss of liquid from a fixture trap due to pressure differential between the inlet and outlet of a trap, often caused by the discharge of another fixture.

**Industrial waste** All liquid or waterborne waste from industrial or commercial processes except domestic sewage.

**Insanitary** A condition that is contrary to sanitary principles or is injurious to health.

**Interceptor** A device designed and installed so as to separate and retain deleterious, hazardous, or undesirable matter from normal wastes and to permit normal sewage or liquid wastes to discharge into the disposal terminal by gravity.

**Invert** Term referring to the lowest point on the interior of a horizontal pipe.

**Labeled** Term describing equipment or materials bearing a label of a listing agency.

**Lateral sewer** A sewer that does not receive sewage from any other common sewer except house connections.

**Leaching well** A pit or receptacle having porous walls that permit the contents to seep into the ground. *Also known as* DRY WELL.

**Leader** The water conductor from the roof to the building (house) storm drain. *Also known as* DOWNSPOUT.

**Liquid waste** The discharge from any fixture, appliance, or appurtenance in connection with a plumbing system that does not receive fecal matter.

**Listed** Term describing equipment and materials included in a list published by an organization acceptable to the authority having jurisdiction and concerned (a listing agency).

**Listing agency** An agency accepted by the administrative authority that lists or labels certain models of a product and maintains a periodic inspection program on the current production of listed models. It makes available a published report of its listing, including information indicating that the products have been tested, comply with generally accepted standards, and are found safe for use in a specified manner.

**Load factor** The percentage of the total connected fixture unit flow that is likely to occur at any point in the drainage system. The load factor represents the ratio of the probable load to the potential load and is determined by the average rates of flow of the various kinds of fixtures, the average frequency of use, the duration of flow during one use, and the number of fixtures installed.

**Loop vent** See VENT, LOOP.

**Main** The principal artery of a system of continuous piping to which branches may be connected.

**Main vent** A vent header to which vent stacks are connected.

**Malleable** Capable of being extended or shaped by beating with a hammer, or by the pressure of rollers. Most metals are malleable. The term "malleable iron" also has the older meaning (still universal in Great Britain) of "wrought iron," abbreviated "Mall."

**Master plumber** An individual who is licensed and authorized to install and assume responsibility for contractual agreements pertaining to plumbing and to secure any required permits. The journeyman plumber is allowed to install plumbing only under the responsibility of a master plumber.

**MSS** Abbreviation for "Manufacturers Standardization Society of the Valve and Fittings Industry, Inc."

**NFPA** Abbreviation for "National Fire Protection Association."

**NSF** Abbreviation for "National Sanitation Foundation Testing Laboratory."

**Offset** A combination of pipe(s) and/or fittings that join two approximately parallel sections of a line of pipe.

**Outfall sewers** Sewers receiving the sewage from a collection system and carrying it to the point of final discharge or treatment. They are usually the largest sewers of an entire system.

**Oxidized sewage** Sewage in which the organic matter has been combined with oxygen and has become stable in nature.

**PB** Abbreviation for "polybutylene."

**PDI** Abbreviation for "Plumbing and Drainage Institute."

**PE** Abbreviation for "polyethylene."

**Percolation** The flow or trickling of a liquid downward through a contact or filtering medium; the liquid may or may not fill the pores of the medium.

**Pitch** The amount of slope or grade given to horizontal piping and expressed in inches or vertically projected drop per foot (mm/m) on a horizontally projected run of pipe.

**Plumbing** The practice, materials, and fixtures used in the installation, maintenance, extension, and alteration of all piping, fixtures, appliances,

# Chapter 1 — Plumbing Formulae, Symbols, and Terminology

and appurtenances in connection with any of the following: sanitary drainage or storm drainage facilities; venting systems and public or private water-supply systems, within or adjacent to any building, structure or convenyance; water supply systems and/or the storm water, liquid waste, or sewage system of any premises to their connection with any point of public disposal or other acceptable terminal.

***Plumbing appliance*** A plumbing fixture that is intended to perform a special plumbing function. Its operation and/or control may be dependent upon one or more energized components, such as a motor, control, heating element, or pressure or temperature-sensing element. Such fixtures may operate automatically through one or more of the following actions: a time cycle, a temperature range, a pressure range, a measured volume or weight; or the fixture may be manually adjusted or controlled by the user or operator.

***Plumbing appurtenances*** A manufactured device, prefabricated assembly, or on-the-job assembly of component parts that is an adjunct to the basic piping system and plumbing fixtures. An appurtenance demands no additional water supply nor does it add any discharge load to a fixture or the drainage system. It is presumed that it performs some useful function in the operation, maintenance, servicing, economy, or safety of the plumbing system.

***Plumbing engineering*** The application of scientific principles to the design, installation, and operation of efficient, economical, ecological, and energy-conserving systems for the transport and distribution of liquids and gases.

***Plumbing fixtures*** Installed receptacles, devices, or appliances that are supplied with water or that receive liquid or liquid-borne wastes and discharge such wastes into the drainage system to which they may be directly or indirectly connected. Industrial or commercial tanks, vats, and similar processing equipment are not plumbing fixtures but may be connected to or discharged into approved traps or plumbing fixtures.

***Plumbing inspector*** Any person who, under the supervision of the department having jurisdiction, is authorized to inspect plumbing and drainage systems as defined in the code for the municipality and complying with the laws of licensing and/or registration of the state, city, or county.

***Plumbing system*** All potable water supply and distribution pipes, plumbing fixtures, and traps, drainage and vent pipe and building (house) drains, including their respective joints and connections, devices, receptacles, and appurtenances within the property lines of the premises and including potable water piping, potable water treating or using equipment, fuel gas piping, water heaters, and vents for same.

***Polymer*** A chemical compound or mixture of compounds formed by polymerization and consisting essentially of repeating structural units.

***Pool*** A water receptacle used for swimming or as a plunge or other bath, designed to accommodate more than one bather at a time.

***Potable water*** Water that is satisfactory for drinking, culinary, and domestic purposes and that meets the requirements of the health authority having jurisdiction.

***Precipitation*** The total measurable supply of water received directly from the clouds as snow, rain, hail, and sleet. It is expressed in inches (mm) per day, month, or year.

***Private sewage disposal system*** A septic tank with the effluent discharging into a subsurface disposal field, one or more seepage pits, or a combination of subsurface disposal field and seepage pit, or of such other facilities as may be permitted under the procedures set forth in a code.

***Private sewer*** A sewer that is privately owned and not directly operated by public authority.

***Private use*** Applies to plumbing fixtures in residences and apartments, private bathrooms in hotels and hospitals, and rest rooms in commercial establishments containing restricted-use single fixtures or groups of single fixtures and similar installations, where the fixtures are intended for the use of a family or an individual.

***Public sewer*** A common sewer directly operated by public authority.

***Public use*** Applies to toilet rooms and bathrooms used by employees, occupants, visitors or patrons, in or about any premises, and locked toilet rooms or bathrooms to which several occupants or employees on the premises possess keys and have access.

***Putrefaction*** Biological decomposition of organic matter with the production of ill-smelling products; usually takes place when there is a deficiency of oxygen.

**PVC** Abbreviation for "polyvinyl chloride."

**PVDF** Abbreviation for "polyvinyl-fluoridine."

**Raw sewage** Untreated sewage.

**Receptor** A plumbing fixture or device of such material, shape, and capacity that it will adequately receive the discharge from indirect waste pipes and so constructed and located that it can be readily cleaned.

**Reduced size vent** Dry vents that are smaller than those allowed by model plumbing codes.

**Reducer** 1. A pipe fitting with inside threads that is larger at one end than at the other. 2. A fitting so shaped at one end that it can receive a larger size pipe in the direction of flow.

**Reflecting pool** A water receptacle used for decorative purposes.

**Relief vent** A vent designed to provide circulation of air between drainage and vent systems or to act as an auxiliary vent.

**Residual pressure** (FP) Pressure less than static that varies with the flow discharged from outlets.

**Return offset** A double offset installed to return the pipe to its original alignment.

**Revent pipe** That part of a vent pipe line that connects directly with an individual waste pipe or group of waste pipes, underneath or at the back of the fixture, and extends either to the main or branch vent pipe. *Also known as* INDIVIDUAL VENT.

**Rim** An unobstructed open edge of a fixture.

**Riser** 1. A water supply pipe that extends vertically one full story or more to convey water to branches or fixtures. 2. (FP) A vertical pipe used to carry water for fire protection to elevations above or below grade, such as a standpipe riser, sprinkler riser, etc.

**Roof drain** A drain installed to remove water collecting on the surface of a roof and discharge it into the leader (downspout).

**Roughing in** The installation of all parts of a plumbing system that can be completed prior to the installation of fixtures. This includes drainage, water supply and vent piping, and the necessary fixture supports.

**Sand filter** A water-treatment device for removing solid or colloidal material with sand as the filter medium.

**Sanitary sewer** A conduit or pipe carrying sanitary sewage. It may include storm water and infiltrated ground water.

**Seepage pit** A lined excavation in the ground that receives the discharge of a septic tank that is designed to permit effluent from the tank to seep through its bottom and sides.

**Septic tank** A watertight receptacle that receives the discharge of a drainage system, or part thereof, and is designed and constructed to separate solids from liquids and digest organic matter over a period of detention.

**Sewage** Any liquid waste containing animal, vegetable, or chemical wastes in suspension or solution.

**Sewage ejector** A mechanical device or pump for lifting sewage.

**Siamese** (FP) A hose fitting for combining the flow from two or more lines into a single stream. *See* FIRE DEPARTMENT CONNECTION.

**Side vent** A vent connected to the drain pipe through a fitting at an angle not greater than 45° to the vertical.

**Sludge** The accumulated, suspended solids of sewage deposited in tanks, beds, or basins, mixed with water to form a semiliquid mass.

**Soil pipe** Any pipe that conveys the discharge of water closets, urinals, or fixtures having similar functions, with or without the discharge from other fixtures, to the building (house) drain or building (house) sewer.

**Special wastes** Wastes that require some special method of handling, such as the use of indirect waste piping and receptors; corrosion resistant piping; sand, oil or grease interceptors; condensers; or other pretreatment facilities.

**Sprinkler system** (FP) An integrated system of underground and overhead piping designed in accordance with fire protection engineering standards. The installation includes one or more automatic water supplies. The portion of the sprinkler system above ground is a network of specially sized or hydraulically designed piping installed in a building, structure, or area, generally overhead, and to which sprinklers are attached in a systematic pattern. The valve controlling each system riser is located in the system riser or its supply piping. Each sprinkler system riser includes a device for actuating an alarm when the system is in operation. The system is activated by heat from a fire and discharges water over the fire area.

# Chapter 1 — Plumbing Formulae, Symbols, and Terminology

## Sprinkler system classification

### Automatic sprinkler system types (FP)

1. Wet-pipe systems.
2. Dry-pipe systems.
3. Pre-action systems.
4. Deluge systems.
5. Combined dry-pipe and pre-action systems.

**Sprinkler systems–special types** Special-purpose systems employing departures from the requirements of standards, such as special water supplies and reduced pipe sizing, shall be installed in accordance with their listings.

**Occupancy classification** Relates to sprinkler installations and their water supplies only, not intended to be a general classification of occupancy hazards.

1. *Extra hazard occupancies* Occupancies or portions of other occupancies where quantity and combustibility of contents is very high, and flammable and combustible liquids, dust, lint, or other materials are present, introducing the probability of rapidly developing fires with high rates of heat release. Extra hazard occupancies involve a wide range of variables that may produce severe fires. The following shall be used to evaluate the severity of extra hazard occupancies:

   A. *Extra hazard group 1* Includes occupancies with little or no flammable or combustible liquids.

   B. *Extra hazard group 2* Includes occupancies with moderate to substantial amounts of flammable or combustible liquids or where shielding of combustibles is extensive.

2. *Ordinary hazard occupancies*

   A. *Ordinary hazard group 1* Occupancies or portions of other occupancies where combustibility is low, quantity of combustibles does not exceed 8 ft (2.4 m), and fires with moderate rates of heat release are expected.

   B. *Ordinary hazard group 2* Occupancies or portions of other occupancies where quantity and combustibility of contents is moderate, stockpiles do not exceed 12 ft (3.7 m), and fires with moderate rates of heat release are expected.

   C. *Ordinary hazard group 3* Occupancies or portions of other occupancies where quantity and/or combustibility of contents is high and fires of high rates of heat release are expected.

3. *Light hazard occupancies* Occupancies or portions of other occupancies where the quantity and/or combustibility of contents is low and fires with relatively low rates of heat release are expected.

### Sprinkler types (FP)

**Concealed sprinklers** Recessed sprinklers with cover plates.

**Corrosion-resistant sprinklers** Sprinklers with special coatings or platings to be used in an atmosphere that would corrode an uncoated sprinkler.

**Dry, pendent sprinklers** Sprinklers for use in a pendent position in a dry-pipe or wet-pipe system with the seal in a heated area.

**Dry, upright sprinklers** Sprinklers designed to be installed in an upright position, on a wet-pipe system, to extend into an unheated area with a seal in a heated area.

**Extended-coverage sidewall sprinklers** Sprinklers with special extended, directional, discharge patterns.

**Flush sprinklers** Sprinklers in which all or part of the body, including the shank thread, is mounted above the lower plane of the ceiling.

**Intermediate-level sprinklers** Sprinklers equipped with integral shields to protect their operating elements from the discharge of sprinklers installed at high elevations.

**Large-drop sprinklers** Listed sprinklers that are characterized by a K factor between 11.0 and 11.5 and a proven ability to meet the prescribed penetration, cooling, and distribution criteria prescribed in the large-drop sprinkler examination requirements. The deflector/discharge characteristics of the large-drop sprinkler generate large drops of such size and velocity as to enable effective penetration of a high-velocity fire plume.

**Nozzles** Devices for use in applications requiring special discharge patterns, directional spray, fine spray, or other unusual discharge characteristics.

**Open sprinklers** Sprinklers from which the actuating elements (fusible links) have been removed.

**Ornamental sprinklers** Sprinklers that have been painted or plated by the manufacturer.

**Pendant sprinklers** Sprinklers designed to be installed in such a way that the water stream is directed downward against the deflector.

**Quick-response sprinklers** A type of sprinkler that is both a fast-response and a spray sprinkler.

**Recessed sprinklers** Sprinklers in which all or a part of the body, other than the shank thread, is mounted within a recessed housing.

**Residential sprinklers** Sprinklers that have been specifically listed for use in residential occupancies.

**Sidewall sprinklers** Sprinklers having special deflectors that are designed to discharge most of the water away from a nearby wall in a pattern resembling a quarter of a sphere, with a small portion of the discharge directed at the wall behind the sprinkler.

**Special sprinklers** Sprinklers that have been tested and listed as having special limitations.

**Upright sprinklers** Sprinklers designed to be installed in such a way that the water spray is directed upward against the deflector.

**Stack** The vertical main of a system of soil, waste, or vent piping extending through one or more stories.

**Stack group** The location of fixtures in relation to the stack so that, by means of proper fittings, vents may be reduced to a minimum.

**Stack vent** The extension of a soil waste stack above the highest horizontal drain connected to the stack. *Also known as* WASTE or SOIL VENT.

**Stack venting** A method of venting a fixture or fixtures through the soil or waste stack.

**Stale sewage** Sewage that contains little or no oxygen and is free from putrefaction.

**Standpipe** A vertical pipe generally used for the storage and distribution of water for fire extinguishing.

**Standpipe system** (FP) An arrangement of piping, valves, hose connections, and allied equipment installed in a building or structure with the hose connections located in such a manner that water can be discharged in streams or spray patterns through attached hose and nozzles, for the purpose of extinguishing a fire and so protecting a building or structure and its contents as well as its occupants. This is accomplished by connections to water supply systems or by pumps, tanks, and other equipment necessary to provide an adequate supply of water to the hose connections.

**Standpipe system class of service** (FP)

*Class I* For use by fire departments and those trained in handling heavy fire streams (2 ½-in. hose).

*Class II* For use primarily by the building occupants until the arrival of the fire department (1 ½-in. hose).

*Class III* For use either by fire departments and those trained in handling heavy hose streams (2 ½-in. hose) or by the building occupants (1 ½-in. hose).

**Standpipe system types** (FP)

*Dry standpipe* A system having no permanent water supply, maybe so arranged through the use of approved devices as to admit water to the system automatically by the opening of a hose valve.

*Wet standpipe* A system having the supply valve open and water pressure maintained in the system at all times.

**Stop valve** A valve used for the control of water supply, usually to a single fixture. Can be a straight or angle configuration.

**Storm sewer** A sewer used for conveying rainwater, surface water, condensate, cooling water, or similar liquid wastes, exclusive of sewage and industrial waste.

**Strain** Change of the shape or size of a body produced by the action of stress.

**Stress** Reactions within a body resisting external forces acting on it.

**Subsoil drain** A drain that receives only subsurface or seepage water and conveys it to an approved place of disposal.

**Submain sewer** A sewer into which the sewage from two or more lateral sewers is discharged. *Also known as* BRANCH SEWER.

**Sump** A tank or pit that receives sewage or liquid waste, is located below the normal grade of the gravity system, and must be emptied by mechanical means.

**Sump pump** A mechanical device for removing liquid waste from a sump.

**Supervisory (tamper) switch** (FP) A device attached to the handle of a valve that, when the valve is closed, annunciates a trouble signal at a remote location.

**Supports** Devices for supporting and securing pipe and fixtures to walls, ceilings, floors, or structural members.

# Chapter 1 — Plumbing Formulae, Symbols, and Terminology

***Swimming pool*** A structure, basin, or tank containing water for swimming, diving, or recreation.

***Tempered water*** Water ranging in temperature from 85 to 110°F (29 to 43°C) thermal efficiency.

***Thermal efficiency*** The ratio of the energy output from the system to energy input to the system.

***Trailer park sewer*** That part of the horizontal piping of a drainage system that begins 2 ft (0.6 m) downstream from the last trailer site connection, receives the discharge of the trailer site, and conveys it to a public sewer, private sewer, individual sewage disposal system, or other approved point of disposal.

***Trap*** A fitting or device designed and constructed to provide, when properly vented, a liquid seal that will prevent the back passage of air without significantly affecting the flow of sewage or waste water through it.

***Trap primer*** A device or system of piping to maintain a water seal in a trap.

***Trap seal*** The maximum vertical depth of liquid that a trap will retain, measured between the crown weir and the top of the dip of the trap.

***Turbulence*** Any deviation from parallel flow in a pipe due to rough inner wall surfaces, obstructions, or directional changes.

***Underground piping*** Piping in contact with the earth below grade.

***Upstream*** Term referring to a location in the direction of flow before reaching a referenced point.

***Vacuum*** Any pressure less than that exerted by the atmosphere. *Also known as* NEGATIVE PRESSURE.

***Vacuum breaker*** *See* BACKFLOW PREVENTER.

***Vacuum relief valve*** A device to prevent excessive vacuum in a pressure vessel.

***Velocity*** Time rate of motion in a given direction and sense.

***Vent, loop*** Any vent connecting a horizontal branch or fixture drain with the stack vent of the originating waste or soil stack.

***Vent stack*** A vertical vent pipe installed primarily for the purpose of providing circulation of air to and from any part of the drainage system.

***Vertical pipe*** Any pipe or fitting that is installed in a vertical position or that makes an angle of not more than 45° with the vertical.

***Vitrified sewer pipe*** Conduit made of fired and glazed earthenware installed to receive waste or sewage or sewerage.

***Waste*** The discharge from any fixture, appliance, area, or appurtenance that does not contain fecal matter.

***Waste pipe*** The discharge pipe from any fixture, appliance, or appurtenance in connection with the plumbing system that does not contain fecal matter.

***Water-conditioning or treating device*** A device that conditions or treats a water supply to change its chemical content or remove suspended solids by filtration.

***Water-distributing pipe*** A pipe that conveys potable water from the building supply pipe to the plumbing fixtures and other water outlets in the building.

***Water hammer*** The forces, pounding noises, and vibration that develop in a piping system when a column of noncompressible liquid flowing through a pipeline at a given pressure and velocity is stopped abruptly.

***Water hammer arrester*** A device, other than an air chamber, designed to provide protection against excessive surge pressure.

***Water main*** The water supply pipe for public or community use. Normally under the jurisdiction of the municipality or water company.

***Water riser*** A water supply pipe that extends vertically one full story or more to convey water to branches or fixtures.

***Water-service pipe*** The pipe from the water main or other source of water supply to the building served.

***Water supply system*** The building supply pipe, the water distributing pipes, and the necessary connecting pipes, fittings, control valves, and all appurtenances carrying or supplying potable water in or adjacent to the building or premises.

***Wet vent*** A vent that also serves as a drain.

***Yoke vent*** A pipe connecting upward from a soil or waste stack to a vent stack for the purpose of preventing pressure changes in the stacks.

# RECOMMENDED PRACTICE FOR CONVERSION TO THE INTERNATIONAL SYSTEM OF UNITS

The International System of Units was developed by the General Conference of Weights and Measures, an international treaty organization, and has been officially abbreviated "SI" from the French term, "Systeme International and d'Unites." The SI system of units is a preferred international measurement system that evolved from earlier decimal metric systems.

When President Ford signed the Metric Conversion Act (Public Law 94-168) on December 23, 1975, a metric system in the United States was declared and a United States Metric Board was established to coordinate the national voluntary conversion effort to the metric system. The Metric Conversion Act specifically defines the metric system of measurement to be used as the International System of Units (SI), established by the General Conference of Weights and Measures and as interpreted and modified by the Secretary of Commerce.

The "recommended practice" section that follows outlines a selection of SI units, including multiples and submultiples, for use in plumbing design and related fields of science and engineering. It is intended to provide the technical basis for a comprehensive and authoritative standard guide for SI units to be used in plumbing design and related fields of science and engineering.

The section also is intended to provide the basic concepts and practices for the conversion of units given in several systems of measurement to the SI system. Rules and recommendations are detailed for the presentation of SI units and their corresponding symbols and numerical values used in conjunction with the SI system.

A selection of conversion factors to SI units for use in plumbing design and related fields of science and engineering is also given. It should be noted that the SI units, rules and recommendations listed herein comply with those provisions set forth in the American National Standard Metric Practice, ANSI Z210.1 (ASTM E380).

## Terminology and Abbreviations

For uniformity in the interpretation of the provisions set forth in this recommended practice section, the following definitions and abbreviations will apply:

***Accuracy*** The degree of conformity of a measured or calculated value to some recognized standard or specified value.

***Approximate value*** A quantity that is nearly but not exactly correct or accurate.

***CGPM*** Abbreviation for the General Conference on Weights and Measures, from the French term, "Conference Generale de Poids et Measures."

***Coherent unit system*** A system in which relations between units contain as numerical factor only the number 1 (or unity). All derived units have a unity relationship to the constituent base or supplementary units.

***Deviation*** The variation from a specified dimension or design requirement, defining the upper and lower limits.

***Digit*** One of the ten arabic numerals (0 to 9).

***Dimension*** A geometric element in a design or the magnitude of such a quantity.

***Feature*** An individual characteristic of a component or part.

***Nominal value*** A value assigned for the purpose of convenient designation, existing in name only.

***Precision*** The degree of mutual agreement between individual measurements, namely, repeatability and reproducibility.

***Significant digit*** Any digit that is necessary to define a value or quantity.

***Tolerance*** The total range of variation permitted; the upper and lower limits between which a dimension must be maintained.

***Unit*** The reference value of a given quantity as defined by CGPM.

## Types of Conversion

**Exact** These conversions denote the precise (or direct) conversion to the SI unit value, accurate to a number of decimal places.

**Soft** These conversions denote the conversion to the SI unit value in the software only. The materials and products remain unchanged and minimal rounding off to the nearest integer is usually applied.

**Hard** These conversions denote that the product or material characteristics are physically changed from existing values to preferred SI unit values.

# Chapter 1 — Plumbing Formulae, Symbols, and Terminology

## SI Units and Symbols[5]

The International System of Units has three types of units, as follows:

**Base units** These units are used for independent quantities. There are seven base units:

| Quantity | Unit | Symbol |
|---|---|---|
| Length | meter | m |
| Mass | kilogram | kg |
| Time | second | s |
| Current (electric) | ampere | A |
| Temperature (thermodynamic) | kelvin | K |
| Substance (amount) | mole | mol |
| Intensity (luminous) | candela | cd |

**Supplementary units** These units are used to denote angles. There are two supplementary units:

| Quantity | Unit | Symbol |
|---|---|---|
| Plane angle | radian | rad |
| Solid angle | steradian | sr |

**Derived units** These units are defined in terms of their derivation from base and supplementary units. Derived units are classified in two categories: (1) derived units with special names and symbols and (2) derived units with generic or complex names, expressed in terms of a base unit, two or more base units, base units and/or derived units with special names, or supplementary units and base and/or derived units.

| Quantity | Unit | Symbol |
|---|---|---|
| Frequency | hertz | Hz |
| Force | newton | N |
| Pressure, stress | pascal | Pa |
| Energy, work, heat (quantity) | joule | J |
| Power | watt | W |
| Electricity (quantity) | coulomb | C |
| Electric potential, electromotive force | volt | V |
| Electric capacitance | farad | F |
| Electric resistance | ohm | $\Omega$ |
| Magnetic flux | weber | Wb |
| Illuminance | lux | lx |
| Electric inductance | henry | H |
| Conductance | siemens | S |
| Magnetic flux density | tesla | T |
| Luminous flux | lumen | lm |

The following are classified as derived units with generic or complex names, expressed in various terms:

| Quantity | Unit | Symbol |
|---|---|---|
| Linear acceleration | meter per second sq. | m/s$^2$ |
| Angular acceleration | radian per second sq. | rad/s$^2$ |
| Area | meter squared | m$^2$ |
| Density | kilogram per cubic meter | kg/m$^3$ |
| Electric charge density | coulomb per cubic meter | C/m$^3$ |
| Electric permittivity | farad per meter | F/m |
| Electric permeability | henry per meter | H/m |
| Electric resistivity | ohm-meter | $\Omega \cdot$m |
| Entropy | joule per kelvin | J/K |
| Luminance | candela per meter sq. | cd/m$^2$ |
| Magnetic field strength | ampere per meter | A/m |
| Mass per unit length | kilogram per meter | kg/m |
| Mass per unit area | kilogram per meter sq. | kg/m$^2$ |
| Mass flow rate | kilogram per second | kg/s |
| Moment of inertia | kilogram-meter sq. | kg·m$^2$ |
| Momentum | kilogram-meter per sec. | kg·m/s |
| Torque | newton-meter | N·m |
| Specific heat | joule per kg per kelvin | J/kg·K |
| Thermal conductivity | watt per meter per kelvin | W/m·K |
| Linear velocity | meter per second | m/s |
| Angular velocity | radian per second | rad/s |
| Dynamic viscosity | pascal-second | Pa·s |
| Kinematic viscosity | meter squared per second | m$^2$/s |
| Volume, capacity | cubic meter | m$^3$ |
| Volume flow rate | cubic meter per second | m$^3$/s |
| Specific volume | cubic meter per kilogram | m$^3$/kg |

---

[5]The following units are the ones most commonly encountered in plumbing engineering and related fields of science and engineering. For additional units, readers are referred to ASNI Z210.1 (ASTM E380), available from the American National Standards Institute, 1430 Broadway, New York, NY 10017, or the American Society for Testing and Materials, 1916 Race Street, Philadelphia, PA 19103.)

## Non-SI Units and Symbols for Use with the SI System

There are several (non-SI) units that are traditional and acceptable for use in the SI system of units due to their significance in specific and general applications. These units are as follows:

| Quantity | Unit | Symbol |
|---|---|---|
| Area | hectare | ha |
| Energy | kilowatt-hour | kW·h |
| Mass | metric ton | t |
| Temperature | degree celsius | C |
| Time | minute, hour, year | min, h, y (respectively) |
| Velocity | kilometer per hour | km/h |
| Volume | liter | L |

## SI Unit Prefixes and Symbols

The SI unit system is based on multiples and submultiples. The following prefixes and corresponding symbols are accepted for use with SI units.

| Factor | Prefix | Symbol |
|---|---|---|
| $10^{18}$ | exa | E |
| $10^{15}$ | peta | P |
| $10^{12}$ | tera | T |
| $10^{9}$ | giga | G |
| $10^{6}$ | mega | M |
| $10^{3}$ | kilo | k |
| $10^{2}$ | hecto[a] | h |
| $10^{1}$ | deka[a] | da |
| $10^{-1}$ | deci[a] | d |
| $10^{-2}$ | centi[a] | c |
| $10^{-3}$ | milli | m |
| $10^{-6}$ | micro | µ |
| $10^{-9}$ | nano | n |
| $10^{-12}$ | pico | p |
| $10^{-15}$ | femto | f |
| $10^{-18}$ | atto | a |

[a]Use of these prefixes should be avoided whenever possible.

## SI Units Style and Use

1. Multiples and submultiples of SI units are to be formed by adding the appropriate SI prefixes to such units.

2. Except for the kilogram, SI prefixes are not to be used in the denominator of compound numbers.

3. Double prefixes are not to be used.

4. Except for exa (E), peta (P), teca (T), giga (G), and mega (M), SI prefixes are not capitalized.

5. The use of units from other systems of measurement is to be avoided.

6. Except when the SI unit is derived from a proper name, the symbol for SI units is not capitalized.

7. SI unit symbols are always denoted in singular form.

8. Except at the end of a sentence, periods are not used after SI units symbols.

9. Digits are placed in groups of three numbers, separated by a space, to the left and to the right of the decimal point. In the case of four digits, spacing is optional.

10. A center dot indicates multiplication and a slash indicates division (to the left of the slash is the numerator and to the right of the slash is the denominator).

11. When equations are used, such equations are to be restated using SI terms.

12. All units are to be denoted by either their symbols or their names written in full. Mixed use of symbols and names is not allowed.

# Chapter 1 — Plumbing Formulae, Symbols, and Terminology

## SI Unit Conversion Factors

To convert from other systems of measurement to SI values, the following conversion factors are to be used. (*Note*: For additional conversion equivalents not shown herein, refer to ANSI Z210.1–also issued as ASTM E380).

### Acceleration, linear

    foot per second squared = 0.3048 m/s$^2$      m/s$^2$ = 3.28 ft/s$^2$

    inch per second squared = 0.0254 m/s$^2$      m/s$^2$ = 39.37 in/s$^2$

### Area

    acre = 4046.9 m$^2$      m$^2$ = 0.0000247 acre

    foot squared = 0.0929 m$^2$      m$^2$ = 10.76 ft$^2$

    inch squared = 0.000645 m$^2$ = 645.16 mm$^2$      m$^2$ = 1550.39 in$^2$

    mile squared = 2 589 988 m$^2$ = 1.59      km$^2$ = 0.39 mi$^2$

    yard squared = 0.836 m$^2$      m$^2$ = 1.2 yd$^2$

### Bending movement (torque)

    pound-force-inch = 0.113 N·m      N·m = 8.85 lb$_f$-in

    pound-force-foot = 1.356 N·m      N·m = 0.74 lb$_f$-ft

### Bending movement (torque) per unit length

    pound-force-inch per inch = 4.448 N·m/m      N·m/m = 0.225 lb$_f$-in/in

    pound-force-foot per inch = 53.379 N·m/m      N·m/m = 0.019 lb$_f$-ft/in

### Electricity and magnetism

    ampere = 1A

    ampere-hour = 3600C

    coulomb = 1C

    farad = 1F

    henry = 1H

    ohm = 1Ω

    volt = 1V

### Energy (work)

    British Thermal Unit (Btu) = 1055 J      J = 0.000948 Btu

    foot-pound-force = 1.356 J      J = 0.74 ft-lb$_f$

    kilowatt-hour = 3 600 000 J      J = 0.000000278 kW-h

### Energy per unit area per unit time

    Btu per foot squared-second = 11 349 W/m$^2$      W/m$^2$ = 0.000088 Btu/ft$^2$-s

### Force

    ounce-force = 0.287 N      N = 3.48 oz$_f$

    pound-force = 4.448 N      N = 0.23 lb$_f$

    kilogram-force = 9.807 N      N = 0.1 kg$_f$

**Force per unit length**

    pound-force per inch = 175.1 N/m          N/m = 0.0057 $lb_f$/in

    pound-force per foot = 14.594 N/m         N/m = 0.069 $lb_f$/ft

**Heat**

    Btu-inch per second-foot squared-F = 519.2 W/m·K      W/m·K = 0.002 BTU-in/s-$ft^2$F

    BTU-inch per hour-foot squared-F = 0.144 W/m·K      W/m·K = 6.94 BTU-in/h-$ft^2$F

    BTU per foot squared = 11 357 J/$m^2$         J/$m^2$ = 0.000088 BTU/$ft^2$

    BTU per hour-foot squared-F = 5.678 W/$m^2$·K      W/$m^2$·K = 0.176 BTU/h-$ft^2$F

    BTU per pound-mass = 2326 J/kg         J/kg = 0.00043 BTU/$lb_m$

    BTU per pound-mass-F = 4186.8 J/kg·K         J/kg·K = 0.000239 BTU/$lb_m$ F

    F-hour-foot squared per BTU = 0.176 K·$m^2$/W      K·$m^2$/W = 5.68 F-h-$ft^2$/BTU

**Length**

    inch = 0.0254 m         m = 39.37 in

    foot = 0.3048 m         m = 3.28 ft

    yard = 0.914 m         m = 1.1 yd

    mile = 1609.3 m         m = 0.000621 mi

**Light (illuminance)**

    footcandle = 10.764 lx         lx = 0.093 ftcd

**Mass**

    ounce-mass = 0.028 kg         kg = 35.7 $oz_m$

    pound-mass = 0.454 kg         kg = 2.2 $lb_m$

**Mass per unit area**

    pound-mass per foot squared = 4.882 kg/$m^2$      kg/$m^2$ = 0.205 $lb_m$/$ft^2$

**Mass per unit length**

    pound-mass per foot = 1.488 kg/m         kg/m = 0.67 $lb_m$/ft

**Mass per unit time (flow)**

    pound-mass per hour = 0.0076 kg/s         kg/s = 131.58 $lb_m$/h

**Mass per unit volume (density)**

    pound-mass per cubic foot = 16.019 kg/$m^3$      kg/$m^3$ = 0.062 $lb_m$/$ft^3$

    pound-mass per cubic inch = 27 680 kg/$m^3$      kg/$m^3$ = 0.000036 $lb_m$/$in^3$

    pound-mass per gallon = 119.8 kg/$m^3$      kg/$m^3$ = 0.008347 $lb_m$/gal

**Moment of inertia**

    pound-foot squared = 0.042 kg·$m^2$         kg·$m^2$ = 23.8 lb-$ft^2$

**Plane angle**

    degree = 17.453 mrad         mrad = 0.057 deg

    minute = 290.89 μrad         μrad = 0.00344 min

    second = 4.848 μrad         μrad = 0.206 s

# Chapter 1 — Plumbing Formulae, Symbols, and Terminology 45

## Power

    Btu per hour = 0.293 W                                    W = 3.41 Btu/h

    foot-pound-force per hour = 0.38 mW            mW = 2.63 ft-$lb_f$/h

    horsepower = 745.7 W                                W = 0.00134 hp

## Pressure (stress), force per unit area

    inches water column = 25.4 mm water           mm water = 0.0394 in. wc

    atmosphere = 101.325 kPa                      kPa = 0.009869 atm

    inch of mercury (at 60°F) = 3.3769 kPa         kPa = 0.296 in. Hg

    inch of water (at 60°F) = 248.8 Pa              Pa = 0.004 in. $H_2O$

    pound-force per foot squared = 47.88 Pa        Pa = 0.02 $lb_f/ft^2$

    pound-force per inch squared = 6.8948 kPa     kPa = 0.145 $lb_f/in.^2$ (psi)

    pounds per square inch = 0.0703 kg/$cm^3$       kg/$cm^3$ = 14.22 psi

    pounds per square inch = 0.069 bars            bars = 14.50 psi

## Temperature equivalent

    $t_k = (t_f + 459.67)/1.8$                             $t_f = 1.8 t_k - 459.67$

    $t_c = (t_f - 32)/1.8$                                  $t_f = 1.8 t_c + 32$

## Velocity (length per unit time)

    foot per hour = 0.085 mm/s                      mm/s = 11.76 ft/h

    foot per minute = 5.08 mm/s                    mm/s = 0.197 ft/min

    foot per second = 0.3048 m/s                  m/s = 3.28 ft/s

    inch per second = 0.0254 m/s                  m/s = 39.37 in./s

    mile per hour = 0.447 m/s                       m/s = 2.24 mi/h

## Volume

    cubic foot = 0.028 $m^3$ = 28.317 L            $m^3$ = 35.71 $ft^3$

    cubic inch = 16 378 mL                          mL = 0.061 $in^3$

    gallon = 3.785 L                                 L = 0.264 gal

    ounce = 29.574 mL                             mL = 0.034 oz

    pint = 473.18 mL                                mL = 0.002 pt

    quart = 946.35 mL                               mL = 0.001 qt

    acre-foot = 1233.49 $m^3$                        $m^3$ = 0.00081 acre-ft

## Volume per unit time (flow)

    cubic foot per minute = 0.472 L/s               L/s = 2.12 $ft^3$/min

    cubic inch per minute = 0.273 mL/s            mL/s = 3.66 $in.^3$/min

    gallon per minute = 0.063 L/s                 L/s = 15.87 gal/min

    cubic feet per hour = 0.0283 $m^3$/h           $m^3$/h = 35.31 $ft^3$/h (cfh)

    cubic feet per hour = 0.007866 L/s           L/s = 127.13 cfh

## Table 1-4  Temperature Conversion Chart, °F – °C

The numbers in the center column refer to the known temperature, in either °F or °C, to be converted to the other scale. If converting from °F to °C, the number in the center column represents the known temperature, in °F, and its equivalent temperature, in °C, will be found in the left column. If converting from °C to °F, the number in the center represents the known temperature, in °C, and its equivalent temperature, in °F, will be found in the right column.

| Known Temp. ||| Known Temp. ||| Known Temp. ||| Known Temp. ||| Known Temp. |||
| --- | --- | --- | --- | --- | --- | --- | --- | --- | --- | --- | --- | --- | --- | --- |
| °C | (°F or °C) | °F | °C | (°F or °C) | °F | °C | (°F or °C) | °F | °C | (°F or °C) | °F | °C | (°F or °C) | °F |
| −59 | −74 | −101 | −34.4 | −30 | −22.0 | −10.0 | 14 | 57.2 | 14.4 | 58 | 136.4 | 43 | 110 | 230 |
| −58 | −73 | −99 | −33.9 | −29 | −20.2 | −9.4 | 15 | 59.0 | 15.0 | 59 | 138.2 | 49 | 120 | 248 |
| −58 | −72 | −98 | −33.3 | −28 | −18.4 | −8.9 | 16 | 60.8 | 15.6 | 60 | 140.0 | 54 | 130 | 266 |
| −57 | −71 | −96 | −32.8 | −27 | −16.6 | −8.3 | 17 | 62.6 | 16.1 | 61 | 141.8 | 60 | 140 | 284 |
| −57 | −70 | −94 | −32.2 | −26 | −14.8 | −7.8 | 18 | 64.4 | 16.7 | 62 | 143.6 | 66 | 150 | 302 |
| −56 | −69 | −92 | −31.6 | −25 | −13.0 | −7.2 | 19 | 66.2 | 17.2 | 63 | 145.4 | 71 | 160 | 320 |
| −56 | −68 | −90 | −31.1 | −24 | −11.2 | −6.7 | 20 | 68.0 | 17.8 | 64 | 147.2 | 77 | 170 | 338 |
| −55 | −67 | −89 | −30.5 | −23 | −9.4 | −6.1 | 21 | 69.8 | 18.3 | 65 | 149.0 | 82 | 180 | 356 |
| −54 | −66 | −87 | −30.0 | −22 | −7.6 | −5.6 | 22 | 71.6 | 18.9 | 66 | 150.8 | 88 | 190 | 374 |
| −54 | −65 | −85 | −29.4 | −21 | −5.8 | −5.0 | 23 | 73.4 | 19.4 | 67 | 152.6 | 93 | 200 | 392 |
| −53 | −64 | −83 | −28.9 | −20 | −4.0 | −4.4 | 24 | 75.2 | 20.0 | 68 | 154.4 | 99 | 210 | 410 |
| −53 | −63 | −81 | −28.3 | −19 | −2.2 | −3.9 | 25 | 77.0 | 20.6 | 69 | 156.2 | 100 | 212 | 414 |
| −52 | −62 | −80 | −27.7 | −18 | −0.4 | −3.3 | 26 | 78.8 | 21.1 | 70 | 158.0 | 104 | 220 | 428 |
| −52 | −61 | −78 | −27.2 | −17 | 1.4 | −2.8 | 27 | 80.6 | 21.7 | 71 | 159.8 | 110 | 230 | 446 |
| −51 | −60 | −76 | −26.6 | −16 | 3.2 | −2.2 | 28 | 82.4 | 22.2 | 72 | 161.6 | 116 | 240 | 464 |
| −51 | −59 | −74 | −26.1 | −15 | 5.0 | −1.7 | 29 | 84.2 | 22.8 | 73 | 163.4 | 121 | 250 | 482 |
| −50 | −58 | −72 | −25.5 | −14 | 6.8 | −1.1 | 30 | 86.0 | 23.3 | 74 | 165.2 | 127 | 260 | 500 |
| −49 | −57 | −71 | −25.0 | −13 | 8.6 | −0.6 | 31 | 87.8 | 23.9 | 75 | 167.0 | 132 | 270 | 518 |
| −49 | −56 | −69 | −24.4 | −12 | 10.4 | 0 | 32 | 89.6 | 24.4 | 76 | 168.8 | 138 | 280 | 536 |
| −48 | −55 | −67 | −23.8 | −11 | 12.2 | 0.6 | 33 | 91.4 | 25.0 | 77 | 170.6 | 143 | 290 | 554 |
| −48 | −54 | −65 | −23.3 | −10 | 14.0 | 1.1 | 34 | 93.2 | 25.6 | 78 | 172.4 | 149 | 300 | 572 |
| −47 | −53 | −63 | −22.7 | −9 | 15.8 | 1.7 | 35 | 95.0 | 26.1 | 79 | 174.2 | 154 | 310 | 590 |
| −47 | −52 | −62 | −22.2 | −8 | 17.6 | 2.2 | 36 | 96.8 | 26.7 | 80 | 176.0 | 160 | 320 | 608 |
| −46 | −51 | −60 | −21.6 | −7 | 19.4 | 2.8 | 37 | 98.6 | 27.2 | 81 | 177.8 | 166 | 330 | 626 |
| −45.6 | −50 | −58.0 | −21.1 | −6 | 21.2 | 3.3 | 38 | 100.4 | 27.8 | 82 | 179.6 | 171 | 340 | 644 |
| −45.0 | −49 | −56.2 | −20.5 | −5 | 23.0 | 3.9 | 39 | 102.2 | 28.3 | 83 | 181.4 | 177 | 350 | 662 |
| −44.4 | −48 | −54.4 | −20.0 | −4 | 24.8 | 4.4 | 40 | 104.0 | 28.9 | 84 | 183.2 | 182 | 360 | 680 |
| −43.9 | −47 | −52.6 | −19.4 | −3 | 26.6 | 5.0 | 41 | 105.8 | 29.4 | 85 | 185.0 | 188 | 370 | 698 |
| −43.3 | −46 | −50.8 | −18.8 | −2 | 28.4 | 5.6 | 42 | 107.6 | 30.0 | 86 | 186.8 | 193 | 380 | 716 |
| −42.8 | −45 | −49.0 | −18.3 | −1 | 30.2 | 6.1 | 43 | 109.4 | 30.6 | 87 | 188.6 | 199 | 390 | 734 |
| −42.2 | −44 | −47.2 | −17.8 | 0 | 32.0 | 6.7 | 44 | 111.2 | 31.1 | 88 | 190.4 | 204 | 400 | 752 |
| −41.7 | −43 | −45.4 | −17.2 | 1 | 33.8 | 7.2 | 45 | 113.0 | 31.7 | 89 | 192.2 | 210 | 410 | 770 |
| −41.1 | −42 | −43.6 | −16.7 | 2 | 35.6 | 7.8 | 46 | 114.8 | 32.2 | 90 | 194.0 | 216 | 420 | 788 |
| −40.6 | −41 | −41.8 | −16.1 | 3 | 37.4 | 8.3 | 47 | 116.6 | 32.8 | 91 | 195.8 | 221 | 430 | 806 |
| −40.0 | −40 | −40.0 | −15.6 | 4 | 39.2 | 8.9 | 48 | 118.4 | 33.3 | 92 | 197.6 | 227 | 440 | 824 |
| −39.4 | −39 | −38.2 | −15.0 | 5 | 41.0 | 9.4 | 49 | 120.2 | 33.9 | 93 | 199.4 | 232 | 450 | 842 |
| −38.9 | −38 | −36.4 | −14.4 | 6 | 42.8 | 10.0 | 50 | 122.0 | 34.4 | 94 | 201.2 | 238 | 460 | 860 |
| −38.3 | −37 | −34.6 | −13.9 | 7 | 44.6 | 10.6 | 51 | 123.8 | 35.0 | 95 | 203.0 | 243 | 470 | 878 |
| −37.8 | −36 | −32.8 | −13.3 | 8 | 46.4 | 11.1 | 52 | 125.6 | 35.6 | 96 | 204.8 | 249 | 480 | 896 |
| −37.2 | −35 | −31.0 | −12.8 | 9 | 48.2 | 11.7 | 53 | 127.4 | 36.1 | 97 | 206.6 | 254 | 490 | 914 |
| −36.7 | −34 | −29.2 | −12.2 | 10 | 50.0 | 12.2 | 54 | 129.2 | 36.7 | 98 | 208.4 | 260 | 500 | 932 |
| −36.1 | −33 | −27.4 | −11.7 | 11 | 51.8 | 12.8 | 55 | 131.0 | 37.2 | 99 | 210.2 | | | |
| −35.5 | −32 | −25.6 | −11.1 | 12 | 53.6 | 13.3 | 56 | 132.8 | 37.8 | 100 | 212.0 | | | |
| −35.0 | −31 | −23.8 | −10.6 | 13 | 55.4 | 13.9 | 57 | 134.6 | 38 | 100 | 212 | | | |

# Chapter 1 — Plumbing Formulae, Symbols, and Terminology

## Table 1-5  Conversion to SI Units

| Multiply | By | To Obtain |
|---|---|---|
| acre | 0.4047 | ha |
| atmosphere (standard) | 101.325[a] | kPa |
| bar | 100[a] | kPa |
| barrel (42 US gal, petroleum) | 159 | L |
|  | 0.159 | $m^3$ |
| Btu (International Table) | 1.055 | kJ |
| Btu/$ft^2$ | 11.36 | kJ/$m^2$ |
| Btu/$ft^3$ | 37.3 | kJ/$m^3$ |
| Btu/gal | 279 | kJ/$m^3$ |
| Btu · ft/h · $ft^2$ · °F | 1.731 | W/(m · K) |
| Btu · in/h · $ft^2$ · °F (thermal conductivity, $k$) | 0.1442 | W/(m · K) |
| Btu/h | 0.2931 | W |
| Btu/h · $ft^2$ | 3.155 | W/$m^2$ |
| Btu/h · $ft^2$ · °F (overall heat transfer coefficient, $U$) | 5.678 | W/($m^2$ · K) |
| Btu/lb | 2.326[a] | kJ/kg |
| Btu/lb · °F (specific heat, $c_p$) | 4.184 | kJ/(kg · K) |
| bushel | 0.03524 | $m^3$ |
| calorie, gram | 4.184 | J |
| calorie, kilogram (kilocalorie) | 4.184 | kJ |
| centipoise (dynamic viscosity, µ) | 1.00[a] | mPa · s |
| centistokes (kinematic viscosity, $v$) | 1.00[a] | $mm^2$/s |
| clo | 0.155 | $m^2$ · K/W |
| dyne/$cm^2$ | 0.100[a] | Pa |
| EDR hot water (150 Btu/h) | 44.0 | W |
| EDR steam (240 Btu/h) | 70.3 | W |
| EER | 0.293 | COP |
| ft | 0.3048[a] | m |
|  | 304.8[a] | mm |
| ft/min, fpm | 0.00508[a] | m/s |

| To Obtain | By | Divide |
|---|---|---|

| Multiply | By | To Obtain |
|---|---|---|
| ft/s, fps | 0.3048[a] | m/s |
| ft of water | 2.99 | kPa |
| ft of water per 100 ft pipe | 0.0981 | kPa/m |
| $ft^2$ | 0.09290 | $m^2$ |
| $ft^2$ · h · °F/Btu (thermal resistance, $R$) | 0.176 | $m^2$ · K/W |
| $ft^2$/s (kinematic viscosity, $v$) | 92.900 | $mm^2$/s |
| $ft^3$ | 28.32 | L |
|  | 0.02832 | $m^3$ |
| $ft^3$/min, cfm | 0.4719 | L/s |
| $ft^3$/s, cfs | 28.32 | L/s |
| ft · $lb_f$ (torque or moment) | 1.356 | N · m |
| ft · $lb_f$ (work) | 1.356 | J |
| ft · $lb_f$/lb (specific energy) | 2.99 | J/kg |
| ft · $lb_f$/min (power) | 0.0226 | W |
| footcandle | 10.76 | lx |
| gallon (US, 231 $in^3$) | 3.7854[a] | L |
| gph | 1.05 | mL/s |
| gpm | 0.0631 | L/s |
| gpm/$ft^2$ | 0.6791 | L/(s · $m^2$) |
| gpm/ton refrigeration | 0.0179 | mL/J |
| grain (1/7000 lb) | 0.0648 | g |
| gr/gal | 17.1 | g/$m^3$ |
| gr/lb | 0.143 | g/kg |
| horsepower (boiler) (33,470 Btu/h) | 9.81 | kW |
| horsepower (550 ft · $lb_f$/s) | 0.746 | kW |
| inch | 25.4[a] | mm |
| in of mercury (60°F) | 3.377 | kPa |
| in of water (60°F) | 249 | Pa |
| in/100 ft, thermal expansion | 0.833 | mm/m |
| in · $lb_f$ (torque or moment) | 113 | mN · m |
| $in^2$ | 645 | $mm^2$ |
| $in^3$ (volume) | 16.4 | mL |

| To Obtain | By | Divide |
|---|---|---|

*(Continued)*

*(Table 1-5 continued)*

| Multiply | By | To Obtain |
|---|---|---|
| in³/min (SCIM) | 0.273 | mL/s |
| in³ (section modulus) | 16,400 | mm³ |
| in⁴ (section moment) | 416,200 | mm⁴ |
| km/h | 0.278 | m/s |
| kWh | 3.60[a] | MJ |
| kW/1000 cfm | 2.12 | kJ/m³ |
| kilopond (kg force) | 9.81 | N |
| kip (1000 lb$_f$) | 4.45 | kN |
| kip/in² (ksi) | 6.895 | MPa |
| litre | 0.001[a] | m³ |
| met | 58.15 | W/m² |
| micron (µm) of mercury (60°F) | 133 | mPa |
| mile | 1.609 | km |
| mile, nautical | 1.852[a] | km |
| mph | 1.609 | km/h |
| | 0.447 | m/s |
| millibar | 0.100[a] | kPa |
| mm of mercury (60°F) | 0.133 | kPa |
| mm of water (60°F) | 9.80 | Pa |
| ounce (mass, avoirdupois) | 28.35 | g |
| ounce (force, thrust) | 0.278 | N |
| ounce (liquid, US) | 29.6 | mL |
| ounce inch (torque, moment) | 7.06 | mN · m |
| ounce (avoirdupois) per gallon | 7.49 | kg/m³ |
| perm (permeance) | 57.45 | ng/(s · m² · Pa) |
| perm inch (permeability) | 1.46 | ng/(s · m · Pa) |
| pint (liquid, US) | 473 | mL |
| pound | | |
|   lb (mass) | 0.4536 | kg |
| | 453.6 | g |
|   lb$_f$ (force, thrust) | 4.45 | N |
|   lb/ft (uniform load) | 1.49 | kg/m |

| To Obtain | By | Divide |
|---|---|---|

| Multiply | By | To Obtain |
|---|---|---|
| lb$_m$/ft · h (dynamic viscosity, µ) | 0.413 | mPa · s |
| lb$_m$/ft · s (dynamic viscosity, µ) | 1490 | mPa · s |
| lb$_f$ · s/ft² (dynamic viscosity, µ) | 47.88 | Pa · s |
| lb/h | 0.126 | g/s |
| lb/min | 0.00756 | kg/s |
| lb/h [steam at 212°F (100°C)] | 0.284 | kW |
| lb$_f$/ft² | 47.9 | Pa |
| lb/ft² | 4.88 | kg/m² |
| lb/ft³ (density, ρ) | 16.0 | kg/m³ |
| lb/gallon | 120 | kg/m³ |
| ppm (by mass) | 1.00[a] | mg/kg |
| psi | 6.895 | kPa |
| quad (10¹⁵ Btu) | 1.055 | EJ |
| quart (liquid, US) | 0.946 | L |
| square (100 ft²) | 9.29 | m² |
| tablespoon (approx.) | 15 | mL |
| teaspoon (approx.) | 5 | mL |
| therm (US) | 105.5 | MJ |
| ton, long (2240 lb) | 1.016 | Mg |
| ton, short (2000 lb) | 0.907 | Mg; t (tonne) |
| ton, refrigeration (12,000 Btu/h) | 3.517 | kW |
| ton (1 mm Hg at 0°C) | 133 | Pa |
| watt per square foot | 10.76 | W/m² |
| yd | 0.9144[a] | m |
| yd² | 0.836 | m² |
| yd³ | 0.7646 | m³ |

| To Obtain | By | Divide |
|---|---|---|

*Notes*: 1. Units are US values unless noted otherwise. 2. Litre is a special name for the cubic decimetre. 1 L = dm³ and 1 mL = 1 cm³.

[a] Conversion factor is exact.

## REFERENCES

1. American Society of Heating, Refrigerating and Air-Conditioning Engineers (ASHRAE). 1997 [1982]. *Handbook of fundamentals.* Atlanta, GA: ASHRAE.

2. Baumeister, Theodore, and Lionel S. Marks. *Standard handbook for mechanical engineers.* New York: McGraw-Hill.

3. Chan, Wen-Yung W., and Milton Meckler. 1983. *Pumps and pump systems.* Sherman Oaks, CA: American Society of Plumbing Engineers.

4. National Fire Protection Association (NFPA). Standard 170.

5. Steele, Alfred. 1982. *Engineered plumbing design.* Elmhurst, IL: Construction Industry Press.

# LOOKING FOR QUALITY PLUMBING PRODUCTS THAT WON'T LET YOU DOWN?

# LET ME INTRODUCE MY FAMILY.

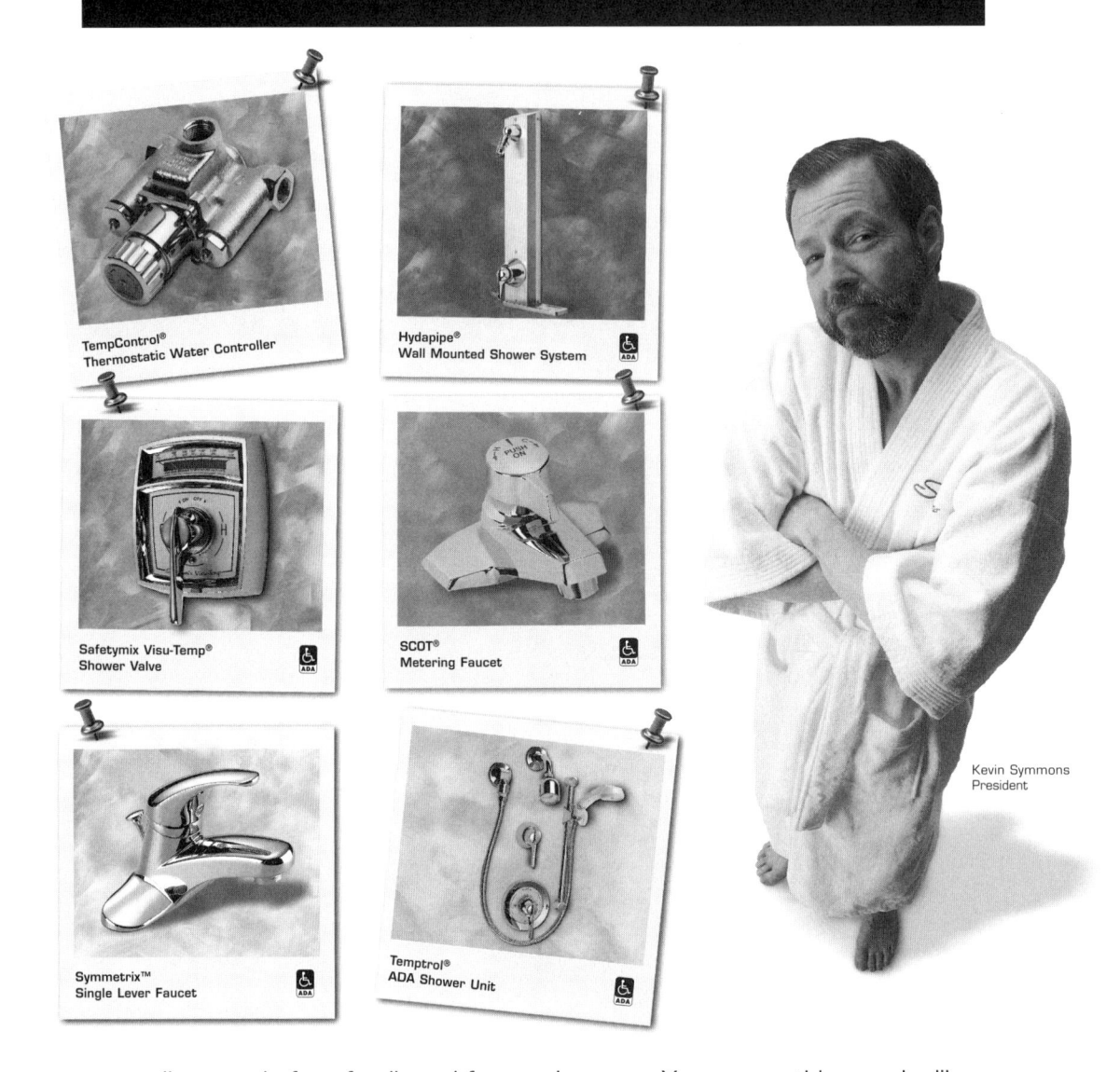

**TempControl®**
Thermostatic Water Controller

**Hydapipe®**
Wall Mounted Shower System

**Safetymix Visu-Temp®**
Shower Valve

**SCOT®**
Metering Faucet

**Symmetrix™**
Single Lever Faucet

**Temptrol®**
ADA Shower Unit

Kevin Symmons
President

I'm proud of my family and for good reason. You see, nothing works like a Symmons product. Meticulously engineered, constructed of the finest materials, rigorously tested to the highest standards of performance and dependability, a Symmons valve, faucet, or shower system is made to work. And made to last.

So, next time you're specifying plumbing products, specify the name you can count on. Symmons…because quality runs in our family.

## Symmons®
*Quality Runs In Our Family*

©1998 Symmons Industries, Inc., 31 Brooks Drive, Braintree, MA 02184-3804  1-800-SYMMONS  Fax: 1-800-961-9621
www.symmons.com

# 2 Standards for Plumbing Materials and Equipment

The codes and standards listed in Table 2-1 represent practices, methods, and standards sponsored by the organizations indicated. They are valuable guides to help the plumbing design engineer determine test methods, ratings, performance requirements, and limits that apply to the equipment in plumbing systems design. Copies can usually be obtained from the organizations whose addresses and abbreviations are listed in Table 2-2. Design engineers must check with local governing codes and authorities to determine which standards (and dates) are accepted.

### Table 2-1 Standards for Plumbing Materials and Equipment

| Description | ANSI | ASTM | FS | Other |
|---|---|---|---|---|
| ***Ferrous Pipe and Fittings*** | | | | |
| Black and galvanized steel pipe, welded and seamless | | A53-90b | WW-P-406d(1)-73 | |
| Buttwelding ends | B16.25-86 | | | |
| Cast-iron fittings for sovent drainage systems | B16.45-87 | | | |
| Cast-iron pipe flanges and flanged fittings | B16.1-89 | | | |
| Cast-iron pressure fittings | A21.10-87 | A377-89 | | AWWA C110-77 |
| Cast-iron pressure pipe | A21.1-82 | A377-89 | | |
| Cast-iron soil pipe and fittings including hubless cast-iron pipe and fittings | A112.5-73 | A74-94 | WW-P-401e-74 | CISPI 301-85 |
| Cast-iron threaded fittings | B16.4 - 85 | A126-93 | WW-P-50lb-74 | |
| Cast-iron threaded drainage fittings | B16.12-91 | | WW-P-491b-67 | IAPMO PS 5-77 |
| Ductile iron pipe | A21.51-76 | A377-89 | WW-P-421c-67 | AWWA C151-81 |
| Ductile iron pipe flanges and flanged fittings, classes 150 and 300 | B16.42-87 | | | |
| Factory-made wrought steel buttwelding fittings | B16.9-86 | | | |
| Ferrous pipe plugs, bushings, and locknuts with pipe threads | B16.14-91 | | WW-P-471b-70 | |
| Forged steel fittings, socket-welding and threaded | B16.11-91 | | | |

*(Continued)*

(Table 2-1 continued)

| Description | ANSI | ASTM | FS | Other |
|---|---|---|---|---|
| Hubless cast-iron soil pipe and fittings (installation) | | | | IAPMO IS 6-89 |
| Malleable iron fittings (threaded) | | A197/A 197M-87 | | |
| Malleable iron threaded fittings | B16.3-85 | | WW-P-521f-77 | |
| Malleable iron threaded pipe unions | B16.39-86 | | WW-U-531c-65 | |
| Nipples—pipe (threaded) | | | WW-P-351b(1)-70 | |
| Orifice flanges | B16.36-88 | | | |
| Pipe flanges and flanged fittings | B16.5-88 | | | |
| Pipe threads (except dryseal) | B1.20.1-83 | | | |
| Pipe threads | B1.20.1-83 | | | |
| Ring-joint gaskets and grooves for steel pipe flanges | B16.20-73 | | | |
| Stainless steel water pipe, grade H | | A268/A 268M-91 | | |
| Subdrains for built-up shower pans | | | | IAPMO PS 16-90 |
| Welded and seamless steel pipe | | A53-90b | WW-P-471b-70 | |
| | | A120-83 | WW-P-406d(1)-73 | |
| Wrought steel buttwelding short radius elbows and returns | B16.28-86 | | | |

***Nonferrous Pipe and Fittings***

| Description | ANSI | ASTM | FS | Other |
|---|---|---|---|---|
| Cast bronze threaded fittings | B16.15-85 | | WW-P-460c-87 | |
| Cast copper alloy fittings for flared copper tubes | B16.26-88 | | | |
| Cast copper alloy pipe flanges and flanged fittings | B16.24-91 | | | |
| Cast copper alloy solder joint drainage fittings (DWV) | B16.23-84 | | | |
| Cast copper alloy solder joint fittings for solvent drainage systems | B16.32-84 | | | |
| Cast copper alloy solder joint pressure fittings | B16.18-84 | | | |
| Copper and copper alloy welded water tubes (installation) | | | | IAPMO IS 21-89 |
| Copper drainage tube (DWV) | | B306-88 | | |
| Copper plumbing tube, pipe and fittings (installation) | | | | IAPMO IS 3-89 |
| Diversion tees and twin waste elbows | | | | IAPMO PS 9-84 |
| Drains for prefabricated and precast showers | | | | IAPMO PS 4-90 |
| Flexible copper water connectors | | | | IAPMO PS 14-89 |
| Seamless brass tube | | B135-91b | WW-T-791a-71 | |
| Seamless copper pipe | | B42-92 | WW-P-377d-76 | |
| Seamless copper tube | | B75-92a | | |
| Seamless copper water tube (K, L, and M) | | B88-93a | WW-T-799f-79 | |

*(Continued)*

# Chapter 2 — Standards for Plumbing Materials and Equipment

*(Table 2-1 continued)*

| Description | ANSI | ASTM | FS | Other |
|---|---|---|---|---|
| Seamless red brass pipe | | B43-91 | WW-P-351a-75 | |
| Seamless and welded copper distribution tube (type D) | | B641-92 | | |
| Threadless copper pipe | | B302-87 | | |
| Tubing trap wall adapters | | | | IAPMO PS 7-84 |
| Welded brass tube | | B587-88 | | |
| Welded copper alloy (C21000) water tube | | B642-86 | | |
| Welded copper and copper-alloy heat exchanger tube | | B543-91 | | |
| Welded copper tube | | B447-92a | | |
| Wrought copper and copper alloy solder joint fittings for solvent drainage systems | B16.43-82 | | | |
| Wrought copper and copper alloy solder joint pressure fittings | B16.22-89 | | | |
| Wrought copper and wrought copper alloy solder joint drainage fittings (DWV) | B16.29-86 | | | |
| Wrought seamless copper and copper-alloy pipe and tube | | B251-88 | | |

### Nonmetallic Pipe

| Description | ANSI | ASTM | FS | Other |
|---|---|---|---|---|
| Acrylonitrile-butadiene-styrene (ABS) DWV pipe and fittings (installations) | | | | IAPMO IS 5-90 |
| Acrylonitrile-butadiene-styrene (ABS), schedule 40 plastic DWV pipe and fittings | B72.18-72 | D2661-93a | L-P-332B-73 | NSF14-90 |
| Acrylonitrile-butadiene-styrene (ABS) sewer pipe and fittings | | D2751-93 | | NSF14-90 |
| Asbestos cement nonpressure sewer pipe | | C428-74 | | |
| Asbestos cement pressure pipe | | C296-91 | | UL 107 |
| Asbestos cement pressure pipe for water and other liquids | | | | AWWA C400-75 |
| Asbestos cement pressure pipe for water service and yard piping (installation) | | | | IAPMO IS 15-82 |
| Bell–end joints for IPS (PVC) pipe using solvent cement | B72.20-71 | D2672-89 | | NSF14-90 |
| Chlorinated poly (vinyl chloride) (CPVC) plastic hot water distribution systems | | D2846-90a | | NSF14-90 |
| Chlorinated poly (vinyl chloride) (CPVC) plastic pipe, schedules 40 and 80 | | F441-82 | | NSF14-90 |
| Clay drain tile | A6.1-72 | C462-75 | SS-P-1299a-68 | |
| Clay pipe, perforated, standard and extra strength | A106.8-78 | C700-78 | SS-P-361E | |
| Concrete drain tile | | C412-90 | | |

*(Continued)*

*(Table 2-1 continued)*

| Description | ANSI | ASTM | FS | Other |
|---|---|---|---|---|
| Concrete sewer, storm drain, and culvert pipe .... | | C14-90 | SS-P-371e-68 | |
| Drain, waste, and vent (DWV) plastic fitting patterns | | D3311-91 | | NSF14-90 |
| Extra-strength and standard-strength clay pipe and perforated clay pipe ................. | | C700-91 | | |
| Fiberglass pipe — centrifugal cast ............. | | D-2997 | | |
| Fiberglass fittings ......................... | | D-3567 | | AWWA C-950 |
| Filament wound fiberglass resin pipe ........... | | D-2996-88 | | |
| Fitting for joining polyethylene pipe for water service and yard piping .................... | | | | IAPMO PS 25-84 |
| Homogeneous bituminized fiber drain and sewer pipe ........................... | A176.1-71 | D1861-88 | SS-P-1540a-70 | |
| Nonmetallic building and house sewers (installation) | | | | IAPMO IS 1-90 |
| Plastic insert fittings for polyethylene (PE) plastic pipe ............................ | | D2609-90a | | |
| Polybutylene (PB) cold water building supply and yard piping and tubing (installation) ......... | | | | IAPMO IS 17-90 |
| Polybutylene (PB) hot and cold water distribution tubing systems using insert fittings (installation) . | | F845-88 | | IAPMO IS 22-90 |
| Polybutylene (PB) plastic hot water distribution system ................................ | | D3309-89a | | NSF14-90 |
| Polybutylene (PB) plastic pipe (SDR-PR) ........ | | D2662-89 | | NSF14-90 |
| Polybutylene (PB) plastic tubing ............... | | D2666-89 | | NSF14-90 |
| Polyethylene (PE) cold water building supply and yard piping (installation) .................. | | | | IAPMO IS 7-90 |
| Polyethylene (PE) plastic pipe (SIDR-PR) ........ | B72.1-75 | D2239-89 | L-P-315c-75 | NSF14-90 |
| Poly (vinyl chloride) (PVC) DWV (installation) .... | | | | IAPMO IS 12-90 |
| Poly (vinyl chloride) (PVC) DWV pipe and fittings . | K65.56-71 | D2665-93a | L-P-320a-66 | IAPMO IS 9-90<br>NSF 14-90 |
| Poly (vinyl chloride) (PVC) natural gas yard piping (installation) ............................ | | | | IAPMO IS-10-90 |
| Poly (vinyl chloride) (PVC) plastic pipe (SDR-PR) . | | D2241-89 | | NSF14-90 |
| Poly (vinyl chloride) (PVC) plastic pipe and fittings, schedules 40, 80, and 120 ................. | B72.7-71 | D1785-91 | L-P-1035A-74 | NSF14-90<br>UL 1285 |
| Poly (vinyl chloride) (PVC) plastic pipe fittings, schedule 40 ........................... | | D2466-90a | | NSF14-90 |
| Primers for solvent cement joints (PVC) ......... | | F656-80 | | |
| PVC cold water building supply and yard piping (installation) ............................ | | | | IAPMO IS 8-89 |
| Rubber rings for asbestos cement pipe ......... | | D1869-78 (1989) | | |

*(Continued)*

# Chapter 2 — Standards for Plumbing Materials and Equipment

*(Table 2-1 continued)*

| Description | ANSI | ASTM | FS | Other |
|---|---|---|---|---|
| Safe handling of solvent cements and primers used for joining thermoplastic pipe and fittings | | F402-93 | | |
| Socket type chlorinated poly (vinyl chloride) (CPVC) plastic pipe fittings, schedule 40 | | F438-90 | | NSF14-90 |
| Socket type chlorinated poly (vinyl chloride) (CPVC) plastic pipe fittings, schedule 80 | | F439-90 | | NSF14-90 |
| Socket type poly (vinyl chloride) (PVC) plastic pipe fittings, schedule 80 | | D2467-92 | | NSF14-90 |
| Solvent cement for acrylonitrite-butadiene-styrene (ABS) plastic pipe and fittings | B72.23-71 | D2235-88 | | NSF14-90 |
| Solvent cement for chlorinated poly (vinyl chloride) (CPVC) plastic pipe and fittings | | F493-89 | | NSF14-90 |
| Solvent cement for poly (vinyl chloride) (PVC) plastic pipe and fittings | B72.16-71 | D2564-93 | | NSF14-90 |
| Solvent cement for styrene rubber plastic pipe and fittings | | D3122-80 | | |
| Styrene-rubber plastic drain pipe and fittings | | D2852-89 | | |
| Thermoplastic accessible and replaceable plastic tube and tubular | | F409-88 | | |
| Thermoplastic gas pressure pipe, tubing, and fittings | | | | NSF14-90 |
| Thread poly (vinyl chloride) (PVC) threaded pipe and fittings, schedule 80 | K65.166-71 | D2464-90 | | NSF14-90 |
| Type PSM poly (vinyl chloride) (PVC) pipe and fittings, sewer (installation) | | D3034-93 | | IAPMO IS 1-90 |
| Type PSP poly (vinyl chloride) (PVC) pipe and fittings, sewer (installation) | | D3033-83 | | IAPMO IS 1-90 |

### *Pipe Joining Materials, Gaskets, and Supports*

| Description | ANSI | ASTM | FS | Other |
|---|---|---|---|---|
| Calking, lead wool and lead pig | | | QQ-C-40 (2) 1970 | |
| Compression joints for vitrified clay bell and spigot pipe | A106.6-77 | C425-92a | | |
| Compression joints, vitrified clay pipe, bell and spigot pipe and fittings | | C425-92a | | |
| Copper sheet, strip, plate and rolled bar | | B152-92 | | |
| Dishwasher drain airgaps | | | | AHAM DW-1<br>ASSE 1021 |
| Flexible elastomeric joints for drain and sewer plastic pipes | | D3212-92 | TT-P-1536a-75 | |
| Flexible elastomeric joints for plastic pressure pipes | | D3139-89 | | NSF14-90 |
| Heat joining polyolefin pipe and fittings | | D2657-90 | | |

*(Continued)*

*(Table 2-1 continued)*

| Description | ANSI | ASTM | FS | Other |
|---|---|---|---|---|
| **Hubless soil pipe couplings** | | | | |
|   Bolts and nuts | B18.2 | | | |
|   Cast-iron couplings | | A126 | | |
|   Clamps | | A48 | | |
|   Gaskets | | C564-93 | | |
|   Stainless steel couplings | | | | CISPI 310-85 |
| Joints for circular concrete sewer and culvert pipe | | C443-85a(1990) | | |
| Neoprene rubber gaskets for hub and spigot cast-iron soil-pipe and fittings | | C564-93 | | |
| Nonmetallic flat gaskets for pipe flanges | B16.21-78 | | | |
| Pipe hangers and supports | | | | MSS-SP-58-93 |
| | | | | MSS-SP-69-91 |
| Rubber gasket joints for ductile iron and cast-iron pressure pipe and fittings | A21.11-1985 | | | AWWA C111-85 |
| | | | | UL 194 |
| Rubber gaskets for cast-iron soil-pipe and fittings | | C564-93 | | CISPI HSN85 |
| Rubber gaskets, molded or extruded, for concrete nonpressure sewer pipe | | C443-85a(1990) | HH-G-160b-68 | |
| Rubber gaskets, sheet | J7.2-71 | D1330-85(1990) | | |
| Rubber rings for asbestos cement pipe | | D1869-78 (1989) | | |
| Sealing compound, preformed plastic, for expansion joints and pipe joints | | | SS-S-00210 (1981) | |
| Sealing compound, sewer, bituminous, two-component, mineral-filled, cold-applied | | | SS-S-168(2)-62 | |
| Sheet lead | | | QQ-L-201 (2) (1970) | |
| Solder metal | | B32-93 | QQ-S-571e (1986) | |
| Solder: tin alloy, tin-lead alloy, and lead alloy | | | QQ-S-571-75 (1989) | |
| ***Plumbing Fixtures*** | | | | |
| Drinking fountains | A112.18.1-73 | | WW-P-541/6a-71 | NSF61-88 |
| Drinking water coolers, self-contained, mechanically refrigerated | A112.11.1-73 | | | (ARI-1010-73) |
| | | | | UL 399 |
| Floor drains | A112.21.1M-91 | | | |
| Hand-held showers | | | | ASSE 1014-79 |
| Individual shower control valves, anti-scald type | | | | ASSE 1016-79 |
| Porcelain enameled cast-iron plumbing fixtures | A112.19.1M-79 | | WW-P-541/3B-81 | |
| | | | WW-P-541/5B-81 | |
| Porcelain enameled formed-steel plumbing fixtures | A112.19.4-84 | | WW-P-541/3B-81 | |
| | | | WW-P-541/4B-81 | |
| | | | WW-P-541/6B-81 | |
| Shower baths and heads and water control valves | A112.181M-79 | | WW-P-541/7B-81 | |

*(Continued)*

# Chapter 2 — Standards for Plumbing Materials and Equipment

*(Table 2-1 continued)*

| Description | ANSI | ASTM | FS | Other |
|---|---|---|---|---|
| Stainless steel fixtures | A112.19.3-87 | | WW-P-541/5B-81 | |
| Thermostatic mixing valves | | | | ASSE 1017-86 |
| Tile-lined roman bathtubs (installation) | | | | IAPMO IS-2-90 |
| Tile-lined shower receptors (installation) | | | | IAPMO IS-4-90 |
| Trim for water closet bowls, tanks, and urinals | A112.19.5-79(90) | | | |
| Vitreous china plumbing fixtures | A112.19.2-90 | | WW-P-541/1B-81<br>WW-P-541/2B-81<br>WW-P-541/4B-81<br>WW-P-541/6B-81 | |

### Valves

| Description | ANSI | ASTM | FS | Other |
|---|---|---|---|---|
| Air admittance valves for DWV systems | | | | ASSE 1051-90 |
| Air gaps in plumbing systems | A112.1.2 | | | |
| Arrester, water hammer | A112.26.1-84 | | | ASSE 1010-82<br>PDI WH-201-77 |
| Backflow preventers with intermediate atmospheric vent | | | | ASSE1012-93 |
| Backflow prevention devices | | | | IAPMO PS 31-90<br>AWWA C506-69 |
| Backwater, sewer | A112.14.1-75 | | | IAPMO PS 8-77 |
| Ball | | | WWU-531c-65 | MMS-SP-72-92 |
| Bronze gate | | | WWV-54D-73 | MSS-SP-80-87 |
| Cast-iron gate | | | WWV-58b-71 | MSS-SP-70-76 |
| Cleanouts | A112.36.2M-91 | | | |
| Double check valve backflow preventer | | | | ASSE 1015-88<br>UL 312 |
| Drain valve, water heater | | | | ASSE 1005-86 |
| Dual check valve type backflow preventers | | | | ASSE 1024-90 |
| Face-to-face and end-to-end dimensions of valves | B16.10-86 | | | |
| Finished and rough brass plumbing fixture fittings | A112.18.1-89 | | | |
| Flanged and buttwelding end-valves | B16.34-88 | | | |
| Globe-type log lighter (gas) | | | | IAPMO PS 10-84 |
| Hydrostatic testing of control valves | B16.37-80 | | | |
| Large metallic valves for gas distribution (manually operated NSP-2 ½ to 12, 125 psig maximum) | B16.38-85 | | | |
| Manually operated gas valves | A21.15.88 | | | UL 125 |
| Manually operated metallic gas valves for use in gas piping systems up to 125 psig | B16.33-90 | | | |

*(Continued)*

*(Table 2-1 continued)*

| Description | ANSI | ASTM | FS | Other |
|---|---|---|---|---|
| Pipe applied atmospheric vacuum breakers | A1001-82 | | | |
| Pressure reducing and regulating valves | A112.26.2-81 | | | UL 1468 |
| Reduced pressure principle backflow preventer | | | | ASSE 1013-93 |
| Relief valves and automatic gas shutoff device for hot water supply systems and addendum | Z21.22-86 | | | UL 132 |
| Swing check, cast iron | | | WWV-35a-65 | MMS-SP-71-76 |
| Thermoplastic gas shutoffs and valves in gas distribution systems (manually operated) | B16.40-85 | | | |
| Trap seal primer valves, water distribution type | | | | ASSE 1044-86 |
| Unions, brass or bronze | | | WWV-35a-65 | MMS-SP-72-70 |
| Unions, pipe, steel or malleable iron | B16.39-77 | | | MMS-SP-83-87 |
| Vacuum breakers, anti-siphon | A112.1.1-82 | | | ASSE 1001-82 |
| Vacuum breakers, hose connection | A112.1.3-82 | | | ASSE 1011-82 |
| Vacuum breaker, pressure type | A112.1.7-81 | | | ASSE 1020-90 |
| Water closet flush tank ballcocks, performance requirements | A112.11-82 | | | ASSE 1002-89 |

**Plumbing Appliances**

| Description | ANSI | ASTM | FS | Other |
|---|---|---|---|---|
| Dishwashing machines, commercial | A197.3-73 | | Qo-D-431c(2)-70 | UL 921-78<br>ASSE 1004-67 |
| Dishwashing machines, household | A197.1-73<br>C33.69-84 | | | ASSE 1006-89<br>UL749 |
| Electric water heaters | | | WH-196j(j)-71<br>WH-201-77 | UL 174, UL 834 |
| Food waste disposer units, household | C33.59-86 | | QQ-G-001513-68 | ASSE 1007<br>UL 430 |
| Gas-fired steam and hot water boilers and addenda | Z21.13-89 | | | |
| Gas water heaters | Z21.10.3-90 | | | |
| Home laundry equipment | C33.13-85 | | | ASSE 1007<br>UL560 |
| Home laundry equipment, plumbing requirements for | A197.2-86 | | | |
| Household dishwashers, plumbing requirements for | A197.1-86 | | | |
| Household food waste disposer units, plumbing requirements for | A197.3-80 | | | ASSE 1008-89 |
| Interceptors, grease | | | | PDI G101 |
| Metal connectors for gas appliances and addenda | Z21.24-87 | | | |
| Oil-fired boilers | | | | UL 726 |
| Oil-fired water heaters | | | | UL 732 |

*(Continued)*

# Chapter 2 — Standards for Plumbing Materials and Equipment

*(Table 2-1 continued)*

| Description | ANSI | ASTM | FS | Other |
|---|---|---|---|---|
| Vents, gas | | | | UL 441 |
| ***Miscellaneous*** | | | | |
| Boiler and pressure vessel code | BPV-1 | | | ASME 16-91 |
| Compound, plumbing fixture setting | | | TT-P-1536 | |
| Liquified petroleum gases, storage and handling | | | | NFPA58(1995) UL 21, UL 25 UL 30, UL 51 UL 79, UL 87 UL 132, UL 144 UL 147, UL 180 UL 495, UL 565 UL 567, UL 569 UL 644, UL 1185, UL 1313, UL1314, UL 1316 |
| Low-pressure air test for building sewers (installation) | | | | IAPMO IS 16-84 |
| Physically challenged people, providing accessiblility | A117.1-86 | | | |
| Power piping (welding and brazing) | ASME B31.1-92 | | | |
| Prefabricated concrete septic tanks | | | | IAPMO PS 1-87 |
| Protective coated pipe (installation of) | | | | IAPMO IS 13-84 |
| Protective pipe coating, plant-applied | | | | IAPMO PS 22-84 |
| Roof drain | A112.21.2-71 | | | |
| Scheme for the identification of piping systems | A13.1-81 | | | |
| Thermal insulation, calcium silicate block and pipe | | C533-90 | | |
| Thermal insulation, cellular glass | | C552-91 | | |
| Thermal insulation, mineral fiber preformed pipe | | C547-77 | | |
| Water chlorination principles and practices | | | | AWWA M-20-73 |
| ***Fire Protection Systems*** | | | | |
| Installation of private fire service mains | | | | NFPA24-1995 |
| Installation of sprinkler systems | | | | NFPA13-1994 |
| Installation of sprinklers systems in one- and two-family dwellings and manufactured homes | | | | NFPA13D-1994 |
| Installation of sprinkler systems in residential occupancies up to 4 stories in height | | | | NFPA13R-1994 |
| Installation of standpipe and hose systems | | | | NFPA14-1996 |

*Note*: The codes and standards listed in Table 2-1 can be obtained from the organizations listed in Table 2-2.

### Table 2-2  Organization Abbreviations, Addresses, and Phone Numbers

| | | | |
|---|---|---|---|
| AGA | American Gas Association, Inc.<br>Laboratories<br>1515 Wilson Blvd.<br>Arlington, VA 22209 | HUD | US Department of Housing and Urban Development<br>451 7th Street SW<br>Washington, DC 20410<br>(202)708-1422 |
| AHAM | Association of Home Appliance Manufacturers<br>20 North Wacker Drive<br>Chicago, IL 60606<br>(312) 984-5800 | IAPMO | International Association of Plumbing and Mechanical Officials<br>20001 South Walnut Drive<br>Walnut, CA 91789<br>(909) 595-8449 |
| ANSI | American National Standards Institute<br>11 West 42nd Street<br>New York, NY 10036<br>(212) 354-3300 | ICBO | International Conference of Building Officials<br>5360 Workman Mill Road<br>Whittier, CA 90601-2298<br>(310) 699-0541 |
| ARI | Air Conditioning and Refrigeration Institute<br>4301 N. Fairfax Drive, Suite 425<br>Arlington, VA 22209<br>(703) 524-8800 | IPC | International Code Council<br>(International Plumbing Code)<br>5360 Workman Mill Road<br>Whittier, CA 90601-2298<br>(310) 908-8182 |
| ASME | American Society of Mechanical Engineers<br>United Engineering Center<br>345 East 47th Street, New York, NY 10017<br>(212) 705-7722 | MSS | Manufacturers Standardization Society of the Valve and Fitting Industry<br>5203 Leesburg Pike, Suite 502<br>Falls Church, VA 22041<br>(703) 281-6613 |
| ASSE | American Society of Sanitary Engineering<br>PO Box 40362<br>Bay Village, OH 44140<br>(216) 835-3040 | NFPA | National Fire Protection Association, Inc.<br>Batterymarch Park<br>Quincy, MA 02269<br>(617) 770-3000 |
| ASTM | American Society for Testing and Materials<br>1916 Race Street<br>Philadelphia, PA 19103-1187<br>(215) 299-5400 | NSF | National Sanitation Foundation<br>PO Box 130140<br>Ann Arbor, MI 48113-0140 |
| AWWA | American Water Works Association<br>6666 W. Quincy Avenue<br>Denver, CO 80235<br>(303) 794-7711 | OSHA | Office of Safety and Health Administration<br>US Department of Labor<br>200 Constitution Avenue NW<br>Washington, DC 20210<br>(202)219-6091 |
| BOCA | Building Officials and Code Administrators International, Inc.<br>4051 West Flossmoor Road<br>Country Club Hills, IL 60478-5795<br>(708) 799-2300 | PDI | Plumbing Drainage Institute<br>1106 West 77th Street, South Drive<br>Indianapolis, IN 46260-3318 |
| CISPI | Cast Iron Soil Pipe Institute<br>5959 Shallowford Road, Suite 419<br>Chattanooga, TN 37421<br>(615) 892-0137 | SBCCI | Southern Building Code Congress International, Inc.<br>900 Montclair Road<br>Birmingham, AL 35213-1206<br>(205) 591-1853 |
| CSA | Canadian Standards Association<br>178 Rexdale Blvd.<br>Rexdale (Toronto), Ontario, CN M9W 1R3 | UL | Underwriters Laboratories, Inc.<br>333 Pfingsten Road<br>Northbrook, IL 60062-2096<br>(847) 272-8800 |
| EPA | US Environmental Protection Agency<br>86 Alexander Drive<br>Research Triangle Park, NC 27711<br>(919)541-2350 | | |
| FS | Federal Specifications, General Services Administration<br>7th and D Streets<br>Specification Section, Room 6039<br>Washington, DC 20402<br>(202) 755-0325 | | |

# ACORN
## Quality Has No Limitations

### Meridian "Designer Series" Fixtures

**Stainless Steel**

**Terrazzo**

**Corterra™ Solid Surface**

*Meridian* Toilets and Urinals combine vandal-resistance and aesthetics. (ADA compliant)

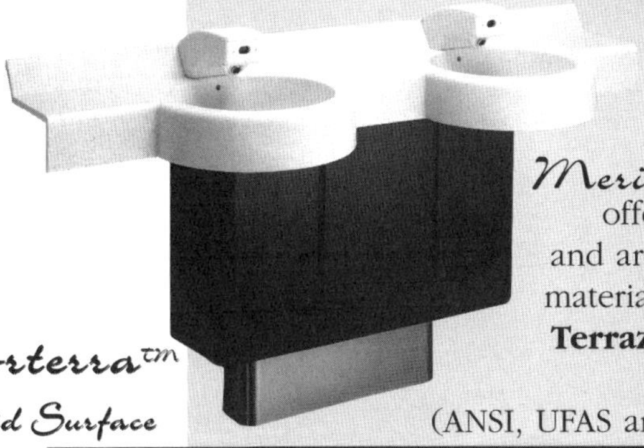

*Meridian* Washbasins offer design versatility and are available in three materials: **Stainless Steel**, **Terrazzo** and **Corterra™ Solid Surface**. (ANSI, UFAS and ADA compliant)

### Terrazzo-Ware™

**Shower Bases**

**Mop Sinks**

Terrazzo-Ware™ mop sinks and shower bases are available in a variety of sizes, configurations and colors. (ADA compliant models are available)

### Wash-Ware®

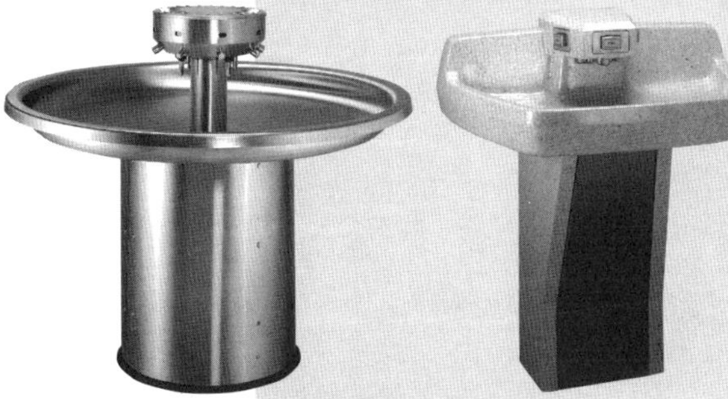

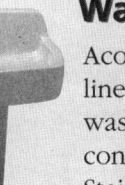

Acorn's Wash-Ware® line offers group washing systems constructed of Stainless Steel and Terrazzo. (ADA compliant)

### Penal-Ware®

Penal-Ware® high-security, vandal resistant products include: toilets, urinals, showers, drinking fountains, individual basins and the "Comby". Electronic software is also available for complete valve control.

### Shower-Ware®

Shower-Ware® includes three basic configurations: wall mounted, built-in panels and freestanding columns. (ANSI, UFAS and ADA compliant models are available)

### Hose Boxes

Stainless Steel and Aluminum Alloy Hose Boxes.

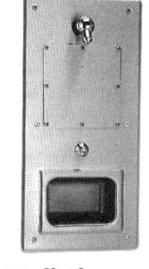

**Comby**

**Wall Shower (Penal-Pack)**

**Acorn ADA COMPLIANCE**

**Acorn Engineering Company**

### "There is no substitute for quality"

For more information call us at (800) 488-8999 or visit our website: www.acorneng.com

P.O. Box 3527 • 15125 Proctor Ave. • City of Industry, CA 91744-0527 U.S.A. • (626) 336-4561 • Fax (626) 961-2200

# EVERYWHERE YOU LOOK

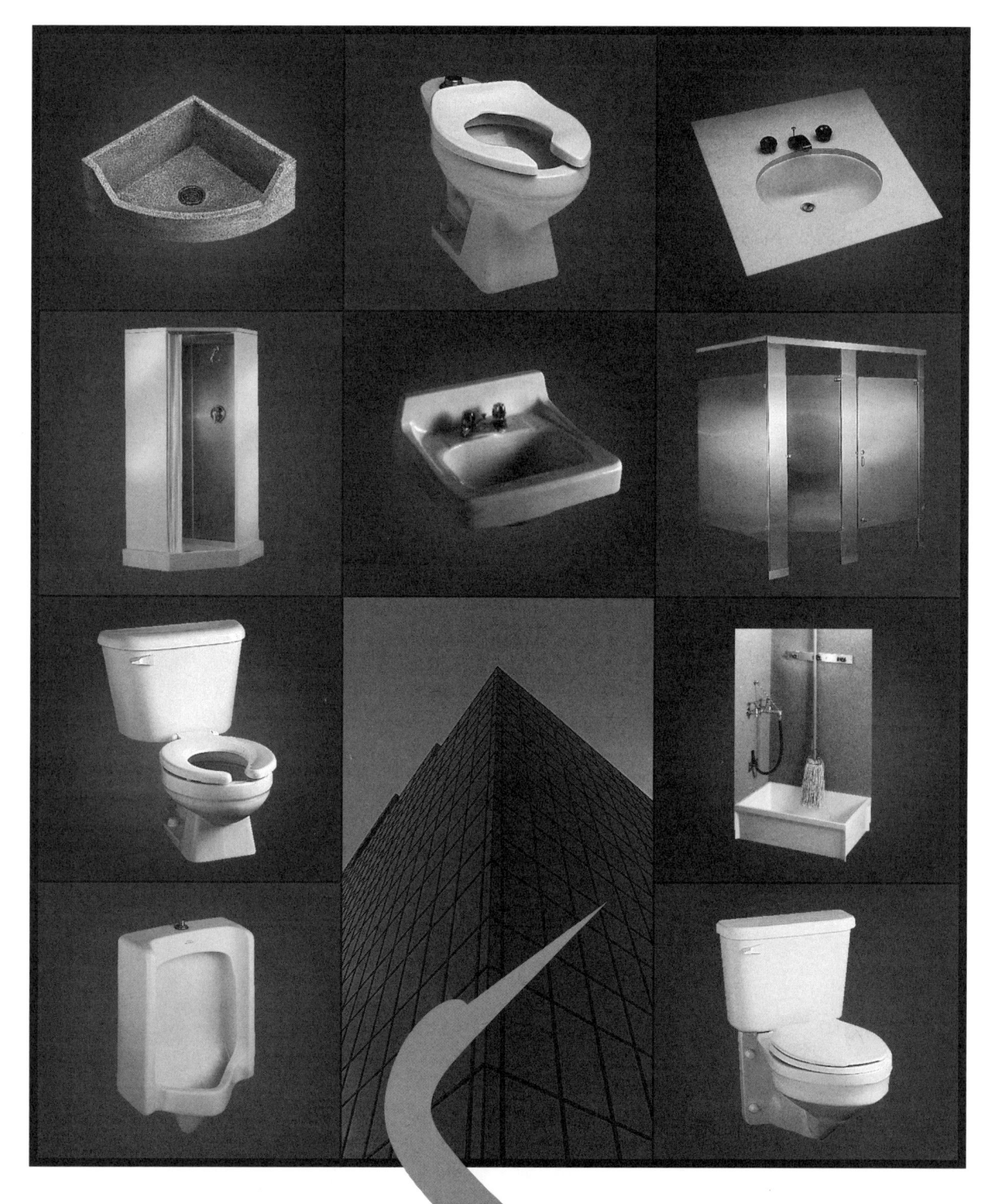

## CRANE PLUMBING IS THERE
HIGH RISE • RESTAURANT • MALL • HOTEL/MOTEL • DAY CARE

# CRANE PLUMBING

FIAT PRODUCTS    1235 HARTREY AVE., EVANSTON, IL 60202 • 847-864-9777    Sanymetal

# 3 Plumbing Specifications

## INTRODUCTION

Plumbing drawings graphically illustrate the scope of plumbing design, showing equipment locations and pipe routings, the quantity of required materials, and the interface of plumbing work with the work of other trades.

Plumbing specifications verbally describe required materials and equipment, quality levels for installation and equipment, and the methods by which materials and equipment are to be assembled and installed and are to interface with each other. They also set requirements for the administration of the plumbing contract.

The plumbing drawings, plumbing specifications, general conditions, special conditions, and addenda comprise the "contract documents" that make up the contract between the building owner and the contractor. None of these items can stand alone: the drawings cannot serve as a contract without the specifications and vice versa. The plumbing designer must, therefore, be familiar with specification writing. If the designer does not actually prepare the plumbing specifications, he/she must be able to coordinate the plumbing drawings with the plumbing specifications.

The writing of specifications for plumbing systems must be precise and exact. The essence of a well-written specification is clarity and brevity as well as correctness and completion.

Specification writers should follow uniform practices to ensure good communication between the design professional and all other segments of the construction industry. The result will be a means by which an engineer in one part of the country can converse meaningfully with a supplier or contractor in another part, using the same language, with the same meanings and in the same way.

## CONSTRUCTION CONTRACT DOCUMENTS

The Construction Specifications Institute (CSI), which was organized in Washington, DC, in 1948, has developed and implemented nationally a set of documents known as the Manual of Practice. At present, there are 16 divisions of construction contract documents, and "plumbing" is included in Division 15–Mechanical. This set of documents is intended to provide an orderly, logical, simple, and flexible format for construction documents, and it has found wide acceptance by engineers, architects, contractors, and suppliers.

It is essential that the plumbing engineer be familiar with and understand all the components that constitute the Manual of Practice to write good specifications for plumbing work.

To discuss these documents effectively, it is necessary to define the terms used in arranging the documents so that one term (and only that term) is used for any one part.

The term "bidder" is used only in connection with the requirements applicable to bidding and awarding the contract. "Contractor" is used only in connection with the requirements applicable to the successful bidder after the contract has been awarded. It should be noted that no instructions to the bidder should appear in the technical specifications; the specifications are instructions addressed to the contractor only.

"Construction contract documents" consist of the drawings and the project manual. Many times, these documents are incorrectly called "plans and specifications." Many items included in the construction contract documents are neither plans nor specifications. Instead of "plans," the correct term is "drawings," and instead of "specifications," the correct term (to describe all documents with the exception of the drawings) is "project manual."

## Project Manual

As previously stated, the project manual is an accurate and descriptive term to define all documents other than the drawings. The manual consists of the following documents:

1. "Pre-bid information" advises prospective bidders about the proposed project. For private work, architects and engineers frequently send invitations through the mail or contact individuals by telephone. For public work, the law usually requires that pre-bid information be published in a newspaper of general circulation.

2. "Instructions to bidders" inform bidders how to prepare the bid so that all bids are in the same format and can be easily and fairly compared.

3. "Bid/tender forms" provide uniform submittals by the bidders and facilitate comparison and evaluation of the bids received.

4. "Bonds and certificates" are legal documents that bind a third party into the contract as a security that the bidder and contractor will perform as agreed. The types of bonds commonly used are (1) a bid bond, (2) a performance bond, (3) a labor and materials payment bond, (4) a guaranty bond, and (5) a maintenance bond.

5. "Form of agreement" is often confused with the contract. This document sets forth the performance agreement with the bidder.

6. "General conditions" serve to define the contractual procedures relative to the project. The American Institute of Architects (AIA) has developed General Conditions of the Contract for Construction, document A201, and a printed copy of this document is often included in the manual. General conditions documents are also available from the National Society of Professional Engineers (NSPE), the American Consulting Engineers Council (ACEC), the American Society of Civil Engineers (ASCE), and the Construction Specifications Institute (CSI).

7. "Supplementary conditions." The "general conditions" portion of the specifications is usually produced as a standard document for use on all projects. Each project, however, has its own unique characteristics, and so it becomes necessary to modify certain items in the general conditions. These modifications comprise the "supplementary conditions."

8. "Specifications" define the quality and type of materials and the installation requirements upon which the contract is based. All work items in the contract should be specified in this document. The drawings show quantities, locations, and routings.

9. "Addenda" are used to modify the contract documents previous to the preparation of the "agreement" and are not used after the contract has been awarded. Change orders, field orders, or some other modifications to the contract are employed when modifications are necessary after the award of the contract.

Each of the foregoing is a separate document; together, they are commonly referred to as the "front-end documents." Although the specifications document comprises the bulk of the project manual, it is only one of the required documents. If the construction project is primarily plumbing work, the plumbing engineer may be responsible for the preparation of the entire project manual. Usually the plumbing engineer works as a member of a project team and is only responsible for

# Chapter 3 — Plumbing Specifications

the plumbing portion of the specifications. It is important, however, that the plumbing engineer be aware of the other portions of the specifications, especially Division 1, in order to coordinate the plumbing work with that of all other parties.

## CSI FORMAT

Typical plumbing specifications may contain thousands of requirements. There would be chaos among the other design disciplines, contractors, and suppliers who have to read and interpret plumbing specifications if the requirements were not organized and expressed in a logical and familiar way. It is possible to pick up specifications prepared by almost any design office and find the pertinent information in it quickly and easily. This is due to widespread acceptance of the specifications format and principles established by the Construction Specifications Institute (CSI).

The CSI format offers many benefits for plumbing specifiers. It is the industry standard that is understood and endorsed by all sectors of the construction industry. Indeed, many architectural firms, building owners, and government agencies require the plumbing consultant to adhere to CSI format. When all design disciplines on a project use CSI format, the resulting project manual has greater consistency and is easier to cross reference than if they don't. Use of CSI format can reduce errors and omissions in plumbing specifications and result in more consistent bidding on projects. Development of computerized product databases and specification writing systems would be set back without CSI format to standardize data organization and specification requirements.

Some plumbing specifiers have resisted CSI formats and principles. They argue that the CSI precepts were developed primarily for architectural applications and are not sufficiently flexible for mechanical engineering. While there may have been a basis for this attitude, CSI has made significant strides during recent years to accommodate the needs of plumbing engineers. Plumbing specifiers who have made the transition and invested the time to understand CSI formats and principles agree that the system does work for plumbing specifications and offers many benefits.

During the construction of the project and after its completion, there are questions raised about various parts or methods which require reference to the specifications. It is important to be able to quickly and efficiently locate the necessary data in a specification. With CSI format providing a consistent and logical pattern for every section, product data and installation requirements are easily pinpointed.

### Document Categories

CSI format works on a hierarchy of levels. At the broadest level, CSI and Engineers Joint Contracts Document Committee (EJCDC) have organized construction documents into categories such as bidding requirements, agreement forms, conditions of the contract, and drawings, in addition to specifications. Each type of document is best suited for a different set of construction requirements. For example, the rights, relationships, and responsibilities of the parties to a construction contract are stated in agreement forms and conditions of the contract. Locations, quantities, and physical relationships between products are best shown in drawings. Specifications should be used to describe the quality of materials and workmanship along with the administrative, procedural, and temporary facility requirements necessary to accomplish the work. The relationship among construction documents is described in the *Uniform Location of Subject Matter* (EJCDC 1910-16).

Each document in the construction contract documents serves a particular and distinct function. Each function is served by one document only. Many times, particulary on smaller projects, specifications appear on drawings. The purpose of a drawing is to define the physical relationship of materials, while the function of a specification is to define quality and types of materials and workmanship.

As previously indicated, specifications are directed to the contractor only. The data, however, should be presented in a form that is of maximum use to the contractor in dealing with subcontractors.

CSI has developed a "three-part section" format. A specification section, when written, is divided into three separate and distinct parts:

A. *Part 1–General* This part includes the scope, necessary references to related work described elsewhere, codes and standards to be followed in the work, qualifications of personnel and manufacturers, required submittals and format, required samples, information on product handling and storage, replacement parts, and other information not specifically included in Parts 2 and 3.

B. *Part 2–Products* All products to be used in the work are included in this part. The products should be described as accurately and as briefly as possible. The remarks should be limited to data about the products. Installation instructions should be included in Part 3.

C. *Part 3–Execution* Detailed instructions for how these products are to be installed and the work performed are included in this part. There should be a detailed description of the installation for each product listed in Part 2. This part should also include tests to be performed, coordination with other trades, acceptance of substrate, and tolerance of installation.

## MASTERFORMAT Organization

The next level of specifications organization is CSI MASTERFORMAT, which organizes all aspects of construction into 16 specification divisions and bidding requirements, contract forms, and conditions of the contract documents. Each division is further divided into sections identified with five-digit section numbers. When used throughout an engineering office, MASTERFORMAT provides an indexing system for organizing product catalogs, cost estimates, drawing key notes, and operation and maintenance manuals.

Most plumbing work falls within Division 15–Mechanical. But MASTERFORMAT is organized by product not by trade, so additional work affecting plumbing can be found throughout the project manual. For example, site utilities are in Division 2–Site Work, concrete pads for plumbing equipment in Division 3–Concrete, flashings in Division 7– Thermal and Moisture Protection, plumbing accessories in Division 10–Specialties, and connections to electrical power and controls in Division 16–Electrical.

Within each section, the CSI section format (CSI *Manual of Practice*, Part II, Chapter 1) provides a three-part structure for organizing product requirements. Part 1 contains general requirements, such as system performance requirements, submittals, and warranties. Part 2 includes specifications for each product and off-site work, such as factory finishing or prefabrication. Part 3 specifies execution of the work, including installation, cleaning and protection of installed products, and balancing and adjusting of equipment and systems.

## Page Formatting

The lowest level within the CSI hierarchy of formats is the page format (CSI *Manual of Practice*, Part III, Chapter 5), which establishes layout standards for paragraph numbering, margins, and headers. Universal adherence to this page format should be welcomed by plumbing engineers, who must constantly revise their specifications to match the inconsistent graphic styles of the architects for whom they consult.

# SPECIFICATIONS

## Writing the Specifications

In addition to espousing formats, CSI has also enunciated principles for effective specification writing. For example, CSI recommends that the specifications be addressed to the contractor and not to subcontractors, such as a plumbing or fire protection subcontractor. This is in accordance with standard forms of general conditions to the construction contract, such as that published by the AIA, which states: "Organization of the Specifications into divisions, sections and articles. . . shall not control the Contractor in dividing the Work among Subcontractors or in establishing the extent of the Work to be performed by any trade."

Well-written specifications are technically correct, complete, clear, concise, and consistent. A general rule of thumb is to assess the quality of the specifications by the range in bid prices. A narrow range usually indicates clarity and consistency of the construction documents. Although this rule of thumb can be applied only after the fact, a careful analysis of the reasons for wide discrepancies should be of assistance when the writer is preparing the next set of construction documents.

Certain phrases are preferable for specification writing; there are also phrases, usages, and styles that should be avoided. It should always be kept in mind that the specifications are a part of the contract documents and, therefore, must be written with extreme care and in a manner that can be easily understood by all concerned individuals, including nontechnical persons. Long words, complex phraseology, and compound sentences should not be used in specifications. Complexity of writing may lend itself to various and costly interpretations. The specification writer should always give complete information and specific instructions regarding the items specified.

The specification writer should select the word that best conveys the exact meaning and repeat the word as often as necessary for a clear and concise statement. Vague sentences often shift responsibility from the office to the field. Bidders may increase the quotation as an insurance against misinterpretations of nonspecific instructions.

Specifications are actually commands to the contractor. The word "shall" is to be used whenever the contractor is told to do something. In specifications, the word "will" is used to impart information and is not a command. The combination "and/or" is not proper usage. In writing a specification, either "and" or "or" is to be used. The word "or" denotes a choice; therefore, extreme caution should be observed when it is used. The phrase "unless otherwise specified" should not be used. The specifications are explicit instructions to the contractor. The appropriate phraseology should be "except as specified in Section _____." "By others" is another indefinite term that should not be used.

The specification writer should keep in mind that the specifications are directed to the contractor and their purpose is to provide clear and concise instructions.

Coordination of plumbing specifications with other design disciplines requires special attention by plumbing specifiers. Each design office should develop a specification coordination checklist to ensure, for example, that pipe penetrations through fire-rated construction are properly sealed and that electrical connections are available for sensor-operated flushometer valves and faucets. Also, some plumbing-related materials, such as insulation and controls, may be specified outside of 15400–Plumbing.

It should be noted that some engineering firms and some owners utilize customized specifications, which they require to be used. Also, numerous manufacturers offer detailed specifications for their products. These should be used with extreme care, with all proprietary verbiage deleted.

**Division 1** Division 1—General Requirements—is an area that requires special coordination between plumbing specifiers and other design disciplines. While the practice of writing a Division 15 section for mechanical general requirements is widespread, such sections too frequently are redundant or conflict with the project's contract conditions and Division 1 specifications. Such conflicts must be resolved sooner or later, and it is clearly preferable to do so before issuing a project manual. Discussion between the plumbing specifier and the architectural or prime engineering specifier can lead to better coordinated Division 1 specifications, which work for the building owner and all members of the project team.

## Master Specifications

Few plumbing specifications are written from scratch. Instead, project specifications are edited versions of office master specifications or guide specifications published by various industry organizations. Preparing project specifications from master specifications is more efficient than writing original specifications for each project. And, while a specifier might be tempted to use specifications from a previous project as the basis for a new project, this practice carries the risk that the specifications will be based upon incomplete or inapplicable data from the previous project. Instead, master specifications fulfill an important quality-assurance function by providing a checklist that ensures that all aspects of a project's design have been considered. Good master specifications help to provide consistency in a firm's project specifications and reflect the lessons learned through the firm's experience.

The plumbing specifier should first select appropriate guide specifications to use as a starting point when developing office master specifications. Engineering firms that subscribe to MASTERSPEC, published by the AIA, or SPECTEXT, published by CSI-affiliated Constructed Science Research Foundation, will find either of these an excellent basis for preparing master specifications. Firms that do a considerable amount of work for the federal government may also consider specifications by the Corp of Engineers, Naval Facilities Command, or another government agency as an appropriate basis for master specifications. As an alternative, plumbing specifiers may also take advantage of guide specifications offered by leading plumbingware manufacturers.

Some specifiers edit master specifications while working directly on a computer. Most plumbing specifiers, however, still prefer marking up a printed copy of the master specifications for final preparation by a word processing operator. In either case, the specifier first must analyze project requirements and select products to be used. The specifier then edits master specifications by filling in blank spaces, deleting and adding text, and selecting among specification options as required by project conditions. Even when master specifications are used, plumbing specifications must be carefully prepared by a qualified plumbing specifier to satisfy the requirements of a particular building or project. It is important to delete all data that do not apply to the particular project.

## Spec Via Computer

The most dramatic trend in plumbing specifications is the rapid development of computerized systems for plumbing design and specifications. Virtually all plumbing specifications are already prepared with computer-based word processors. With new desktop publishing systems, it is even feasible to include product illustrations in specifications.

New programs such as SuperSpec and SPECSystem go beyond word processing to provide expert system capabilities for specification writing. These systems prepare specifications based on a checklist or interactive computerized input from a plumbing designer.

CD-ROM laser disks are now available containing over 600 megabytes each of specification information, such as guide specifications, regulatory requirements, and product data. Among the programs available on CD-ROM are the Construction Criteria Base (CCB) with government specifications; AIA MASTERSPEC and construction regulations; Sweet Search electronic index to products in Sweet's catalog; and Eclat Electronic Catalog Library. The fast access to a vast amount of plumbing data made possible by these machines will have a profound effect on plumbing design and specification writing. The cost of CD-ROM drives has fallen significantly recently. They are now economical enough to be considered standard equipment for any computerized plumbing specifier.

New systems are emerging that will integrate computer-aided design and drafting (CADD) with specifications automation. Other systems will conduct automated searches of plumbing and building codes. And a number of plumbingware manufacturers have started providing product information in computerized form.

## Administration of Specifications

The job of the specification writer is not finished when the specifications are written, printed, and distributed. Administration of the specifications is as important as the document itself. Proper administration can eliminate many of the inadequacies of specifications, whereas improper administration can ruin specifications.

# Chapter 3 — Plumbing Specifications

## PLUMBING SPECIFIERS

Despite advances in specifications automation, plumbing specifications still depend on the judgment and competence of a professional plumbing engineer. Behind the computerized number crunching of design and the word crunching of specifications, engineering remains fundamentally a creative approach to solving human needs. The essence of plumbing specifications is communication among people, whether they are designers, plumbers, suppliers, contractors, or regulatory authorities. Plumbing specifications are intended to be used for design, bidding, and installation, not for litigation.

Writing effective specifications requires broad experience as a plumbing systems designer. In most engineering offices, specifications are considered such an important part of product selection and system design that plumbing specifications are written by the project engineer or team leader. This is different from the situation in an architectural office, where specifications are frequently written by a specification writer who does not have design or project management responsibilities.

Most specifiers prefer to list a minimum of three manufacturers for each product or piece of equipment to foster competition. There are times, however, when only a single choice will best serve the needs of the owner. In these cases, the specifier has the authority to limit acceptable products.

Plumbing specifiers must have strong communication skills to communicate project requirements to the contractor. Even though specifications seldom rise to the level of poetry or drama, their prose must be clear, precise, and easy to read. Stylistically, plumbing specifications require that a specifier be both an effective technical writer and a precise legal writer.

Compared to drafting, where several draftspeople may collaborate to produce a set of drawings, specification writing tends to be an individual activity. Plumbing specifiers do not work in isolation, however. They must interact with other members of a design team to determine project options and coordinate plumbing work with electrical, mechanical, structural, and architectural requirements. The time required to write plumbing specifications will vary according to the complexity of a project and the method of bidding and contracting. On a significant commercial, institutional, or industrial project, often two or three days are budgeted for specification writing. Hospitals and other complex projects can require a full week or more to specify.

Like most plumbing engineering, specification writing is learned on the job. University-level classes in plumbing specifications are rare. There are, however, continuing education programs in specification writing, including programs offered by CSI at both the national and local level. Plumbing designers who have at least five years of specification writing experience can also demonstrate their proficiency and understanding of specifications formats and principles by taking the CSI Certified Construction Specifier (CCS) examination.

As plumbing specifications continue to evolve, plumbing engineers who enjoy the challenges of specification writing will continue to be a valuable asset to their firms and to the project design teams on which they serve.

# WE OFFER YOU MANY CHOICES BUT ONLY ONE STANDARD.

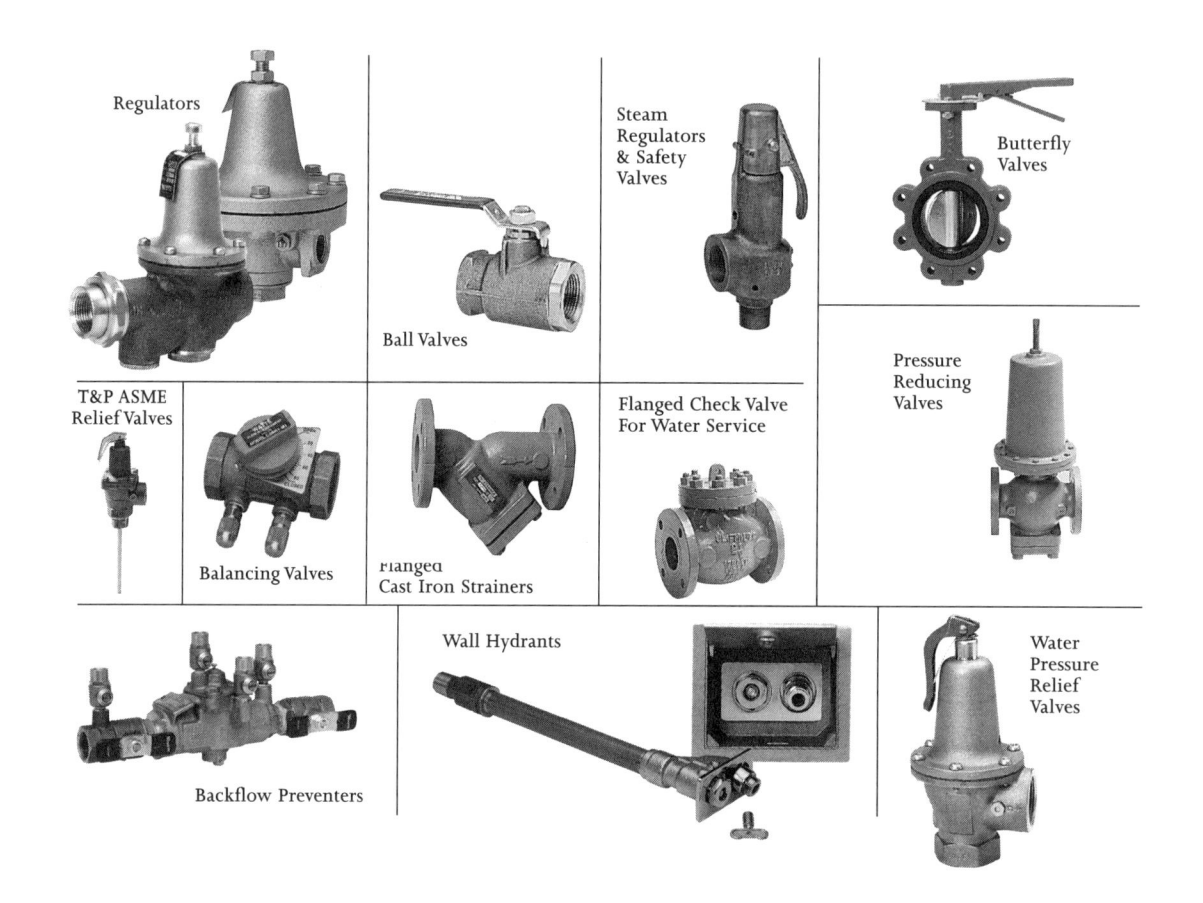

Regulators

Ball Valves

Steam Regulators & Safety Valves

Butterfly Valves

T&P ASME Relief Valves

Balancing Valves

Flanged Cast Iron Strainers

Flanged Check Valve For Water Service

Pressure Reducing Valves

Backflow Preventers

Wall Hydrants

Water Pressure Relief Valves

Whenever you see the Watts name on a valve, you know it signifies the highest standards of engineering, design and manufacturing. From the beginning, Watts products have helped set the performance standards of our industry. So whatever project you're working on, make sure you include Watts.

We offer more valves than anyone else, but only one standard of excellence. For more information, please call 1-800-617-3274 or contact your local Watts representative.

**WATTS REGULATOR** | **ISO 9001 CERTIFIED**

# Chapter 3 — Plumbing Specifications

## APPENDIX 3-A1

CSI MASTERFORMAT–Broad-scope section titles (1988 Edition)

BIDDING REQUIREMENTS, CONTRACT FORMS, AND CONDITIONS OF THE CONTRACT

00010  PRE-BID INFORMATION
00100  INSTRUCTIONS TO BIDDERS
00200  INFORMATION AVAILABLE TO BIDDERS
00300  BID FORMS
00400  SUPPLEMENTS TO BID FORMS
00500  AGREEMENT FORMS
00600  BONDS AND CERTIFICATES
00700  GENERAL CONDITIONS
00800  SUPPLEMENTARY CONDITIONS
00900  ADDENDA

*Note*: The items listed above are not specification sections and are referred to as "documents" rather than "sections" in the master list of section titles, numbers, and broad-scope section explanations.

### Specifications

#### DIVISION 1—GENERAL REQUIREMENTS

01010  SUMMARY OF WORK
01020  ALLOWANCES
01025  MEASUREMENT AND PAYMENT
01030  ALTERNATES/ALTERNATIVES
01035  MODIFICATION PROCEDURES
01040  COORDINATION
01050  FIELD ENGINEERING
01060  REGULATORY REQUIREMENTS
01070  IDENTIFICATION SYSTEMS
01090  REFERENCES
01100  SPECIAL PROJECT PROCEDURES
01200  PROJECT MEETINGS
01300  SUBMITTALS
01400  QUALITY CONTROL
01500  CONSTRUCTION FACILITIES AND TEMPORARY CONTROLS
01600  MATERIAL AND EQUIPMENT
01650  FACILITY STARTUP/COMMISSIONING
01700  CONTRACT CLOSEOUT
01800  MAINTENANCE

#### DIVISION 2—SITE WORK

02010  SUBSURFACE INVESTIGATION
02050  DEMOLITION
02100  SITE PREPARATION
02140  DEWATERING
02150  SHORING AND UNDERPINNING
02160  EXCAVATION SUPPORT SYSTEMS
02170  COFFER DAMS
02200  EARTHWORK
02300  TUNNELING
02350  PILES AND CAISSONS
02450  RAILROAD WORK
02480  MARINE WORK
02500  PAVING AND SURFACING
02600  UTILITY PIPING MATERIALS
02660  WATER DISTRIBUTION
02680  FUEL AND STEAM DISTRIBUTION
02700  SEWERAGE AND DRAINAGE
02760  RESTORATION OF UNDERGROUND PIPE
02770  PONDS AND RESERVOIRS
02780  POWER AND COMMUNICATIONS
02800  SITE IMPROVEMENTS
02900  LANDSCAPING

#### DIVISION 3—CONCRETE

03100  CONCRETE FORMWORK
03200  CONCRETE REINFORCEMENT
03250  CONCRETE ACCESSORIES
03300  CAST-PLACE CONCRETE
03370  CONCRETE CURING
03400  PRECAST CONCRETE
03500  CEMENTITIOUS DECKS AND TOPPINGS
03600  GROUT
03700  CONCRETE RESTORATION AND CLEANING
03800  MASS CONCRETE

#### DIVISION 4—MASONRY

04100  MORTAR AND MASONRY GROUT
04150  MASONRY ACCESSORIES
04200  UNIT MASONRY
04400  STONE
04500  MASONRY RESTORATION AND CLEANING
04550  REFRACTORIES
04600  CORROSION RESISTANT MASONRY
04700  SIMULATED MASONRY

#### DIVISION 5—METALS

05010  METAL MATERIALS
05030  METAL COATINGS
05050  METAL FASTENING
05100  STRUCTURAL METAL FRAMING
05200  METAL JOISTS
05300  METAL DECKING
05400  COLD-FORMED METAL FRAMING
05500  METAL FABRICATIONS
05580  SHEET METAL FABRICATIONS
05700  ORNAMENTAL METAL
05800  EXPANSION CONTROL
05900  HYDRAULIC STRUCTURES

## DIVISION 6—WOOD AND PLASTICS

- 06050 FASTENERS AND ADHESIVES
- 06100 ROUGH CARPENTRY
- 06130 HEAVY TIMBER CONSTRUCTION
- 06150 WOOD AND METAL SYSTEMS
- 06170 PREFABRICATED STRUCTURAL WOOD
- 06200 FINISH CARPENTRY
- 06300 WOOD TREATMENT
- 06400 ARCHITECTURAL WOODWORK
- 06500 STRUCTURAL PLASTICS
- 06600 PLASTIC FABRICATIONS
- 06650 SOLID POLYMER FABRICATIONS

## DIVISION 7—THERMAL AND MOISTURE PROTECTION

- 07100 WATERPROOFING
- 07150 DAMP PROOFING
- 07180 WATER REPELLENTS
- 07190 VAPOR RETARDERS
- 07195 AIR BARRIERS
- 07200 INSULATION
- 07240 EXTERIOR INSULATION AND FINISH SYSTEMS
- 07250 FIREPROOFING
- 07270 FIRESTOPPING
- 07300 SHINGLES AND ROOFING TILES
- 07400 MANUFACTURED ROOFING AND SIDING
- 07480 EXTERIOR WALL ASSEMBLIES
- 07500 MEMBRANE ROOFING
- 07570 TRAFFIC COATINGS
- 07600 FLASHING AND SHEET METAL
- 07700 ROOF SPECIALTIES AND ACCESSORIES
- 07800 SKYLIGHTS
- 07900 JOINT SEALERS

## DIVISION 8—DOORS AND WINDOWS

- 08100 METAL DOORS AND FRAMES
- 08200 WOOD AND PLASTIC DOORS
- 08250 DOOR OPENING ASSEMBLIES
- 08300 SPECIAL DOORS
- 08400 ENTRANCES AND STOREFRONTS
- 08500 METAL WINDOWS
- 08600 WOOD AND PLASTIC WINDOWS
- 08650 SPECIAL WINDOWS
- 08700 HARDWARE
- 08800 GLAZING
- 08900 GLAZED CURTAIN WALLS

## DIVISION 9—FINISHES

- 09100 METAL SUPPORT SYSTEMS
- 09200 LATH AND PLASTER
- 09250 GYPSUM BOARD
- 09300 TILE
- 09400 TERRAZZO
- 09450 STONE FACING
- 09500 ACOUSTICAL TREATMENT
- 09540 SPECIAL WALL SURFACES
- 09545 SPECIAL CEILING SURFACES
- 09550 WOOD FLOORING
- 09600 STONE FLOORING
- 09630 UNIT MASONRY FLOORING
- 09650 RESILIENT FLOORING
- 09680 CARPET
- 09700 SPECIAL FLOORING
- 09780 FLOOR TREATMENT
- 09800 SPECIAL COATINGS
- 09900 PAINTING
- 09950 WALL COVERINGS

## DIVISION 10—SPECIALTIES

- 10100 VISUAL DISPLAY BOARDS
- 10150 COMPARTMENTS AND CUBICLES
- 10200 LOUVERS AND VENTS
- 10240 GRILLES AND SCREENS
- 10250 SERVICE WALL SYSTEMS
- 10260 WALL AND CORNER GUARDS
- 10270 ACCESS FLOORING
- 10290 PEST CONTROL
- 10300 FIREPLACES AND STOVES
- 10340 MANUFACTURED EXTERIOR SPECIALTIES
- 10350 FLAGPOLES
- 10400 IDENTIFYING DEVICES
- 10450 PEDESTRIAN CONTROL DEVICES
- 10500 LOCKERS
- 10520 FIRE PROTECTION SPECIALTIES
- 10530 PROTECTIVE COVERS
- 10550 POSTAL SPECIALTIES
- 10600 PARTITIONS
- 10650 OPERABLE PARTITIONS
- 10670 STORAGE SHELVING
- 10700 EXTERIOR PROTECTION DEVICES FOR OPENINGS
- 10750 TELEPHONE SPECIALTIES
- 10800 TOILET AND BATH ACCESSORIES
- 10880 SCALES
- 10900 WARDROBE AND CLOSED SPECIALTIES

## DIVISION 11—EQUIPMENT

- 11010 MAINTENANCE EQUIPMENT
- 11020 SECURITY AND VAULT EQUIPMENT
- 11030 TELLER AND SERVICE EQUIPMENT
- 11040 ECCLESIASTICAL EQUIPMENT
- 11050 LIBRARY EQUIPMENT
- 11060 THEATER AND STAGE EQUIPMENT

# Chapter 3 — Plumbing Specifications

| | |
|---|---|
| 11070 | INSTRUMENTAL EQUIPMENT |
| 11080 | REGISTRATION EQUIPMENT |
| 11090 | CHECKROOM EQUIPMENT |
| 11100 | MERCANTILE EQUIPMENT |
| 11110 | COMMERCIAL LAUNDRY AND DRY CLEANING EQUIPMENT |
| 11120 | VENDING EQUIPMENT |
| 11130 | AUDIO VISUAL EQUIPMENT |
| 11140 | VEHICLE SERVICE EQUIPMENT |
| 11150 | PARKING CONTROL EQUIPMENT |
| 11160 | LOADING DOCK EQUIPMENT |
| 11170 | SOLID WASTE HANDLING EQUIPMENT |
| 11190 | DETENTION EQUIPMENT |
| 11200 | WATER SUPPLY AND TREATMENT EQUIPMENT |
| 11280 | HYDRAULIC GATES AND VALVES |
| 11300 | FLUID WASTE TREATMENT AND DISPOSAL EQUIPMENT |
| 11400 | FOOD SERVICE EQUIPMENT |
| 11450 | RESIDENTIAL EQUIPMENT |
| 11460 | UNIT KITCHENS |
| 11470 | DARKROOM EQUIPMENT |
| 11480 | ATHLETIC, RECREATIONAL, AND THERAPEUTIC EQUIPMENT |
| 11500 | INDUSTRIAL AND PROCESS EQUIPMENT |
| 11600 | LABORATORY EQUIPMENT |
| 11650 | PLANETARIUM EQUIPMENT |
| 11660 | OBSERVATORY EQUIPMENT |
| 11680 | OFFICE EQUIPMENT |
| 11700 | MEDICAL EQUIPMENT |
| 11780 | MORTUARY EQUIPMENT |
| 11850 | NAVIGATION EQUIPMENT |
| 11870 | AGRICULTURAL EQUIPMENT |

## *DIVISION 12—FURNISHINGS*

| | |
|---|---|
| 12050 | FABRICS |
| 12100 | ARTWORK |
| 12300 | MANUFACTURED CASEWORK |
| 12500 | WINDOW TREATMENT |
| 12600 | FURNITURE AND ACCESSORIES |
| 12670 | RUGS AND MATS |
| 12700 | MULTIPLE SEATING |
| 12800 | INTERIOR PLANTS AND PLANTERS |

## *DIVISION 13—SPECIAL CONSTRUCTION*

| | |
|---|---|
| 13010 | AIR-SUPPORTED STRUCTURES |
| 13020 | INTEGRATED ASSEMBLIES |
| 13003 | SPECIAL PURPOSE ROOMS |
| 13080 | SOUND, VIBRATION, AND SEISMIC CONTROL |
| 13090 | RADIATION PROTECTION |
| 13100 | NUCLEAR REACTORS |
| 13120 | PRE-ENGINEERED STRUCTURES |
| 13150 | AQUATIC FACILITIES |
| 13175 | ICE RINKS |
| 13180 | SITE CONSTRUCTED INCINERATORS |
| 13185 | KENNELS AND ANIMAL SHELTERS |
| 13200 | LIQUID AND GAS STORAGE TANKS |
| 13220 | FILTER UNDERDRAINS AND MEDIA |
| 13230 | DIGESTER COVERS AND APPURTENANCES |
| 13240 | OXYGENATION SYSTEMS |
| 13260 | SLUDGE CONDITIONING SYSTEMS |
| 13300 | UTILITY CONTROL SYSTEMS |
| 13400 | INDUSTRIAL AND PROCESS CONTROL SYSTEMS |
| 13500 | RECORDING INSTRUMENTATION |
| 13550 | TRANSPORTATION CONTROL INSTRUMENTATION |
| 13600 | SOLAR ENERGY SYSTEMS |
| 13700 | WIND ENERGY SYSTEMS |
| 13750 | COGENERATION SYSTEMS |
| 13800 | BUILDING AUTOMATION SYSTEMS |
| 13900 | FIRE SUPPRESSION AND SUPERVISORY SYSTEMS |
| 13950 | SPECIAL SECURITY CONSTRUCTION |

## *DIVISION 14—CONVEYING SYSTEMS*

| | |
|---|---|
| 14100 | DUMBWAITERS |
| 14200 | ELEVATORS |
| 14300 | ESCALATORS AND MOVING WALKS |
| 14400 | LIFTS |
| 14500 | MATERIAL HANDLING SYSTEMS |
| 14600 | HOISTS AND CRANES |
| 14700 | TURNTABLES |
| 14800 | SCAFFOLDING |
| 14900 | TRANSPORTATION SYSTEMS |

## *DIVISION 15—MECHANICAL*

| | |
|---|---|
| 15050 | BASIC MECHANICAL MATERIALS AND METHODS |
| 15250 | MECHANICAL INSULATION |
| 15400 | PLUMBING |
| 15500 | HEATING, VENTILATING, AND AIR CONDITIONING |
| 15550 | HEAT GENERATION |
| 15650 | REFRIGERATION |
| 15750 | HEAT TRANSFER |
| 15850 | AIR HANDLING |
| 15880 | AIR DISTRIBUTION |
| 15950 | CONTROLS |
| 15990 | TESTING, ADJUSTING, AND BALANCING |

## *DIVISION 16—ELECTRICAL*

| | |
|---|---|
| 16050 | BASIC ELECTRICAL MATERIALS AND METHODS |

| 16200 | POWER GENERATION—BUILD-UP SYSTEMS |
| 16300 | MEDIUM VOLTAGE DISTRIBUTION |
| 16400 | SERVICE AND DISTRIBUTION |
| 16500 | LIGHTING |
| 16600 | SPECIAL SYSTEMS |
| 16700 | COMMUNICATIONS |
| 16850 | ELECTRIC RESISTANCE HEATING |
| 16900 | CONTROLS |
| 16950 | TESTING |

## APPENDIX 3-A2

CSI MASTERFORMAT–Broad-scope and narrow-scope section titles used for plumbing work (from 1988 Edition)

### DIVISION 2—SITE WORK

Section        Number        Title

#### 02600 UTILITY PIPING MATERIALS

- -605  Utility Structures
- -610  Pipes and Fittings
- -640  Valves and Cocks
- -645  Hydrants

#### 02660 WATER DISTRIBUTION

- -665  Water Systems
- -670  Water Wells
- -675  Disinfection of Water Distribution Systems

#### 02680 FUEL AND STEAM DISTRIBUTION

- -685  Gas Distribution Systems
- -690  Oil Distribution Systems

#### 02700 SEWERAGE AND DRAINAGE

- -710  Subdrainage Systems
- -720  Storm Sewerage
- -730  Sanitary Sewerage
- -735  Combined Wastewater Systems
- -740  Septic Systems

#### 02760 RESTORATION OF UNDERGROUND PIPE

- -762  Inspection of Underground Pipelines
- -764  Sealing Underground Pipelines
- -766  Relining Underground Pipelines

#### 02800 SITE IMPROVEMENTS

- -810  Irrigation Systems
- -820  Fountains

### DIVISION 15—MECHANICAL

Section        Number        Title

#### 15050 BASIC MECHANICAL MATERIALS AND METHODS

- -060  Pipes and Pipe Fittings
- -100  Valves
    - Manual Control Valves
    - Self-Actuated Valves
- -120  Piping Specialties
- -130  Gages
- -140  Supports and Anchors
- -150  Meters
- -160  Pumps
- -170  Motors
- -175  Tanks
- -190  Mechanical Identification
- -240  Mechanical Sound, Vibration, and Seismic Control

#### 15250 MECHANICAL INSULATION

- -260  Piping Insulation
- -280  Equipment Insulation

#### 15300 FIRE PROTECTION

- -310  Fire Protection Piping
- -320  Fire Pumps
- -330  Wet Pipe Sprinkler Systems
- -335  Dry Pipe Sprinkler Systems
- -340  Pre-Action Sprinkler Systems
- -345  Combination Dry Pipe and Pre-Action Sprinkler Systems
- -350  Deluge Sprinkler Systems
- -355  Foam Extinguishing Systems
- -360  Carbon Dioxide Extinguishing Systems
- -365  Halogen Agent Extinguishing Systems
- -370  Dry Chemical Extinguishing Systems
- -375  Standpipe and Hose Systems

#### 15400 PLUMBING

- -410  Plumbing Piping
- -430  Plumbing Specialties
- -440  Plumbing Fixtures
- -450  Plumbing Equipment
- -475  Pool and Fountain Equipment
- -480  Special Systems

## APPENDIX 3-B
## SECTION SHELL OUTLINE

This shell outline has been developed by the American Institute of Architects conforming to the CSI *Manual of Practice*.

## SECTION XXXXX
## XXXXXXXXXXXXXXXXXXXX

### PART 1—GENERAL

1.1 SUMMARY

A. This section includes [description of essential unit of work included in section].

B. Products furnished but not installed under this section include [description].

C. Products installed but not furnished under this section include [description].

D. Related Sections: The following sections contain requirements that relate to this section:

1. Division [#] Section ["Title"] for [description of related unit of work].
2. Division [#] Section ["Title"] for [description of related unit of work].
3. Division [#] Section ["Title"] for [description of related unit of work].
4. Division [#] Section ["Title"] for [description of related unit of work].

E. Allowances:

F. Unit Prices:

G. Alternates:

1.2 REFERENCES

1.3 DEFINITIONS

1.4 SYSTEM DESCRIPTION

1.5 SYSTEM PERFORMANCE REQUIREMENTS

A. Performance Requirements: Provide [system] complying with performance requirements specified.

1.6 SUBMITTALS

A. General: Submit the following:

B. Product data for each type of [products] specified, including details of construction relative to materials, dimensions of individual components, profiles, and finishes.

C. Product data for the following products:

1. [Product].
2. [Product].
3. [Product].
4. [Product].

D. Shop drawings from manufacturer detailing equipment assemblies and indicating dimensions, weights, loadings, required clearances, method of field assembly, components, utility requirements, and location and size of each field connection.

E. Include setting drawings, templates, and directions for installation of anchor bolts and other anchorages to be installed as unit of work of other sections.

F. Coordination drawings for [unit of work].

G. Coordination drawings for reflected ceiling plans drawn accurately to scale and coordinating penetrations and ceiling-mounted items, including sprinklers, diffusers, grilles, light fixtures, speakers, and access panels.

H. Wiring diagrams from manufacturer for electrically operated equipment.

I. Wiring diagrams detailing wiring for power, signal, and control systems, differentiating between manufacturer and field-installed wiring.

J. Material certificates signed by manufacturer certifying that each material item complies with requirements, in lieu of laboratory test reports, when permitted by architect.

K. Product certificates signed by manufacturers of [products] certifying that their products comply with requirements.

L. Welder certificates signed by contractor certifying that welders comply with requirements of "quality-assurance" article.

M. Qualifications data for firms and persons specified in "quality-assurance" article to demonstrate their capabilities and experience. Include list of completed projects with project name, addresses, name of architects and owners, plus other information specified.

N. Test reports from, and based on tests performed by, qualified independent testing laboratory evidencing compliance of [product] with requirements based on comprehensive testing.

O. Maintenance data for [materials and products], for inclusion in operating and maintenance manuals.

1.7 QUALITY ASSURANCE

A. Installer Qualifications: Engage an experienced installer who has successfully completed [unit of work] similar in material, design, and extent to that indicated for project.

B. Installer's Field Supervision: Require installer to maintain an experienced full-time supervisor who is on jobsite during times that [unit of work] is in progress.

C. Testing Laboratory Qualifications: Demonstrate experience and capability to conduct testing indicated without delaying progress of the work based on evaluation of laboratory-submitted criteria conforming to ASTM E 699.

D. Qualify welding process and welding operators in accordance with ASME "Boiler and Pressure Vessel Code," Section IX, "Welding and Brazing Qualifications."

E. Regulatory Requirements: Fabricate and stamp [product] to comply with [code].

F. Regulatory requirements: Comply with following codes.
   1. [Itemize codes in form of separate subparagraphs under above].

G. UL Standard: Provide [products] complying with UL [designation, title].

H. Electrical Component Standard: Provide components complying with NFPA 70 "National Electrical Code" and which are listed and labeled by UL where available.

I. UL and NEMA Compliance: Provide [components] required as part of [product or system] which are listed and labeled by UL and comply with applicable NEMA standards.

J. ASME Compliance: Fabricate and stamp [product] to comply with ASME Boiler and Pressure Vessel Code, Section VIII, Division 1.

K. Single Source Responsibility: Obtain [system] components from single source having responsibility and accountability to answer and resolve problems regarding proper installation, compatibility, performance, and acceptance.

L. Manufacturer and Product Selection: The drawings indicate sizes, profiles, and dimensional requirements of [product or system]. A [product or system] having equal performance characteristics with deviations from indicated dimensions and profiles may be considered, provided deviations do not change the design concept or intended performance. The burden of proof of equality is on the proposer.

1.8 DELIVERY, STORAGE, AND HANDLING

A. Deliver materials and equipment to site in such quantities and at such times to ensure continuity of installation. Store them at site to prevent cracking, distortion, staining, and other physical damage and so that markings are visible.

B. Lift and support equipment only at designated lifting or supporting points as shown on final shop drawings.

C. Deliver [product] as a factory assembled unit with protective crating and covering.

D. Store [products] on elevated platforms in a dry location.

E. Coordinate delivery of [product] in sufficient time to allow movement into building.

# Chapter 3 — Plumbing Specifications

1.9 PROJECT CONDITIONS

    A. Site Information: Data on indicated subsurface conditions are not intended as representations or warranties of accuracy or continuity of these conditions {between soil borings}. It is expressly understood that owner and engineer will not be responsible for interpretations or conclusions drawn therefrom by contractor. Data are made available for convenience of contractor (and are not guaranteed to represent conditions that may be encountered).

    B. Field Measurements: Verify dimensions by field measurements. Verify that [name of system, product, or equipment] may be installed in compliance with the original design and referenced standards.

1.10 SEQUENCING AND SCHEDULING

    A. Coordinate the size and location of concrete equipment pads. Cast anchor bolt inserts into pad. Concrete reinforcement and formwork requirements are specified in Division 3.

    B. Coordinate the installation of roof penetrations. Roof specialties are specified in Division 7.

1.11 WARRANTY

    A. Special Project Warrant: Submit written warranty, executed by manufacturer, agreeing to repair or replace [product] which fails in materials or workmanship within specified warranty period. This warranty shall be in addition to, and not limitation of, other rights the owner may have against the contractor under the contract documents.

        1. Warranty period is 1 year after date of substantial completion.

1.12 MAINTENANCE

1.13 EXTRA MATERIALS

    A. Deliver extra materials to owner. Furnish extra materials described below matching products installed, packaged with protective covering for storage and identified with labels clearly describing contents.

## PART 2—PRODUCTS

2.1 MANUFACTURERS

    A. *Available Manufacturers:* Subject to compliance with requirements, manufacturers offering products which may be incorporated in the work include, but are not limited to, the following:

    B. *Manufacturers:* Subject to compliance with requirements, provide products by one of the following:

        1. [Name of Product]:

            a. [Manufacturer's Name].

            b. [Manufacturer's Name].

            c. [Manufacturer's Name].

        2. [Name of Product]:

            a. [Manufacturers Name].

            b. [Manufacturer's Name].

        3. [Name of Product]:

            a. [Manufacturer's Names].

        4. [Name of Product]:

            a. [Manufacturer's Names].

    C. *Available Products:* Subject to compliance with requirements, products which may be incorporated in the work include, but are not limited to, the following:

    D. *Products:* Subject to compliance with requirements, provide one of the following:

    E. *Manufacturer:* Subject to compliance with requirements, provide product by [Manufacturer's Name].

2.2 MATERIALS [PRODUCT NAME]

    A. [Material or Product Name]: [Nonproprietary description of material] complying with [standard designation] (for type, grade, etc.).

    B. [Material or Product Name]: [Nonproprietary description of material] complying with [standard designation] (for type, grade, etc.).

    C. [Material or Product Name]: [Standard designation], [type, grade, etc. as applicable to referenced standard].

D. [Material or Product Name]: [Standard designation], [type, grade, etc. as applicable to referenced standard].

## 2.3 MATERIALS, GENERAL [PRODUCTS, GENERAL]

A. [Description] Standard: Provide [product or material] which complies with [standard designation].

B. [Description] Standard: Provide [product or material] which complies with [standard designation].

C. [Kind of Performance] Characteristics: [Insert requirements for kind of performance involved and test method as applicable unless requirements included under Part 1 Article ("System Description).]

D. [Kind of Performance] Characteristics: [Insert requirements for kind of performance involved and test method as applicable unless requirements included under Part 1 Article ("System Description").]

## 2.4 EQUIPMENT [NAME OF MANUFACTURED UNIT]

A. [Equipment or Unit Name]: [Nonproprietary description of . . .] complying with [standard designation] (for type, grade, etc.).

B. [Equipment or Unit Name]: [Nonproprietary description of . . .] complying with [standard designation] (for type, grade, etc.).

C. [Equipment, Unit, or Product Name]: [standard designation], (type, grade, etc. as applicable to referenced standard).

D. [Equipment, Unit, or Product Name]: [standard designation], (type, grade, etc. as applicable to referenced standard).

## 2.5 COMPONENTS

A. [Component Name]: [Nonproprietary description of . . .] complying with [standard designation] (for type, grade, etc.).

B. [Component Name]: [Nonproprietary description of . . .] complying with [standard designation] (for type, grade, etc.).

## 2.6 ACCESSORIES

A. Manufacturer's standard factory finish.

## 2.7 MIXES

## 2.8 FABRICATION

## 2.9 SOURCE OF QUALITY CONTROL

# PART 3—EXECUTION

## 3.1 EXAMINATION

A. Examine [substrates] [areas] [and] [conditions] [with Installer present] for compliance with requirements for [maximum moisture content], installation tolerances, [other specific conditions], and other conditions affecting performance of [unit of work of this section]. Do not proceed with installation until unsatisfactory conditions have been corrected.

B. Examine rough-in drawings for [name] piping systems to verify actual locations of piping connections prior to installation.

C. Examine walls, floors, roof, and [description] for suitable conditions where [name of products or system] are to be installed.

D. Do not proceed until unsatisfactory conditions have been corrected.

## 3.2 PREPARATION

A. Protection:

## 3.3 INSTALLATION, GENERAL [APPLICATION, GENERAL]

A. [Description] Standard: Install [name of product, material, or system] to comply with [standard designation].

## 3.4 INSTALLATION {OF [NAME]} {APPLICATION OF [NAME]}

A. Install [name of unit of work] level and plumb, in accordance with manufacturer's written instructions,

# Chapter 3 — Plumbing Specifications

rough-in drawings, the original design, and referenced standards.

## 3.5 CONNECTIONS (NOT A CSI ARTICLE—BUT USEFUL FOR DIVISION 15)

A. Piping installation requirements are specified in other sections. Drawings indicate general arrangement of piping, fittings, and specialties. The following are specific connection requirements:

B. Install piping adjacent to equipment to allow servicing and maintenance.

## 3.6 FIELD QUALITY CONTROL

A. Testing Laboratory: Owner will employ and pay an independent testing laboratory to perform field quality control testing.

B. Testing Laboratory: Provide the services of an independent testing laboratory experienced in the testing of [unit of work] and acceptable to the engineer, to perform field quality control testing.

C. Extent and Testing Methodology: Arrange for testing of completed [unit of work] in successive stages in areas of extent described below; do not proceed with [unit of work] of next area until test results for previously completed work verify compliance with requirements.

D. Testing laboratory shall report test results promptly and in writing to contractor and engineer.

E. Repair or replace [unit of work] within areas where test results indicate [unit of work] does not comply with requirements.

F. Manufacturer's Field Service: Provide services of a factory-authorized service representative to supervise field assembly of components, installation of [products] including piping and electrical connections, and to report results in writing.

## 3.7 ADJUSTING [CLEANING] [ADJUSTING AND CLEANING]

## 3.8 COMMISSIONING (NOT A CSI ARTICLE—BUT USEFUL FOR DIVISION 15 [DEMONSTRATION])

A. Start-Up Services, General: Provide services of a factory-authorized service representative to provide start-up service and to demonstrate and train owner's maintenance personnel as specified below.

B. Test and adjust controls and safeties. Replace damaged or malfunctioning controls and equipment.

C. Train owner's maintenance personnel on procedures and schedules related to start-up and shut-down, troubleshooting, servicing, and preventative maintenance.

D. Review data in operating and maintenance manuals. Refer to Division 1, Section ["Project Closeouts"] ["Operating and Maintenance Manuals"].

E. Schedule training with owner through architect, with at least 7 days advance notice.

## 3.9 PROTECTION

## 3.10 SCHEDULES

## REFERENCES

1. Chusid, Michael. 1991. Trends in plumbing specifications. *Plumbing Engineer* (March).

2. Construction Specifications Institute (CSI). *Manual of Practice*. Alexandria, VA.

3. CSI. 1985. *The function of reference standards in construction specifications*. CSI Monograph Series no. 01M091.

4. CSI. 1988. MASTERFORMAT. Alexandria, VA.

5. Guzey, Onkal K., and James N. Freehof. 1989. *ConDoc: The new system for formatting and integrating construction documentation*. 2d ed. A Professional Development Program of the American Institute of Architects (AIA).

6. Harris, Cyril M., ed. Specifications. Chap. 10 in *Handbook of utilities and services for buildings, planning, design, and installation*. New York: McGraw-Hill.

7. Massey, Howard C. Appendix A in *Estimating plumbing costs*. Carlsbad, CA: Craftsman Book Co.

8. Maybeck, Edward M. 1988. Plumbing specifica-

tions. *Plumbing Engineer* (April).

9. Rosen, Harold J. *Constructions specification writing, principles and procedures.* 2d ed. New York: John Wiley & Sons.

10. Steele, Alfred. 1978. Spec is a four-letter word. *Plumbing Engineer* November/December.

## BIBLIOGRAPHY

1. Hartman, Robert. 1991. MASTERFORMAT numbers are for more than specs. *The Construction Specifier* (January).

    Describes ConDoc numbering system.

2. Heineman, Tom. 1990. Less-than-full specifications: Part I. The Construction Specifier (November).

    Describes outline specifications.

3. Heineman, Tom. 1990. Less-full specifications: Part II. *The Construction Specifier* (December).

    Describes short-form specifications.

4. Meier, Hans W. 1989. Short-form specs: A hit or a myth? *The Construction Specifier* (February).

    Discusses short-form specs.

5. Missing link. *The Construction Specifier* (June 1991).

    ARCOM's LincSpec enhances the links between automated drawing and specification systems.

6. Raeber, John A. 1990. Guides and manual as reference standards. *The Construction Specifier* (February).

    Suggested rules for reference standards.

7. Shrive, Charles A. 1991. Sum of the parts: Complementary documents. *The Construction Specifier* (June).

    Describes proper specifying using MASTERFORMAT.

8. Smith, Dana K. 1987. Computerizing spec practices. *The Construction Specifier* (April).

    New microcomputer systems are changing the way specifications are being produced.

9. Trends in formats. *The Construction Specifier* (June 1991).

10. Wright, Victor E. 1988. Spec writing made easier. *Design Graphics World* (November).

    Spec-Writer, by Pinkerton/Galewsky Partnership, is an effective productivity enhancer for the A/E office.

## SOURCES OF ADDITIONAL INFORMATION

For additional information on selected plumbing specification resources, contact the following organizations:

American Institute of Architects (AIA)
1735 New York Ave.
Washington, DC 20006
(202) 626-1472

> AIA publishes the MASTERSPEC guide specification system, which is available in either hardcopy or electronic form.

CAP/Electronic Sweet's
169 Monroe Avenue NW
Grand Rapids, MI 49503-2651
(616) 454-0000

> The Electronic Sweet's program includes SweetSearch, a database index to products in Sweet's catalog file.

Construction Specification Institute
601 Madison
Alexandria VA 22314
(703) 684-0300

> Contact CSI for their Services and Publications Catalog, which describes their extensive resources for plumbing specifiers.

Eclat: The Information Automation Company
7041 Koll Center Parkway, Suite 220
Pleasanton, CA 94566
(415) 484-8400, (800) 533-2528

> Eclat's Electronic Catalog Library is a CD-ROM database of Divisions 15 and 16 products.

National Institute of Building Sciences
1201 L. Street N.W., Suite 400
Washington, DC 20005
(202) 289-7800

> NIBS has assembled government specifications and building regulations onto a CD-ROM.

SPECSystem
999 Peachtree Street NE
Atlanta, GA 30367-5401
(404) 881-9880

> SPECSystem is an automated specification writing system in question-and-answer input format.

SuperSpec
PO Box 47440
Jacksonville, FL 32247
(904) 399-5996

> With SuperSpec's specification processing system, a specifier completes a checklist of project requirements that directs the automated processing of a project specification.

# 4 Plumbing Cost Estimation

This chapter describes a method for estimating plumbing work to aid the engineer in determining the cost of installing the components of a plumbing system. The system described, which here has been used successfully by plumbing contractors throughout the United States, is intended to provide a basis for estimating plumbing projects.

With this method, the cost of labor is brought to an average point. Consideration must then be given to the following[1]:

- Location.
- Height.
- Depth.
- Weather.
- Congestion of people.
- Handling equipment.
- Distance.
- Equipment rooms.
- Laboratory equipment.
- Kitchen equipment.
- Distribution.
- Material procurement.
- Length of construction.
- Required tools and/or equipment.
- Site office/trailer and utilities.
- Owner-furnished equipment.
- Special handling.
- Unions.

Estimating is essentially an accumulation of details, with all the costs calculated. The system of applying labor costs to each piece of material has long been recognized as the most accurate method of compiling labor costs.

The labor units shown here represent an average arrived at by the canvassing of 150 plumbing contractors from all areas of the United States. They were developed and prepared by the National Association of Plumbing–Heating–Cooling Contractors (NAPHCC). These labor units are for land labor only and do not include equipment or operators' time for the placing of heavy material. It is advised that equipment time be allotted for material that takes more than two (2) men to handle, assuming that two men can handle 150 lb (68.04 kg) or less.

When preparing an estimate, it is necessary to make an accurate takeoff list, that is, to list all the pipe fittings and valves in the system. When the list is complete, time for each piece of material is allocated. The total number of hours is the basis for determining the final cost.

To apply man-hours to material with this system, each piece of material is evaluated in terms of "joints." For instance, an elbow is a two-joint fitting, a valve is a two-joint fitting, and a tee is a three-joint fitting. Thus, a joint count would be determined for the following list of materials:

| 4 | elbows | = | 8 joints |
| 2 | tees | = | 6 joints |
| 1 | valve | = | 2 joints |
|   | Total | = | 16 joints |

Add one (1) joint for each 5 ft (1.5 m) of pipe. (This will take care of nipples.)

---

[1] It should be noted that the costs of materials are not included. Current costs for these may be obtained from local vendors or from national cost guides.

***Example 4-1*** The material for a given project contains the fittings given below. Determine the total man-hours required for the project.

| Length of Pipe, ft (m) | Dia. of Pipe, in. (mm) | Total Joints | Joint Allowance (h) | Man-Hours |
|---|---|---|---|---|
| 260 (79.25) | 3 (76.2) thread —pipe | 52 | 0.95 | 49.40 |
| 110 (33.50) | 2 (50.8) thread —pipe | 22 | 0.40 | 8.80 |
| 14 (4.27) | 3 (76.2) thread 90° elbows | 28 | 0.95 | 26.60 |
| 8 (2.44) | 2 (50.8) thread 90° elbows | 16 | 0.40 | 6.40 |
| 2 (0.61) | 3 (76.2) thread —tees | 6 | 0.95 | 5.70 |
| 1 (0.30) | 2 (50.8) thread —tees | 3 | 0.40 | 1.20 |
|  |  |  | Total: | 98.10 |

For a more complete, detailed list of fittings and the required joints, see Table 4-1. Tables 4-2, 4-3, and 4-4 show the man-hours allowed per joint, based on the use of standard materials and normal working conditions.

When estimating, it is necessary to figure overhead and profit before arriving at a final cost. A rule of thumb for overhead is to double the cost of labor, which includes fringe benefits. Another approach is to add a straight 15% markup after marking up the labor cost by 30% to include benefits and burdens (workmen's compensation, etc.). As for profit, 10% is generally acceptable.

If we assume that wages and overhead amount to $50.00 per hour, the total labor cost for Example 4-1 would be $4,905.00 plus profit.

It is most important to know what is, and is not, included in the hours indicated. See Table 4-5.

Tables 4-6 and 4-7 show the man-hours required to perform excavation work. It should be noted that some figures indicate hours required for equipment. The cost of equipment rental can be obtained from a local rental yard.

Example 4-2 is a typical sample of estimating man-hours for trenching by both the hand digging and hand mechanical tamping backfill, and the machine digging and backfill methods.

**Table 4-1  Fittings and Their Joint Numbers**

### Pressure Piping

*Valves*  All hand-operated
Valves: gate, globe, angle, butterfly
Cocks: gas, plug, balancing, lubricated
Checks, strainers
CONSIDER ALL AS 2 JOINT FITTINGS.

*Treated separately*
Pressure-reducing, relief, motor-operated, automatic temperature-regulating, solenoid, expansion, quick-opening

*Tees*  Regular or reducing
Tees, crosses, unions, flanges (2)
CONSIDER ALL AS 3 JOINT FITTINGS.

*Treated separately*
Nozzles—welded tee branches (two sizes smaller than main)

*Ells*  Regular or reducing
Ells: 90°, 45°, 22½°, adapters IPS
Adapters (male and female)
Couplings: reducers, flanges, caps
CONSIDER ALL AS 2 JOINT FITTINGS.

*Pipe*  Nearest 10 linear ft (3.05 linear m) (includes nipples)
CONSIDER AS 2 JOINTS PER 10 LINEAR FT (3.05 LINEAR M).

### Drainage Piping[a]

*Tees*  Regular or reducing
Tees: sanitary, straight, cleanout, tapped, vent
Wyes: inverted, upright, cleanout
Combinations, heel outlet ¼ bends
CONSIDER ALL AS 3 JOINT FITTINGS.

*Bends*  Regular or long
Bends: sweeps, ¼, ⅕, ⅙, ⅛, ¹⁄₁₆
Traps: P, S, running, vented, unvented
Reducers, increasers, offsets, double hubs
CONSIDER ALL AS 2 JOINT FITTINGS.

*Crosses*  Regular or reducing
Crosses: sanitary, straight, tapped, side outlet
Double combinations, double wyes
Tees: Side outlet, double vertical
CONSIDER ALL AS 4 JOINT FITTINGS.

*Pipe*  Nearest 10 linear ft (3.05 linear m), 5 ft (1.52 m), 10 ft. (3.05 m) lengths—single and double joint.
CONSIDER ALL AS 2 JOINTS PER 10 LINEAR FT (3.05 LINEAR M).

[a]Spigot end of fitting counted as a joint when connected.

# Chapter 4 — Plumbing Cost Estimation

### Table 4-2  Man-Hours for Installation of Pipes, Hangers, and Fixtures

| Material | ½ (12.7) | ¾ (19.05) | 1 (25.4) | 1¼ (31.75) | 1½ (38.1) | 2 (50.8) | 2½ (63.5) | 3 (76.2) | 3½ (88.9) | 4 (101.6) | 5 (127.0) | 6 (152.4) | 8 (203.32) | 10 (254.0) | 12 (304.8) | Multipliers |
|---|---|---|---|---|---|---|---|---|---|---|---|---|---|---|---|---|
| Thread | 0.25 | 0.27 | 0.30 | 0.36 | 0.38 | 0.40 | 0.90 | 0.95 | 1.00 | 1.00 | 1.45 | 1.50 | 2.00 | — | — | Base: T&C steel pipe—power drive |
| Copper | 0.20 | 0.21 | 0.25 | 0.27 | 0.30 | 0.32 | 0.63 | 0.75 | — | 0.85 | 1.23 | 1.27 | 1.70 | — | — | Base: 50–50 Type "L"[a]  DWV 1.2  1100° alloy braze 1.3 |
| Groove | — | 0.27 | 0.30 | 0.36 | 0.38 | 0.40 | 0.72 | 0.76 | — | 0.80 | 1.16 | 1.20 | 1.60 | 1.84 | 2.08 | Base: groove-end steel pipe, placing gasket & bolting clamp—power drive |
| Weld | 0.40 | 0.40 | 0.50 | 0.55 | 0.60 | 0.80 | 1.00 | 1.15 | — | 1.60 | 2.00 | 2.40 | 3.20 | 4.00 | 4.80 | Base: BE steel pipe SCH 40  SCH 80 1.3 SKT. weld 0.8 nozzle 2.5 |
| Plastic (PR. PPG.) | 0.20 | 0.21 | 0.25 | 0.26 | 0.27 | 0.28 | 0.40 | 0.50 | — | 0.60 | 0.98 | 1.01 | 1.36 | 1.62 | 2.16 | Base: solvent joint  thermoseal 1.5 |
| V. C. compression | — | — | — | — | — | — | — | 0.30 | — | 0.40 | 0.80[b] | 0.90[b] | 1.00[b] | | | Base: 300 LF pipe  Add crane with operator[b] |
| V. C. mortar | — | — | — | — | — | — | — | 0.50 | — | 0.60 | 1.20[b] | 1.40[b] | 1.50[b] | | | Base: 300 LF pipe  Add crane with operator[b] |
| Wtr. main m. j. | — | — | — | — | — | — | 0.60 | — | 0.62 | — | 0.70[b] | 0.72[b] | 0.80[b] | 0.82[b] | | Base: 300 LF pipe  Add crane with operator[b] |
| Wtr. main compr. | — | — | — | — | — | — | 0.47 | — | 0.48 | — | 0.50[b] | 0.52[b] | 0.54[b] | 0.56[b] | | Base: 300 LF pipe  Add crane with operator[b] |
| Wtr. main caulk | — | — | — | — | — | — | 1.10 | — | 1.15 | — | 1.20[b] | 1.60[b] | 2.00[b] | 2.40[b] | | Base: 300 LF pipe  Add crane with operator[b] |
| Soil waste caulk | — | — | — | — | 0.50 | — | 0.55 | — | 0.60 | 0.65 | 0.70 | 1.20 | 1.30 | 1.50 | | Base: B&S SV cast-iron soil  extra heavy 1.02  silicon 1.20 |
| Soil waste compr. | — | — | — | — | 0.40 | — | 0.45 | — | 0.50 | 0.55 | 0.60 | 1.00 | 1.10 | 1.25 | | Base: B&S SV cast-iron soil  extra heavy 1.02 |
| Soil waste hubless | — | — | — | 0.30 | 0.30 | — | 0.35 | — | 0.40 | 0.45 | 0.50 | 0.80 | 0.90 | 1.00 | | Base: hubless SV cast-iron soil  silicon 1.20  glass 1.30 |
| Stainless—weld | 0.60 | 0.70 | 0.75 | 0.80 | 0.85 | 1.00 | 1.90 | 2.35 | — | 3.10 | — | 4.00 | 6.20 | — | — | Base: Heliarc SCH 40 BE  SCH 10  pipe 0.70  aluminum 1.05 |
| Hangers—ring[c] | — | 0.50 | 0.50 | 0.50 | 0.50 | 0.50 | 0.60 | 0.60 | — | 0.70 | 0.70 | 0.80 | 1.00 | 1.00 | 1.00 | |
| Hangers—roller | — | — | — | — | — | — | 1.40 | 1.40 | — | 1.60 | 1.60 | 1.80 | 2.20 | 2.20 | 2.20 | |

[a]Check local conditions.    [b]Material weighing more than 150 lb (68.2 kg).    [c]Included in hanger, rod, and insert.

***Example 4-2***  Using Tables 4-6 and 4-7, estimate the man-hours for trenching by both the hand digging and hand mechanical tamping backfill, and the machine digging and backfill methods.

*Step 1*  Prepare a short table on trenching man-hours per linear foot.

**Trenching (h/linear ft)**

| Depth (ft) | Yd³/ L Ft | Ditch Width, in. | Hand Trenching Sandy | Hand Trenching Med. | Hand Trenching Clay | Backhoe Short Ditch | Backhoe Long Ditch | Cleveland per h | Grading | Hand Grading |
|---|---|---|---|---|---|---|---|---|---|---|
| 6 | 0.45 | 24 | 0.57 | 0.90 | 1.24 | 0.07 | 0.06 | — | 0.05 | 0.04 |

*Note:* 1 in. = 25.4 mm, 1 ft = 0.3048 m, 1 yd³ = 0.7646 m³.

### Table 4-3  Man-Hours for Installation of Plumbing Fixtures

| | | | |
|---|---|---|---|
| Toilet—floor | 1.80 | Laundry tray | 2.50 |
| Toilet—wall | 2.70 | Urinal—wall | 2.80 |
| Lavatory—wall | 2.00 | Urinal—stall | 3.80 |
| Lavatory—counter | 2.50 | Service sink | 3.00 |
| Bathtub | 3.00 | Drinking fountain | 2.00 |
| Shower | 1.00 | Water cooler—floor | 2.00 |
| Sink—single | 2.00 | Water cooler—wall | 2.50 |
| Sink—double | 2.50 | | |

*Step 2* Prepare a short table on backfilling man-hours per linear foot.

**Backfilling (h/linear ft)**

| Depth, (ft) | Yd³/ LFt | Hand Backfill Flood Sandy | Med. | Clay. | Mechanical Hand Tamper Sandy | Med. | Hard | Dozer S | Stomper T | U |
|---|---|---|---|---|---|---|---|---|---|---|
| 6 | 0.45 | 0.14 | 0.18 | 0.23 | 0.36 | 0.45 | 0.59 | S2 | 0.02A | 0.02 |

*Note:* 1 in. = 25.4 mm, 1 ft = 0.3048 m, 1 yd³ = 0.7646 m³.

*Step 3* Estimate hand digging and hand mechanical tamping backfill.

1. Medium hardness type digging.
2. Average depth = 6 ft.
3. Length = 180 ft.
4. Labor unit for hand trenching = 0.90/ft.
5. Total hours of trenching = 180 x 0.90 = 162 h.
6. Labor unit for mechanical hand tamper backfilling = 0.45.
7. Total hours of backfilling = 180 x 0.45 = 81 h.
   Total man-hours = 243 h.
   S  = Hand tamp first portion.
   S2 = 1-ft-deep hand tamp around pipe.
   T  = Dozer.
   U  = Stomper.
   A  = Average man-hour per linear foot.

*Step 4* Estimate machine digging and backfill.

1. Medium hardness type digging.
2. Average depth = 6 ft.
3. Length = 180 ft.
4. Labor unit for backhoe digging = 0.06/ft (long ditch).
5. Total hours trenching = 180 x 0.06 = 10.8 h.
6. Hand grading allowance = 0.04/ft = 7.2 h.
7. Dozer-stomper backfill
   S = Hand tamp first 1 ft = 0.08/ft = 14.40 h.
   T = Dozer = 0.02/ft = 3.6 h.
   U = Stomper = 0.02/ft = 3.6 h.
   A = Allowance for laborer = 3.6 h.

To recap:

Item #5: 10.8 h x cost of backhoe/h.
Item #6: 7.2 h x cost of laborer.
Item #7 (S): 14.40 h x cost of laborer.
Item #7 (T): 3.6 h x cost of dozer.
Item #7 (U): 3.6 h x cost of stomper.
Item #7 (A): 3.6 h x cost of laborer.

Total cost = sum of all recap items.

**Table 4-4  Man-Hour Tables (Miscellaneous)**

| Installation | | Multiplier |
|---|---|---|
| ***Overhead piping and duct*** | | |
| (Base) | 8 ft (2.44 m) ladder | 1.00 |
| | 10 ft (3.05 m) ladder | 1.03 |
| | 12 ft (3.66 m) scaffold | 1.25 |
| | 16 ft (4.88 m) scaffold | 1.35 |
| | 20 ft (6.10 m) scaffold | 1.50 |
| ***In a crawl space and tunnel*** | | |
| | 3 ft (0.91 m) high | 1.50 |
| ***In a ditch*** | | |
| (Base) | 3 ft (0.91 m) deep | 1.00 |
| | 5 ft (1.52 m) deep | 1.10 |
| | 8 ft (2.44 m) deep—shored | 2.00 |
| | 10 ft (3.05 m) deep—shored | 2.25 |
| | 12 ft (3.66 m) deep—shored | 2.50 |
| ***Handling metallic pipe*** | | |
| | Schedule 10 | 0.93 |
| | Schedule 20 | 0.95 |
| | Schedule 30 | 0.97 |
| (Base) | Schedule 40 | 1.00 |
| | Schedule 60 | 1.03 |
| | Schedule 80 | 1.05 |
| | Schedule 120 | 1.07 |
| ***Distribution of material*** | | |
| | *Distance from stock pile:* | |
| (Base) | 100 ft (30.48 m) | 1.00 |
| | 300 ft (91.44 m) | 1.03 |
| | 500 ft (152.40 m) | 1.04 |
| | 1000 ft (304.80 m) | 1.05 |
| ***Piping of*** | Equipment rooms | 1.20 |
| | Lab equipment | 1.10 |
| | Kitchen equipment | 1.10 |

**Table 4-5  Labor Units for Estimating**

*Include time required for planning, layout, measuring, and the personal needs of the mechanics, plus the following:*

**Pipe and Fittings**

| Handle | Fabricate | Install | Test |
|---|---|---|---|
| Unload | Mark | Join | |
| Assort | Cut | | |
| Stockpile | Ream | | |
| Warehouse (bins) | Put on fitting | | |
| Distribute | | | |

*Not included in the labor units is the time required for the following:*

| Hoisting | Identification | Preparing Plans |
|---|---|---|
| Rigging | Pipe markers | Isometrics |
| Operator | Valve tags | Shop drawings |
| Crane setup | | Detail drawings |
| | | Wiring diagrams |

# Chapter 4 — Plumbing Cost Estimation

## Table 4-6  Excavation Table

### Trenching Man-Hours Per Linear Foot

| Depth (ft) | Yd³/LFt | Ditch Width (in.) | Hand Trenching Sandy | Hand Trenching Med. | Hand Trenching Clay | Backhoe Short Ditch | Backhoe Long Ditch | Cleveland (ft/h) | Grading | Hand Grading[b] | Chain Type[a] Short Ditch | Chain Type[a] Long Ditch |
|---|---|---|---|---|---|---|---|---|---|---|---|---|
| 1 | 0.06 | 18 | 0.07 | 0.11 | 0.16 | 0.01 | 0.01 | 150 | 0.01 | 0.03 | 0.02 | 0.02 |
| 1½ | 0.09 | 18 | 0.11 | 0.17 | 0.24 | 0.02 | 0.01 | 150 | 0.01 | 0.03 | 0.03 | 0.02 |
| 2 | 0.12 | 18 | 0.14 | 0.22 | 0.32 | 0.02 | 0.01 | 125 | 0.01 | 0.03 | 0.03 | 0.02 |
| 2½ | 0.14 | 18 | 0.18 | 0.28 | 0.40 | 0.03 | 0.02 | 125 | 0.01 | 0.04 | 0.04 | 0.03 |
| 3 | 0.17 | 18 | 0.21 | 0.34 | 0.48 | 0.03 | 0.02 | 125 | 0.02 | 0.04 | 0.04 | 0.03 |
| 3½ | 0.20 | 18 | 0.25 | 0.40 | 0.56 | 0.03 | 0.02 | 115 | 0.02 | 0.04 | — | — |
| 4 | 0.22 | 18 | 0.28 | 0.44 | 0.64 | 0.04 | 0.03 | 100 | 0.02 | 0.04 | — | — |
| 4½ | 0.25 | 18 | 0.32 | 0.50 | 0.72 | 0.04 | 0.03 | 100 | 0.03 | 0.04 | — | — |
| 5 | 0.38 | 24 | 0.48 | 0.76 | 1.05 | 0.06 | 0.05 | 100 | 0.04 | 0.04 | — | — |
| 5½ | 0.42 | 24 | 0.53 | 0.84 | 1.16 | 0.06 | 0.05 | — | 0.04 | 0.04 | — | — |
| 6 | 0.46 | 24 | 0.57 | 0.90 | 1.24 | 0.07 | 0.06 | — | 0.05 | 0.04 | — | — |
| 6½ | 0.49 | 24 | 0.62 | 1.00 | 1.38 | 0.07 | 0.06 | — | 0.05 | 0.06 | — | — |
| 7 | 0.52 | 24 | 0.91 | 1.30 | 2.08 | 0.08 | 0.07 | — | 0.05 | 0.06 | — | — |
| 7½ | 0.56 | 24 | 1.00 | 1.43 | 2.28 | 0.08 | 0.07 | — | 0.06 | 0.06 | — | — |
| 8 | 0.60 | 24 | 1.06 | 1.50 | 2.40 | 0.09 | 0.08 | — | 0.06 | 0.06 | — | — |
| 8½ | 0.64 | 24 | 1.13 | 1.60 | 2.56 | 0.09 | 0.08 | — | 0.06 | 0.06 | — | — |
| 9 | 0.68 | 24 | 1.20 | 1.70 | 2.72 | 0.10 | 0.09 | — | 0.07 | 0.06 | — | — |
| 9½ | 0.72 | 24 | 1.27 | 1.80 | 2.88 | 0.10 | 0.09 | — | 0.07 | 0.08 | — | — |
| 10 | 0.72 | 24 | 1.34 | 1.88 | 3.00 | 0.11 | 0.10 | — | 0.08 | 0.08 | — | — |
| 10½ | 0.78 | 24 | 1.41 | 1.95 | 3.12 | 0.11 | 0.10 | — | 0.08 | 0.08 | — | — |
| 11 | 0.82 | 24 | 1.48 | 2.08 | 3.32 | 0.12 | 0.11 | — | 0.08 | 0.08 | — | — |
| 11½ | 0.86 | 24 | 1.56 | 2.18 | 3.48 | 0.13 | 0.11 | — | 0.09 | 0.08 | — | — |
| 12 | 0.90 | 24 | 1.63 | 2.28 | 3.64 | 0.13 | 0.12 | — | 0.09 | 0.08 | — | — |
| 12½ | 0.93 | 24 | 1.70 | 2.41 | 3.78 | 0.14 | 0.12 | — | 0.10 | 0.08 | — | — |
| 13 | 0.96 | 24 | 2.75 | 4.55 | 6.75 | 0.14 | 0.13 | — | 0.10 | 0.08 | — | — |
| 13½ | 1.00 | 24 | 2.86 | 4.73 | 7.10 | 0.15 | 0.13 | — | 0.10 | 0.08 | — | — |
| 14 | 1.04 | 24 | 3.03 | 5.00 | 7.51 | 0.16 | 0.14 | — | 0.11 | 0.08 | — | — |

| | | | | |
|---|---|---|---|---|
| Cleveland | $ |
| Stomper | $ |
| Dozer | $ |
| Chain type machine[a] | $ |
| Backhoe | $ |
| Dirt haul – dozer and truck | $ |

### Concrete Sawing

| Depth of cut (in.): | 1 | 1½ | 2 | 2½ | 3 | 3½ | 4 | 5 |
|---|---|---|---|---|---|---|---|---|
| Per ft of saw cut: | 0.03 | 0.03 | 0.04 | 0.04 | 0.05 | 0.05 | 0.06 | 0.07 |

*Notes:* 1. Conversion factors: 1 in. = 25.4 mm, 1 ft = 0.3048 m, 1 yd³ = 0.7646 m³.  2. Concrete breaking w/ comp.: 10ft/h per linear ft. approx. 2 ft wide (does not include removal). 3. Asphalt breaking w/ comp.: 25 ft/h per linear ft approx. 2 ft wide (does not include removal). 4. Asphalt breaking w/ stomper: 100 ft/h per linear ft approx. 2 ft wide (does not include removal).

[a] "Chain type" refers to a gasoline-driven trenching machine, which digs a maximum of 10 in. wide x 3 ½ ft.

[b] Add hand grading for mechanical trenching only if required.

## Table 4-7 Backfill Man-Hours Per Linear Foot

| Depth (ft) | Yd³/L Ft | Hand Backfill Flood Sandy | Hand Backfill Flood Med. | Hand Backfill Flood Clay | Mechanical Hand Tamper Sandy | Mechanical Hand Tamper Med. | Mechanical Hand Tamper Hard | Dozer-Stomper S | Dozer-Stomper T | Dozer-Stomper U | Chain Type, 10-In. Ditch Width V | Chain Type, 10-In. Ditch Width W | Chain Type, 10-In. Ditch Width X |
|---|---|---|---|---|---|---|---|---|---|---|---|---|---|
| 1 | 0.06 | 0.02 | 0.03 | 0.03 | 0.05 | 0.06 | 0.07 | S1 | 0.003[a] | S1 | 0.02 | 0.003[a] | 0.005[b] |
| 1½ | 0.08 | 0.03 | 0.04 | 0.04 | 0.07 | 0.08 | 0.11 | S2 | 0.004[a] | 0.005 | 0.02 | 0.005[c] | |
| 2 | 0.11 | 0.04 | 0.05 | 0.06 | 0.09 | 0.11 | 0.15 | S2 | 0.005[a] | 0.005 | 0.03 | 0.003[a] | 0.005[d] |
| 2½ | 0.14 | 0.05 | 0.06 | 0.07 | 0.11 | 0.14 | 0.20 | S2 | 0.006[a] | 0.005 | 0.04 | 0.004[a] | 0.01[d] |
| 3 | 0.17 | 0.05 | 0.07 | 0.09 | 0.14 | 0.17 | 0.24 | S2 | 0.008[a] | 0.01 | 0.05 | 0.005[a] | 0.01[e] |
| 3½ | 0.20 | 0.06 | 0.08 | 0.10 | 0.16 | 0.20 | 0.28 | S2 | 0.009[a] | 0.01 | | | |
| 4 | 0.22 | 0.07 | 0.09 | 0.11 | 0.18 | 0.22 | 0.29 | S2 | 0.01[a] | 0.01 | | | |
| 4½ | 0.25 | 0.08 | 0.10 | 0.13 | 0.20 | 0.25 | 0.33 | S2 | 0.02[a] | 0.01 | | | |
| 5 | 0.38 | 0.12 | 0.15 | 0.19 | 0.31 | 0.38 | 0.50 | S2 | 0.02[a] | 0.01 | | | |
| 5½ | 0.42 | 0.13 | 0.17 | 0.21 | 0.34 | 0.42 | 0.55 | S2 | 0.02[a] | 0.01 | | | |
| 6 | 0.45 | 0.14 | 0.18 | 0.23 | 0.36 | 0.45 | 0.59 | S2 | 0.02[a] | 0.02 | | | |
| 6½ | 0.49 | 0.15 | 0.20 | 0.25 | 0.40 | 0.49 | 0.63 | S2 | 0.03[a] | 0.02 | | | |
| 7 | 0.52 | 0.16 | 0.21 | 0.26 | 0.42 | 0.52 | 0.68 | S2 | 0.03[a] | 0.02 | | | |
| 7½ | 0.57 | 0.17 | 0.23 | 0.29 | 0.46 | 0.57 | 0.74 | S2 | 0.03[a] | 0.02 | | | |
| 8 | 0.60 | 0.18 | 0.24 | 0.30 | 0.48 | 0.60 | 0.78 | S2 | 0.03[a] | 0.03 | | | |
| 8½ | 0.64 | 0.19 | 0.26 | 0.32 | 0.51 | 0.64 | 0.83 | S2 | 0.03[a] | 0.04 | | | |
| 9 | 0.68 | 0.21 | 0.27 | 0.34 | 0.55 | 0.68 | 0.89 | S2 | 0.04[a] | 0.04 | | | |
| 9½ | 0.72 | 0.22 | 0.29 | 0.36 | 0.58 | 0.72 | 0.94 | S2 | 0.04[a] | 0.05 | | | |
| 10 | 0.75 | 0.23 | 0.30 | 0.38 | 0.60 | 0.75 | 0.98 | S2 | 0.04[a] | 0.05 | | | |
| 10½ | 0.78 | 0.24 | 0.32 | 0.39 | 0.63 | 0.78 | 1.02 | S2 | 0.04[a] | [f] | | | |
| 11 | 0.83 | 0.25 | 0.33 | 0.42 | 0.67 | 0.83 | 1.08 | S2 | 0.05[a] | [f] | | | |
| 11½ | 0.87 | 0.26 | 0.35 | 0.44 | 0.70 | 0.87 | 1.13 | S2 | 0.05[a] | [f] | | | |
| 12 | 0.91 | 0.28 | 0.37 | 0.46 | 0.73 | 0.91 | 1.19 | S2 | 0.05[a] | [f] | | | |
| 12½ | 0.95 | 0.29 | 0.38 | 0.48 | 0.77 | 0.95 | 1.24 | S2 | 0.05[a] | [f] | | | |
| 13 | 0.99 | 0.30 | 0.40 | 0.50 | 0.80 | 0.99 | 1.30 | S2 | 0.06[a] | [f] | | | |
| 13½ | 1.04 | 0.32 | 0.42 | 0.52 | 0.83 | 1.04 | 1.35 | S2 | 0.06[a] | [f] | | | |
| 14 | 1.10 | 0.33 | 0.44 | 0.55 | 0.88 | 1.10 | 1.43 | S2 | 0.06[a] | [f] | | | |

*Notes*: 1. Conversion factors: 1 in. = 25.4 mm, 1 ft = 0.3048 m, 1 yd$^3$ = 0.7646 m$^3$. 2. Sand: 1 yd = 1.3 ton. 3. S: Hand tamp first portion. S1: Must hand tamp all the way. S2: 1 ft hand tamp up to 3-in. pipe = 0.8/ft. 1½ ft hand tamp 4-10-in. pipe = 0.12/ft. T: Dozer. U: Stomper. V: Hand flood. W: Backhoe—flood. X: Backhoe—stomper. 4. Always figure a minimum of 4 h for all equipment. Also, ditches must be at least as long as is indicated in notes b, c, d, and e before equipment may be used. If ditches are shorter, hand laborer must be used.

[a] Must add for stand-by of hand laborer.

[b] If ditch is over 250 ft long.

[c] If ditch is over 200 ft long.

[d] If ditch is over 150 ft long.

[e] If ditch is over 100 ft long.

[f] Call equipment company for price.

# Chapter 4 — Plumbing Cost Estimation

**APPENDIX — WORKSHEETS**

## WATER PIPING—SAMPLE ESTIMATING FORM

| Size | 90° EL | 45° EL | Cplg. | Gate Valve | Check Valve | Union | Hose Bibb | Back-Flow Prev. | PRV | Tee | Reducing Tee | "K" Pipe Length | "L" Pipe Length | Misc. |
|------|--------|--------|-------|------------|-------------|-------|-----------|-----------------|-----|-----|--------------|-----------------|-----------------|-------|
|      |        |        |       |            |             |       |           |                 |     |     |              |                 |                 |       |
|      |        |        |       |            |             |       |           |                 |     |     |              |                 |                 |       |
|      |        |        |       |            |             |       |           |                 |     |     |              |                 |                 |       |
|      |        |        |       |            |             |       |           |                 |     |     |              |                 |                 |       |
|      |        |        |       |            |             |       |           |                 |     |     |              |                 |                 |       |
|      |        |        |       |            |             |       |           |                 |     |     |              |                 |                 |       |
|      |        |        |       |            |             |       |           |                 |     |     |              |                 |                 |       |
|      |        |        |       |            |             |       |           |                 |     |     |              |                 |                 |       |
|      |        |        |       |            |             |       |           |                 |     |     |              |                 |                 |       |
|      |        |        |       |            |             |       |           |                 |     |     |              |                 |                 |       |

# ESTIMATE WORKSHEET

Ind. Dept. _____  Prepared By _____
Type of Estimate _____  Sheet ___ of ___

| Labor Category | Rate |
|---|---|
| LABORER | $ ___ /HR. |
| CARPENTER | $ ___ /HR. |
| CONCRETE FINISHER | $ ___ /HR. |
| IRONWORKER | $ ___ /HR. |
| MILLWRIGHT | $ ___ /HR. |
| ELECTRICIAN | $ ___ /HR. |
| PIPEFITTER | $ ___ /HR. |
| MASON | $ ___ /HR. |
| RIGGER | $ ___ /HR. |
| PAINTER | $ ___ /HR. |

| ITEM NO. | DESCRIPTION | OWNER FURNISHED ||  CONTRACT WORK ||||||| GRAND TOTAL |
|---|---|---|---|---|---|---|---|---|---|---|
| | | | | MATERIAL & EQUIPMENT ||| LABOR ||| |
| | | QUAN. | TOTAL | QUAN. | PRICE | TOTAL | M.H. | RATE | TOTAL | |

**CHECK ITEMS INCLUDED:**
STRUCTURAL   PAINTING   SPRINKLER CHANGES
MECHANICAL   CUTTING & PATCHING   GUARDS
PIPING   UNLOADING   INTERFERENCES
ELECTRICAL   SHEET METAL   SAFETY EQUIPMENT
INSTRUMENTATION   DEMOLITION   TAXES
INSULATION   SPECIAL HANDLING   FREIGHT
PERMITS

JOB NO. ☐☐☐☐☐   DOC. NO. ☐☐☐   IND. DEPT. ☐☐   MAJOR CLASS. ☐☐☐☐   DATE ☐☐

# UNDERGROUND PIPING
# BILL OF MATERIAL

Job No. _____  Project: _____
Estimator: _____  Location: _____
Date: _____  Page: ____ of ____
Dwg: _____
Rev: _____

| Rev. | Date | By | Reqn. |
|------|------|----|----|
|      |      |    |    |
|      |      |    |    |

**Legend: Spec**
(1) Cast-Iron Soil Pipe – Svc. Wt.
(2) Cast-Iron Soil Pipe – Ex. Hvy.
(3) Cast-Iron Soil Pipe – No Hub
(4) Vitrified Clay Pipe – Ex. Hvy.
(5) PVC – DWV
(6)

**Legend: Joint Type**
(A) Push-On (Rubber O-Ring)
(B) Caulked – Lead and Oakum
(C) Mechanical Joint
(D) Caulked – Corrosion Res.
(E)
(F)

Service Description / Line No. / Joint / Spec / Size

Columns: Pipe L.F. × 10' (S.H., D.H.) | Pipe L.F. × 5' (S.H., D.H.) | Sweep ¼ Bends (Long, Sht.) | ¼ Bend 90° | ⅛ Bend 45° | Bend | Y-Branch (Single, Double, Reduc'g) | Combination Y + ⅛ Bend (Single, Reduc'g Single, Double, Reduc'g Double) | Sanitary T-Branch (Single, Reduc'g, Tapped Single, Tapped T-Branch)

Service Description / Line No. / Joint / Spec / Size

Columns: Floor Drain | Cleanout | Increaser | Reducer | S-Trap (No Vent, With Vent) | ½ S-Trap (No Vent, With Vent) | Deep Seal Trap | Running Trap | Plug for Hub | Iron Ferrule w/ Brass Plug | Vent Branch | Seals Rubber O-Ring

# 5 Job Preparation, Plumbing Drawing, and Field Checklists

## GENERAL

1. Determine the scope of the project (systems to be used, cost restrictions, code restrictions, etc.) prior to project start-up.

2. A plumbing job file should be started, properly identified with job name and number. All data, memorandums, calculations, etc. related to the job should be filed in this folder or three-ring binder. All calculations and design data should be placed on standard reproducible sheets.

3. In preparing and issuing drawings, a sheet number should be obtained, and the title should be clearly indicated.

4. Neither sketches nor drawings should be issued without being checked and approved by the department head.

5. Office instructions, checklists, design criteria, calculations, etc. should be filed in the job folder and kept up to date.

6. Compile a team directory, including owner, architect, engineers, consultants, contractors, etc. with telephone and fax numbers, addresses, and e-mail addresses.

## SUGGESTED ITEMS TO BE CHECKED AND INFORMATION TO BE OBTAINED AT THE START OF A JOB

1. Legal description and location of the building(s) on the site, including property lines and address.

2. Size, location, and depth of all adjacent, available sanitary and storm sewer, water, and gas mains.

3. Available water flow and pressure (static and residual) at a given elevation. Obtain water-quality analysis.

4. If job is an alteration or connects to an existing building, existing services and existing loads should be checked, as well as water-heating equipment, pumps, ejectors, backflow preventers, etc.

5. A check with the electrical engineer should be made regarding current characteristics for motors and controls and the maximum size of motors that can be started across the line.

6. A check should be made with the mechanical engineer regarding the availability of steam or boiler feed water for heating domestic water. Boiler water temperature and steam pressure as well as condensate requirements from HVAC or other condensate-producing equipment should also be checked.

7. All codes and ordinances that will be enforced should be checked, and it should be determined if any amendments have been made to the code.

8. A check should be made of fire protection requirements, including fire hose, automatic sprinklers, fire department connections, Halon systems, etc.

9. The need for other systems, such as natural gas, compressed air, vacuum, distilled wa-

ter, deionized water, oxygen, nitrous oxide, acid waste, fuel oil, and steam, should be determined.

10. Detailed layouts for all equipment requiring plumbing connections should be made.

11. A preview should be made, together with the mechanical and electrical engineer, of the mechanical room layout, looking specifically for the locations of the floor drains and connected loads for water, sewer, and gas to equipment.

## SUGGESTED PROCEDURE FOR THE DESIGN OF PLUMBING WORK AND THE PREPARATION OF PLUMBING DRAWINGS

1. Count the plumbing fixtures and estimate the following:

    A. Size of building drain.

    B. Size of water service.

    C. Size of gas main.

    D. Hot water load (tank and heater size).

2. Determine if sump pumps or sewage ejectors are required and properly size. (Does relief port from backflow preventers discharge to sump pump?)

3. Determine if gravity, pneumatic tanks or tankless systems are required.

4. Locate roof drains, leaders, fire standpipe risers, main stacks, sprinkler alarm valves, risers, etc. Verify space requirements, plumbing wall thickness, beams under fixtures, etc.

5. In conjunction with the architect, determine if there are sufficient numbers of plumbing fixtures and the requirements regarding drinking fountains with water, drinking fountains with central chilled water, or self-contained water coolers. Determine requirements for the physically challenged.

6. Determine proper pipe spaces and wall thickness for plumbing work and coordinate with the architect. Check structural drawings for footings, grade beams, beams under plumbing walls, etc.

7. Notify the electrical engineer of the electrical requirements of the plumbing work, such as electrically heat-traced piping, motors, electric water heaters, water coolers, controls, etc.

8. Notify the mechanical engineer of the heating requirements (domestic water heating requirements, if steam or high-temperature water is used) of the plumbing work.

9. Ascertain from the mechanical engineer the drain, water, and gas requirements of the mechanical equipment.

10. Check with mechanical and electrical engineers regarding space conditions for risers, fire hose cabinets, sprinklers, etc.

11. Check the structural drawings for space conditions affecting the plumbing work.

12. All plumbing work should be designed in accordance with good engineering practice and meet the requirements of all applicable codes.

13. All fire standpipe and sprinkler work should be designed in accordance with the authority having jurisdiction, the building code, and the requirements of the National Fire Protection Association (NFPA), the fire marshall, and the owner's insurance company. Determine if a fire pump is required.

14. No piping shall be installed in transformer vaults, switchboard rooms, telephone rooms, or elevator equipment rooms.

15. Floor drains should be provided in rooms with the following equipment:

    A. Pumps, refrigeration compressors, air compressors, vacuum pumps, boilers, water heaters, air conditioning equipment, water softeners, etc.

    B. In commercial kitchens, near dishwashers, steam kettles, large refrigerators, etc.

    C. In toilet rooms, particularly those with more than one water closet.

    D. Provide trap primers on infrequently ulitized floor drains.

16. In general, pressure-reducing valves shall be provided for domestic water lines where static pressure exceeds 80 psi (551.2 kPa). (Refer to the local code.)

17. Water service should be provided with a main valve, and where a building has two or more services, each should be provided with a backflow prevention device. All fire standpipe and sprinkler services shall be provided with a check valve. Verify the need and type of

# Chapter 5 — Job Preparation, Plumbing Drawing, and Field Checklists

approved backflow preventers and their location (check with the local plumbing inspector).

18. Hot water systems should be provided with a circulating return, unless the distance between the heater and farthest fixture is relatively short or as determined by code. Hot water systems with long horizontal runs should be provided with a temperature maintenance system. Specifying automatic flow-control devices will eliminate the very difficult task of balancing the system. The character of the building's use should determine the use of duplex water heating equipment and circulating equipment. Hot water systems should be insulated throughout.

19. Plumbing fixtures on floors below grade, when drained to the sewer, should be provided with backwater valves. Sewage ejectors should be used only where absolutely necessary.

20. Wherever possible, all water supplies to mechanical equipment should be provided with an air gap connection. All direct connections to refrigerating compressors, boilers, blow-off tanks, cooling towers, etc. shall be provided with backflow preventers.

21. House pumps should be provided with spring-loaded check valves in discharge instead of the swing checks.

22. All risers should be provided with control valves and drains.

23. Fire extinguishers should be provided in all fire hose cabinets and elsewhere as directed by the NFPA. ABC rated fire extinguishers should be provided in boiler rooms, kitchens, and all rooms containing electrical switchboards and motors.

24. Riser diagrams shall show water service, water heaters, sump pumps, ejectors, house pumps, house tanks, etc.

25. Coordinate ceiling heights (from the architectural drawings) with the plumbing work.

26. Hose bibbs should be installed in all machinery rooms, kitchens, incinerator rooms and rooms containing floor drains. Provide with vacuum breaker or backflow preventer, as directed by code.

27. Frost-proof hose bibbs should be provided where freezing temperatures are possible.

28. In general, piping should be run as directly as possible. However, except for piping buried underground, all piping should be run parallel to or at right angles to the walls, partitions, etc. and should be neatly grouped in parallel lines. Cold water lines should be insulated where condensation could present a problem.

29. Pipes rising within a story should be noted as "Rise." Pipes rising to another story should be noted as "Up." Pipes dropping to another story should be noted as "Dn." Pipes at ceiling should be noted as "At Ceiling," when exposed and as "Abv Ceiling" when concealed. Pipes under the floor, other than obvious fixture drain lines, should be noted as "Blw Floor," "At Ceiling Below," or "Abv Ceiling Below."

30. Roof drains and stacks through the roof should be kept 12 to 18 in. (0.3 to 0.5 m) away from all parapet walls, building offsets, roof openings, etc. to allow for proper flashing.

31. Unless the architect or owner has a requested list of symbols to use, standard symbols and abbreviations should be utilized on all plumbing drawings.

32. When attending meetings related to the job, detailed notes should be taken on all items of discussion pertaining, either directly or indirectly, to the plumbing work. (After the meeting, a confirming memorandum should be written, with copies sent to those present at the meeting.)

33. In general, piping should be run to clear steel and concrete beams. Where absolutely necessary, piping may penetrate beams. Where it is necessary to clip beam flanges or run pipes through the web of steel beams or concrete beams, permission from the structural engineer should be obtained and confirmed, and all such special conditions should be clearly noted on the drawings and confirmed in writing to the structural engineer.

34. All motor controllers, remote control stations, alarm panels, remote alarms, flow switches, and electrode control units should be located and noted on the drawings or details.

35. All information received from or given to the architect, owner, etc., either directly or on the telephone, should be immediately confirmed with a memorandum and copies sent to the person and other interested or affected parties.

36. The following information should be obtained from the utilities and authorities having jurisdiction, with the necessary approvals:

A. *Fire department* Street fire hydrants, on-site hydrants, flow demands, type of hydrant preferred and meter requirements, approval of detector check valve, fire department connection, and fire main.

B. *Building department* Location, size, depth, etc. of sanitary and storm sewers, sewer laterals, points of connection, grease traps, etc.

C. *Water department* Water pressures, location and size of water mains, hardness, backflow preventer requirements.

D. *Gas company* Meter location, pressures, size of meters, etc.

37. Provide a room or space for all plumbing-related equipment, such as water heaters, water softeners, pumps, tanks, backflow preventers, etc.

# Chapter 5 — Job Preparation, Plumbing Drawing, and Field Checklists

## PLUMBING DRAWING CHECKLIST

JOB NAME _____

LOCATION _____

DATE _____

ENGINEER _____

*Note: Mark off each item after it is checked!*

1. Completeness of titles and title block, with all information filled in.
2. North arrow, company name, column numbers and room names and/or room numbers on all plans.
3. General notes, legends, and symbols.
4. Floor elevations on all riser diagrams or on floor plans (if no risers).
5. Reference notes between plans and details.
6. Show fixture units on all waste and vent stacks, water risers, and water mains.
7. Check with all disciplines, particularly structural, for clearances, etc.
8. Verify that the electrical engineer has provided for plumbing electrical work, including fire protection systems (fire sprinkler alarm), connection to the building's fire alarm system, emergency requirements (fuel oil, tanks, pumps) booster pumps, and electric water heater.
9. Protection against freezing, for wet standpipes, valved outlets, stop and drain valve to roof, exposed pipes against exterior walls, in attic spaces, hose bibbs, dry barrel fire hydrants, etc.
10. Send coordination drawings to:
    Electrical engineer_____Dated_____
11. Send coordination drawings to:
    Architect engineer_____Dated_____
12. Send coordination drawings to:
    Structural engineer_____Dated_____
13. Indicate elevation of house (building) drain leaving the building. Check if lowest gravity line is below curb/property line to see if backwater valve is required (only on those fixtures actually below floor sinks/floor drains, etc.).
14. Reduced pressure backflow devices and/or double check backflow preventers, where required, to protect domestic water systems.
15. Acid waste system (acid vents, acid floor sinks, etc.).
16. Indicate hose bibbs etc. around building, landscape watering, shut-off valves, and frost-proof hose bibbs, if required.
17. Provide expansion loops, bends, joints, guides, anchors, etc. in piping, where needed.
18. Sectional shut-off valves and access panels for water, gas, air, and steam piping systems are provided, where needed.
19. Water softeners with bypass and backwash piping. Verify capacity of drain for backwash flow rate.
20. Adequate wet standpipes (fire hose cabinet) and dry standpipes coverage.
21. Housekeeping pads under all floor-mounted equipment.
22. Combustion air for gas-fired equipment, water heaters boilers, etc.
23. Make-up water for air-conditioning system, boilers, cooling towers, etc.
24. Coordinate connections for food preparation equipment.
25. Adequate drinking fountains.
26. Cooling tower/evaporative condenser bleed and overflow to sanitary, storm, or catch basin.
27. Adequate floor drains are provided in public toilet rooms and at pieces of equipment that discharge to drain.
28. Computer-assisted drawings (CADs) require established procedures that minimize labor requirements and maximize drawing production and efficiency. Some of these procedures include the use of blocks, dimensioning, stamp dating, etc.

*Note: This checklist should be regarded only as a starting reference. Users of the checklist are urged to modify such list to fit the particular application.*

# FIELD CHECKLIST

Most plumbing inspections can be broken down into three classifications: (1) underground, (2) finished rough and topout, and (3) setting of fixtures and finish, including adjustment. In general, the engineer should make a minimum of three visits to the job site. The following is a list of important items that should not be overlooked by the engineer when visiting the job site.

## Underground Inspection

The first inspection is usually when the underground system, or portion thereof, is being tested and made ready for backfill. Systems and materials that should be inspected at this time include:

### Building drain, storm sewer

1. General alignment and conformity to layout, with particular attention to deviations, close or congested areas, and fittings that could cause a stoppage.
2. Alignment and workmanship of joints, with particular attention to method of prefill and proper pipe support to prevent sagging.
3. Slope or pitch of lines and how determined (string level, instrument, etc.). Questionable portions should be checked with surveyor's instruments and rod.
4. Spacing and accessibility of cleanouts.
5. Direction, slope, and termination of vent connections.
6. Sleeves for material and centering of pipes to allow for expansion and contraction–also caulking and waterproofing, where required.
7. Pipe size increases at specified points. Invert elevation at termination.
8. Sumps, sand traps, grease or oil interceptors, manholes, and other structures for size, level, and proper rim and invert elevations.
9. Trap primer connections in place and terminated above floor. All temporary terminations of piping plugged or capped to prevent entry of foreign material.
10. Acid or industrial wastes that require special materials, dilution tanks, settling tanks, or basins should receive special attention to ensure that they have been installed to perform safely and properly.

### Water and gas

1. Compliance with utility company's regulations on connections to mains, locations and installation of shut-off valves or cocks, meters, pressure-reducing valves, and bypasses, and accessibility for replacement or repair.
2. Water mains for depth of installation, alignment, location of stubs and shut-off valves. Lawn sprinklers or future connections (if required) and installation, size, and placement of thrust blocks.

## Above-Ground Inspection

1. Installation and placement of expansion joints or loops and anchors, with attention to types of materials used.
2. Specified thickness, continuity, and waterproofing on insulations of fittings and piping, including cold water lines where condensation might occur.
3. Condensate drains, relief valve drains, and indirect wastes in place, properly anchored, supported, and terminated, with air gaps to their receptacles.
4. Proper location and elevation of drains, such that they can perform the intended function. Adequate slope of floors and areas to drains are important. Rims of receptacles intended for other use should be set at an elevation to preclude their use as floor drains.
5. Locations and weather tightness of flashings of piping through walls and roof, for sanitary vents, clearance to openings and outside air intakes of air conditioning equipment.
6. Fireproofing of pipe penetrations at fire separations and access panels for valves.
7. Connection of elements of the plumbing system to those of other trades in boiler rooms, mechanical areas, and on rooftop equipment with regard to proper use of material, arrangement of piping, unions, and valves for repair or replacement, backflow prevention, trap priming, and drainage.
8. Proper hanging of all pipes at ceiling. Check for type of hangers being used and if they meet the intent of the specifications and/or industry standards.

9. Check piping configuration of all pieces of major equipment (i.e., boosters, pumps, water heaters, etc.), that they meet the intent of the drawings and/or specifications.

## Setting of Fixtures and Finish Inspection

The engineer should make a final inspection after all piping has been covered and fixtures are in place.

Water systems should be checked for compliance with sterilization procedures, temperature, and balance. All fixtures and trims should be properly aligned at specified heights and solidly mounted. All pieces of special equipment should be fully tested and owners or operators familiarized with their operation. Finally, the engineer should make sure that the necessary manuals, parts lists, control diagrams, and other data have been properly assembled for the owner's or operator's files.

# Nature created bacteria, mold, and mildew for specific purposes.

**(None of which include bathing.)** Aqua Glass gelcoat tubs and showers now have built-in Microban® antibacterial protection. Microban® protection inhibits the growth of bacteria, mold, and mildew that cause odors and stains, so Aqua Glass bathing fixtures are much easier to keep clean. Microban® is also engineered to last the life of the unit. It's everything your customers have been clamoring for. So when you install an Aqua Glass tub or shower in your customers' homes, you help create an indulgent atmosphere.

*Call 1-800-238-3940 or visit www.aquaglass.com.*

### AQUA GLASS®
With built-in Microban® antibacterial protection.
A Masco Company

Microban® antibacterial protection

*Pure Indulgence.*™

Microban and the Microban symbol are trademarks of the Microban Products Company, Huntersville, NC.

# 6 Plumbing for Physically Challenged Individuals

## INTRODUCTION

The plumbing engineer must be prepared to provide adequate facilities for physically challenged persons, whether or not the requirements for these facilities are covered specifically in the local jurisdiction's applicable code. Most US plumbing codes today include some type of provision for the physically challenged. Also, the Americans with Disabilities Act (ADA) of 1990 includes plumbing provisions. The plumbing engineer must determine which codes are applicable to the project he or she is designing and incorporate any provisions these codes require in addition to ADA requirements.

This chapter presents background information on past and current legislation affecting plumbing for the physically challenged and design requirements for compliance with ANSI A117.1-1998 and the Americans with Disabilities Act Accessibility Guidelines for Buildings and Facilities (ADAAG), July 26, 1991. Throughout this chapter, there are references to standards and guidelines giving dates of issue. The reader must be sure to review and reference the latest editions of these documents, in accordance with those documents listed and referred to in local codes.

## BACKGROUND

Many design and construction features of buildings and facilities caused problems for individuals with physical impairments. These architectural barriers make it difficult for physically challenged persons to participate in educational, employment, and recreational activities.

In 1959, a general conference was called and those groups vitally interested in the problem of accessibility were invited to participate and be represented. The attendees recommended the initiation of a standards-development project to study the cases and to prepare a national document.

In 1961, the American Standards Association (now the American National Standards Institute) issued the American Standard *Specifications for Making Buildings and Facilities Usable by the Physically Handicapped,* ASA A117.1-1961. This document was reaffirmed in 1971 with no changes and redesignated as A117.1-1961 (R1971). In 1998, the standard was renamed *Accessible and Usable Buildings and Facilities.*

The US Department of Housing and Urban Development, along with the National Easter Seal Society and the President's Committee on Employment of the Handicapped (the original co-secretariat of the A117 standards committee), sponsored two (2) years of research and development to revise the A117.1 standard in 1974. This work (extended to include residential environments) resulted in the 1980 version of this standard. The scope of ANSI A117.1-1980 was greatly expanded. Curb ramps, accessible bathrooms and kitchens, and other elements of housing were included in the standard; an appendix was added in order to assist the designer in understanding the standard's minimum requirements; and more illustrations were incorporated.

The standard was also upgraded in 1985, in compliance with ANSI standard practice, which requires a review every five years at the minimum. The standard, issued as ANSI A117.1-1986, further reinforced the concept that the standard is basically a resource for design specifications and leaves to the adopting, enforcing agency application criteria such as where, when, and to what extent such specifications will apply. Clarification of this "how-to" function of ANSI A117.1-1986 facilitated its referencing in building codes and federal design standards—a major step toward achieving uniformity in design specifications.

The technical data contained in ANSI A117.1-1986 were expanded greatly to incorporate additional elevator and plumbing data as well as, for the first time, specifications for alarm and communications systems for use by individuals with visual or hearing impairments.

The technical data contained in the 1986 issue have been used as the basis of most state and local codes, as well as the Uniform Federal Accessibility Standard (UFAS) and the US Architectural and Transportation Barriers Compliance Board (ATBCB) requirements.

As part of an ongoing review process, the A117.1 committee was reconvened in 1989 with the intention of reissuing the standard in 1990. The magnitude of the changes, both in technical data and in format, resulted in a delay in publication of the standard until December 15, 1992. This standard was the most comprehensive to date and the involvement from disability advocates and interested parties was remarkable. The 1992 standard is now referenced in several model codes and has resulted in improved accessibility in many regards.

In 1995 the A117.1 committee was called again and charged with the task of reviewing the standard for changes. The makeup of the committee had grown to include many disability advocacy groups, model code representatives, associations—including the American Society of Plumbing Engineers (ASPE)—and design professionals. The committee worked for more than three (3) years, through three (3) public reviews, examining over 1000 proposed changes, during 23 days of meetings to produce the 1998 ANSI A117.1-1998 standard. The 1998 standard has been developed to work in harmony with federal accessibility laws, including the current *Fair Housing Accessibility Guidelines* and the proposed *Americans with Disabilities Act Accessibility Guidelines* (ADAAG).

New provisions for type B dwelling units are intended to provide technical requirements consistent with the *Fair Housing Accessibility Guidelines* of the US Department of Housing and Urban Development (HUD). HUD is currently in the process of reviewing the 1998 standard to determine equivalency with the guidelines. In addition, the A117.1 committee worked closely with the ADAAG Review Federal Advisory Committee to harmonize the 1998 edition with proposed revisions to ADAAG. The US Architectural and Transportation Barriers Compliance Board (Access Board) is expected to publish a "notice of proposed rule making" later this year.

## LEGISLATION

In 1969, Public Law 90-480 (known as the Architectural Barriers Act of 1968) was signed by US President Lyndon B. Johnson. The main thrust of this legislation was that any building constructed, in whole or in part, with federal funds must be made accessible to, and usable by, the physically challenged. Public Law 93-112, known as the Rehabilitation Act of 1973, was passed by the federal government in 1973.

State and municipal governments also began issuing their own ordinances regarding architectural barriers. These legislative acts were usually modified versions of the ANSI A117.1 document. At the present time, just about every state has adopted some legislation covering this subject; however, there are major differences from one ordinance to another. Like the federal government, the original legislation usually applied to government-owned or government-financed structures, but now the requirements generally apply to all public accommodations.

The Americans with Disabilities Act (ADA or "the Act") was enacted by the US Congress and signed by US President George Bush on July 26, 1990. The ADA prohibits discrimination based on physical or mental disabilities in private places of employment and public accommodation, in addition to requiring transportation systems and communication systems to facilitate access by the disabled. The Act is modeled to a considerable extent on the Rehabilitation Act of 1973, which applies to federal grantees and contractors.

The ADA is essentially civil rights legislation, but its implementation has a major impact on the construction industry. In order to clarify construction requirements, the US attorney general's office commissioned the US Architectural and Transportation Barrier Compliance Board (ATBCB) to prepare architectural guidelines to ensure that the construction industry understood what was required in order to comply with the Act. The ATBCB, which is represented on the A117.1 committee, used much of the completed how-to data that were available from A117.1, and where-to data from the ongoing scoping work being done by the Board for Coordination of Model Codes (BCMC), their governmental experiences, and public comments to produce the guidelines commonly referred to as "ADAAG."

After incorporating public comments, the "final rule" was issued on July 26, 1991, in the federal register (28 CFR Part 36) as "Nondiscrimination on the Basis of Disability by Public Accommodations and in Commercial Facilities." The Act became effective on January 26, 1992, and applies to all construction with application for permit after January 26, 1992. This "final rule" preempted state and local laws affecting entities subject to the ADA, to the extent that those laws directly conflict with the statutory requirements of the ADA. The attorney general's office established as a procedure for the certification of state and local accessibility codes or ordinances that they meet or exceed the requirements of the ADA. It was hoped that, with such a certified code enforced by local inspectors, compliance with ADA would not be decided in the courts.

In 1994, the ATBCB commissioned a new committee to make recommendations for an improved document to replace the current *Americans with Disabilities Act Accessibility Guidelines* (ADAAG). This committee met for more than two (2) years to review proposed changes to the document and to remove the ambiguities that have been a cause of contention to designers as well as code enforcement officials. This new committee included 22 members representing: advocacy groups (American Council of the Blind; Disability Rights Education and Defense Fund, Inc; Eastern Paralyzed Veterans Association; Maryland Association of the Deaf; World Institute on Disability), code enforcement officials [Virginia Building and Code Officials Association; Texas Department of Licensing and Regulation; Southern Building Code Congress International, Inc. (SBCCI); National Fire Protection Association (NFPA); National Conference of States on Building Codes and Standards; International Conference of Building Officials (ICBO); Council of American Building Officials (CABO); Building Officials and Code Administrators International, Inc. (BOCA)], and designers (AIA, American Society of Interior Designers). The document that came from this committee's work was presented to the ATBCB on October 10, 1996, and it is expected that the ATBCB will issue a "notice of proposed rule making" late in 1998. Design professionals must continue to review the ADA in its entirety, and the forthcoming revisions, as well as state and local codes for application to their projects. Some states require preapproval of accessible plumbing fixtures. Approval of the fixtures is the responsibility of the fixture manufacturers, but the specifier must specify and approve only those fixtures that have received approval.

There are still a number of concerns regarding whether the established standards properly address the specific needs of children and the elderly. Children cannot necessarily reach fixtures set at established heights for the physically challenged. Also, the elderly may have trouble accessing fixtures set low to meet established height requirements for the physically challenged.

## DESIGN

Although plumbing is only a small portion of the overall effort to create a totally barrier-free environment, it is one of the most important areas to be dealt with by the engineers.

The following are the various classifications of disabilities:

- *Nonambulatory disabilities* Those that confine individuals to wheelchairs.
- *Semiambulatory disabilities* Those that necessitate individuals to require the aid of braces, crutches, walkers, or some other type of device in order to walk.
- *Sight disabilities* Total blindness and other types of impairment affecting an individual's sight.
- *Hearing disabilities* Total deafness and other types of impairment affecting an individual's hearing.

**Table 6-1 Graphic Conventions**

| Convention | Description |
|---|---|
| 36 / 915 | Typical dimension line showing US customary units (in in.) above the line and SI units (in mm) below. |
| 9 / 230 | Dimensions for short distances indicated on extended line. |
| 9  36 / 230  915 | Dimension line showing alternate dimensions required. |
| ⇐ | Direction of approach. |
| max | Maximum. |
| min | Minimum. |
| – – – – – – – | Boundary of clear floor area. |
| ——— – ——— – ——— ₵ | Centerline. |

*Note:* Dimensions that are not marked "minumum" or "maximum" are absolute, unless indicated otherwise in text or captions.

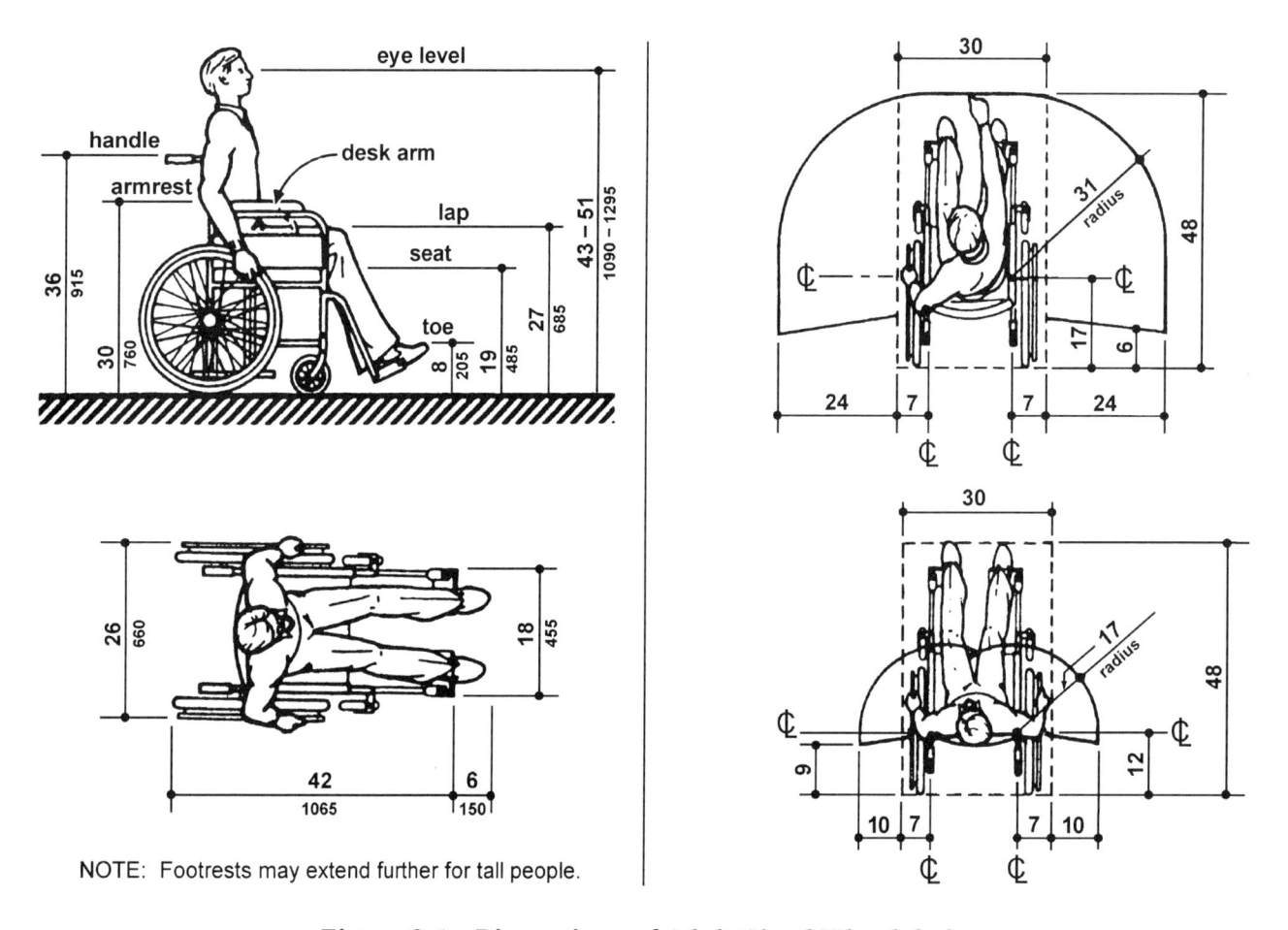

NOTE: Footrests may extend further for tall people.

**Figure 6-1  Dimensions of Adult-Sized Wheelchairs**

# Chapter 6 — Plumbing for Physically Challenged Individuals

- *Coordination disabilities* Those caused by palsy due to cerebral, spinal, or peripheral nerve injury.
- *Aging disabilities* Those brought on by the natural process of aging, which reduces mobility, flexibility, coordination, and perceptiveness in individuals. (*Note*: To some extent, various national standards—e.g., US Dept. of Housing and Urban Development's *Minimum Property Standards*—differentiate the elderly from the "physically handicapped.")

The disability classifications that affect the plumbing engineer the most, in terms of design, are the nonambulatory and the semiambulatory groups. Adequate plumbing facilities must be provided for these individuals. The plumbing engineer should analyze the needs of a person confined to a wheelchair and those forced to use walking aids such as crutches and braces. The plumbing designer should become familiar with the characteristics of the wheelchair and various associated types of equipment to make a complete analysis of the prevailing conditions. At the present time, there are many variations in wheelchair design available on the market. The specifications in these guidelines are based on adult dimensions and anthropometrics. An illustration of a typical wheelchair design is shown in Figure 6-1.

In addition to the dimensions of the wheelchair, the plumbing engineer must take into consideration how wheelchairs are employed and how the person in a wheelchair utilizes plumbing fixtures.

The following information on fixture requirements for the use of the physically challenged is based on the recommended design criteria contained in the proposed ANSI A117.1-1998. For convenient reference to the ANSI A117.1 text, the corresponding ANSI article numbers have been used (e.g., "601.1" and "602.5"). Illustrations, in most cases, are the same as or similar to those in ANSI A117.1. Therefore, A117.1 figure numbers (such as "Figure B4.15.2.1" and "Figure B4.20.3.1") have been included with the *Data Book* figure numbers.

Explanatory notes have been added after the recommendations for each fixture, where deemed of value. Where there are differences between A117.1 and ADAAG other than of an editorial nature, it is also noted.

## Clear Floor or Ground Space for Wheelchairs

The minimum clear floor or ground space required to accommodate a single, stationary wheelchair and occupant is 30 in. × 48 in. (760 mm × 1220 mm). (See Figure 6-2—B4.2.4.1.) The minimum clear floor or ground space for wheelchairs may be positioned for forward or parallel approach to an object (see Figure 6-3—B4.2.4.2). Clear floor or ground space for wheelchairs may be part of the knee space required under some objects. One full, unobstructed side of the clear floor or ground space for a wheelchair shall adjoin another wheelchair clear floor space. If a clear floor space is located in an alcove or otherwise confined on all or part of three sides, additional maneuvering clearances shall be provided as shown in Figure 6-4 (B4.2.4.4).

## Anthropometrics

**Forward reach** If the clear floor space only allows forward approach to an object, the maximum high forward reach allowed shall be 48 in. (1220 mm). (See Figure 6-5—B4.2.5.1.) The minimum low forward reach is 15 in. (380 mm). If the high forward reach is over an obstruction, reach and clearances shall be as shown in Figure 6-6 (B4.2.5.2).

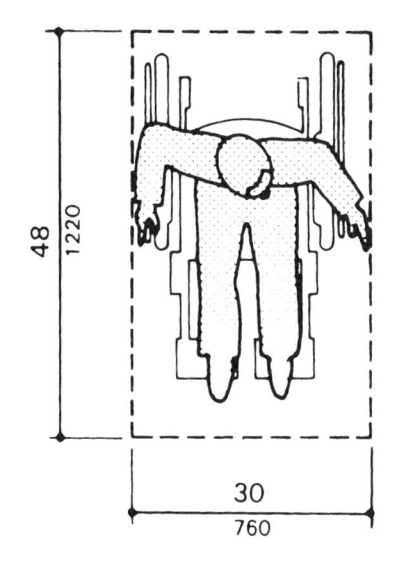

**Figure 6-2 Clear Floor Space for Wheelchairs**

*Source:* CABO/ANSI A117.1-1992. Reprinted with permission.

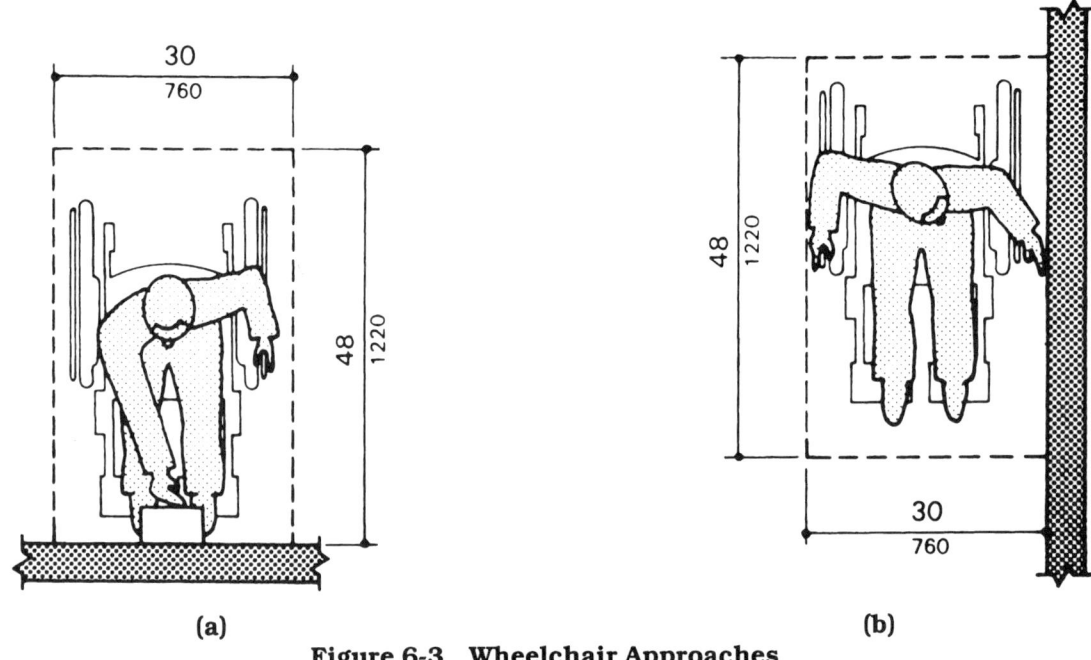

**Figure 6-3 Wheelchair Approaches**
**(a) Forward Approach  (b) Parallel Approach**

*Source:* CABO/ANSI A117.1-1992. Reprinted with permission.

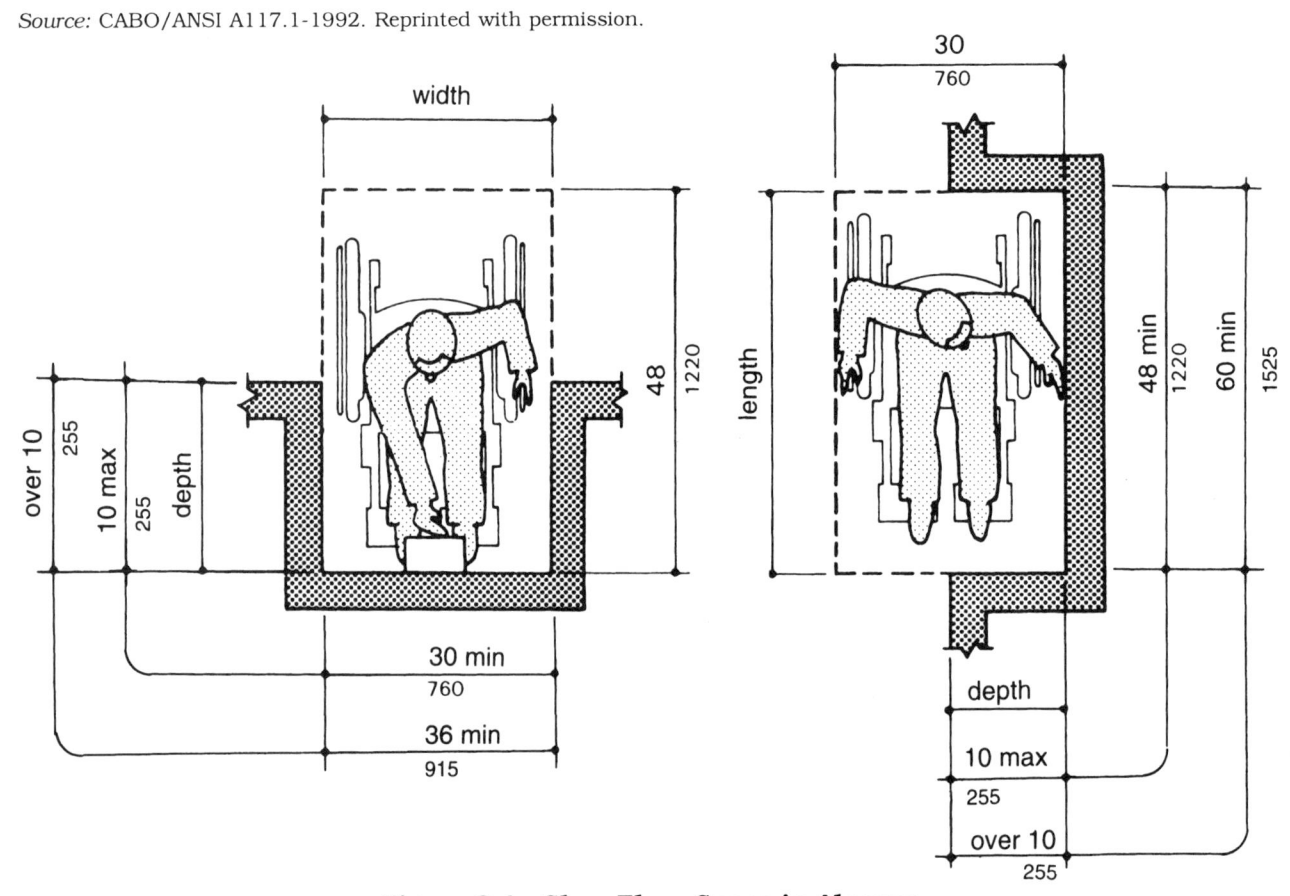

**Figure 6-4  Clear Floor Space in Alcoves**

*Source:* CABO/ANSI A117.1-1992. Reprinted with permission.

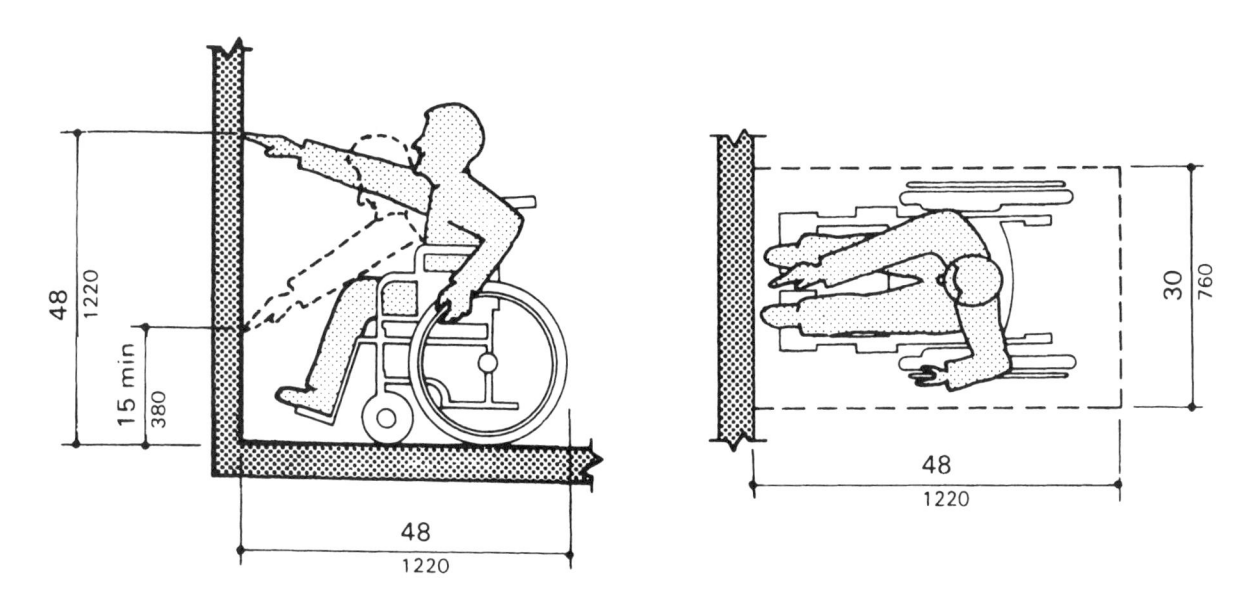

# Chapter 6 — Plumbing for Physically Challenged Individuals

**Figure 6-5 Unobstructed Forward Reach Limit**

*Source:* CABO/ANSI A117.1-1992. Reprinted with permission.

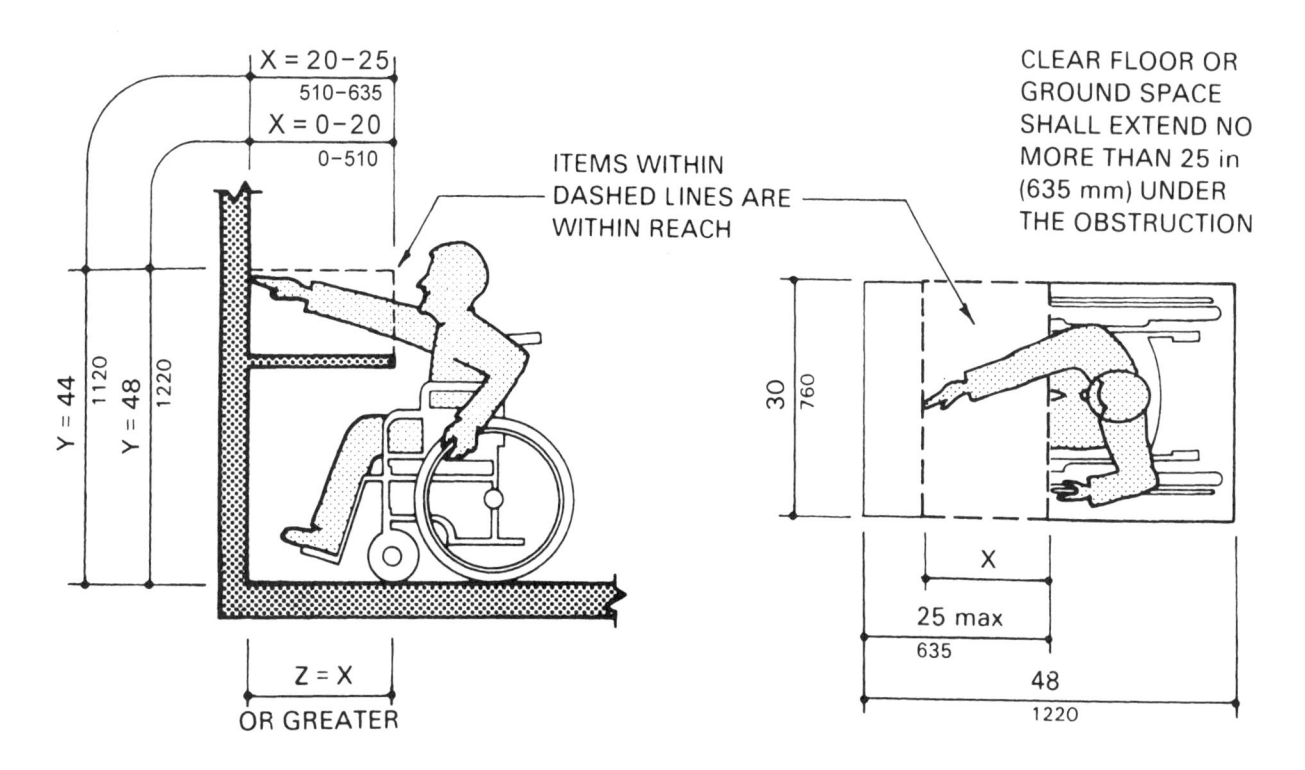

**Figure 6-6 Forward Reach Over an Obstruction**

*Source:* CABO/ANSI A117.1-1992. Reprinted with permission.

*Note:* X = reach depth, Y = reach height, Z = clear knee space. Z is the clear space below the obstruction, which shall be at least as deep as the reach distance, X.

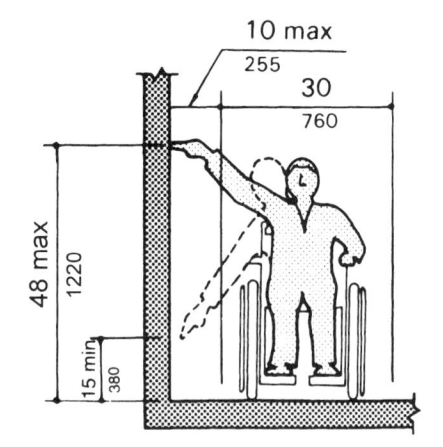

**Figure 6-7  Unobstructed Side Reach Limit**

Source: CABO/ANSI A117.1-1992. Reprinted with permission.

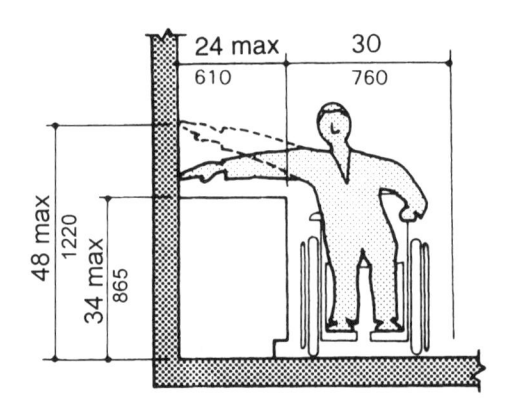

**Figure 6-8  Obstructed Side Reach Limit**

Source: CABO/ANSI A117.1-1992. Reprinted with permission.

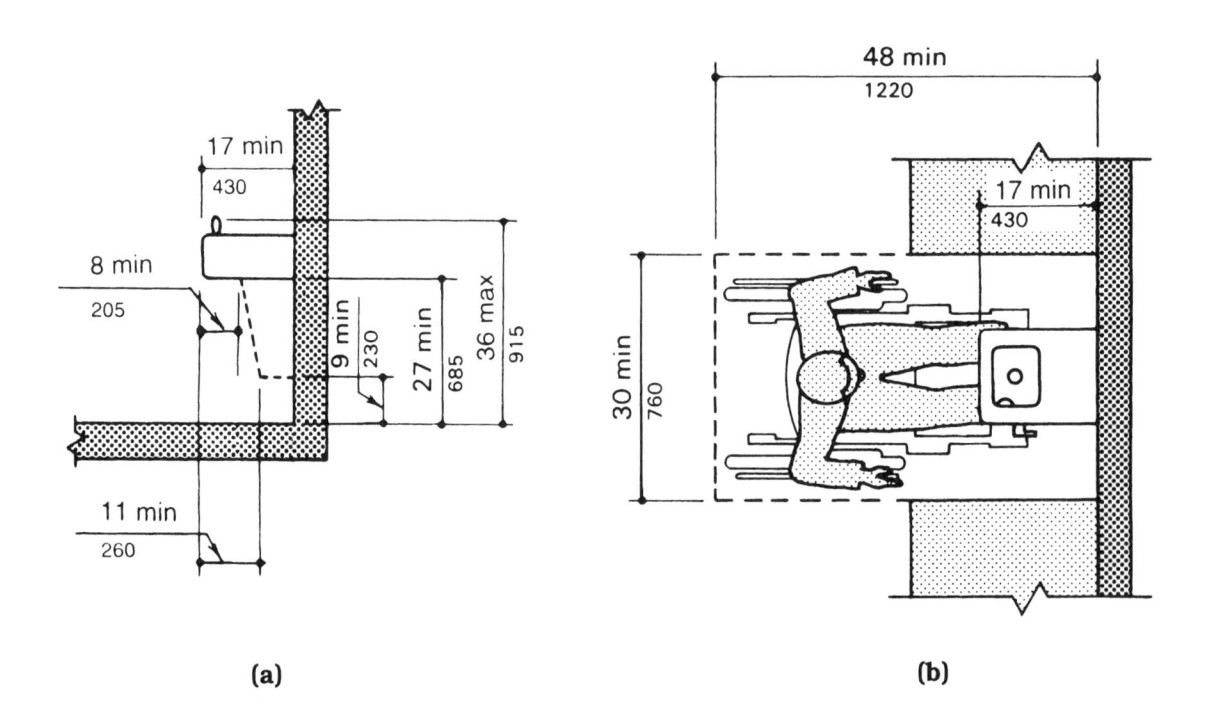

(a)

(b)

**Figure 6-9  Cantilevered Drinking Fountains and Water Coolers
(a) Spout Height and Leg Clearance  (b) Clear Floor Space**

Source: CABO/ANSI A117.1-1992. Reprinted with permission.

Note, Figure 6-9a only: Equipment permitted within dashed lines if mounted below apron.

# Chapter 6 — Plumbing for Physically Challenged Individuals

**Side reach** If the clear floor space allows parallel approach by a person in a wheelchair, the maximum high side reach allowed shall be 48 in. (1220 mm) and the low side reach shall be 15 in. (380 mm) (see Figure 6-7—B4.2.6.1). If the side reach is over an obstruction, the reach and clearances shall be as shown in Figure 6-8 (B4.2.6.2).

## PLUMBING ELEMENTS AND FACILITIES[1]

### 601 General

**601.1 Scope** Plumbing elements and facilities required to be accessible by scoping provisions adopted by the administrative authority shall comply with the applicable provisions of this chapter.

### 602 Drinking Fountains and Water Coolers

**602.1 General** Accessible fixed drinking fountains and water coolers shall comply with 602.

**602.2 Clear floor or ground space** A clear floor or ground space complying with 305 shall be provided.[2]

*602.2.1 Forward approach* Where a forward approach is provided, the clear floor or ground space shall be centered on the unit and shall include knee and toe clearance complying with 306.[2]

*602.2.2 Parallel approach* Where a parallel approach is provided, the clear floor or ground space shall be centered on the unit.

**602.3 Operable parts** Operable parts shall comply with 309.[2]

**602.4 Spout height** Spout outlets shall be 36 in. (915 mm) maximum above the floor or ground. (See Figure 6-9A—B4.15.2.1A.)

**602.5 Spout location** Units with a parallel approach shall have the spout 3½ in. (89 mm) maximum from the front edge of the unit, including bumpers. Units with a forward approach shall have the spout 15 in. (380 mm) minimum from the vertical support and 5 in. (125 mm)

[1]Text source: CABO/ANSI A117.1-1998.
[2]See CABO/ANSI A117.1-1998.

maximum from the front edge of the unit, including bumpers.

**602.6 Water flow** The spout shall provide a flow of water 4 in. (100 mm) high minimum to allow the insertion of a cup or glass under the flow of water. The angle of the water stream from spouts within 3 in. (75 mm) of the front of the unit shall be 30° maximum. The angle of the water stream from spouts between 3 in. (75 mm) and 5 in. (125 mm) from the front of the unit shall be 15° maximum. The angle of the water stream shall be measured horizontally, relative to the front face of the unit. (See Figure 6-10—B4.15.2.3.)

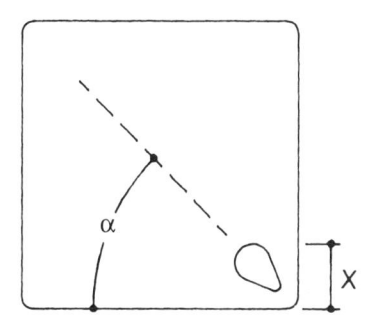

When: x = 3 in    α = 30° max
3 < x < 5 in α = 15° max

**Figure 6-10  Horizontal Angle of Water Stream — Plan View**

*Source:* CABO/ANSI A117.1-1992. Reprinted with permission.

**602.7 Protruding objects** Units shall comply with 307.[2]

*Drinking fountain note:*

*The easiest way for someone in a wheelchair to use a drinking fountain is to approach it from the front and bend down to the spout. The plumbing engineer should therefore specify a fountain or cooler with a spout located as close to the front edge and as low as possible. There are self-contained units available that can be mounted so that spout heights of 33 to 34 in. (839 to 864 mm) can be obtained, without interfering with required leg clearances.*

*Parallel approach units are more difficult to use than the cantilevered type and should be avoided if possible. If used, the spout should be mounted as close to 30 in. (762 mm) as the fountain will permit.*

It is desirable to provide some water coolers or fountains with spout heights of approximately 42 in. (1067 mm) to serve semiambulatory users who can have difficulty bending to lower elevations.

Drinking fountains must be provided not only for wheelchair-bound individuals but also for back-disabled individuals (ADAAG section 4.1.3, item no. 10, and appendix A4.15.2). Where only one (1) fountain is required by code, it must be an accessible bi-level unit, or two (2) separate accessible units mounted at different heights must be provided. Where more than one (1) fountain is required by code, 50% must be installed for wheelchair-bound individuals.

## 603 Toilet and Bathing Rooms

**603.1 General** Accessible toilet and bathing rooms shall comply with 603.

**603.2 Clearances**

***603.2.1 Wheelchair turning space*** A wheelchair turning space complying with 304 shall be provided within the room.[2]

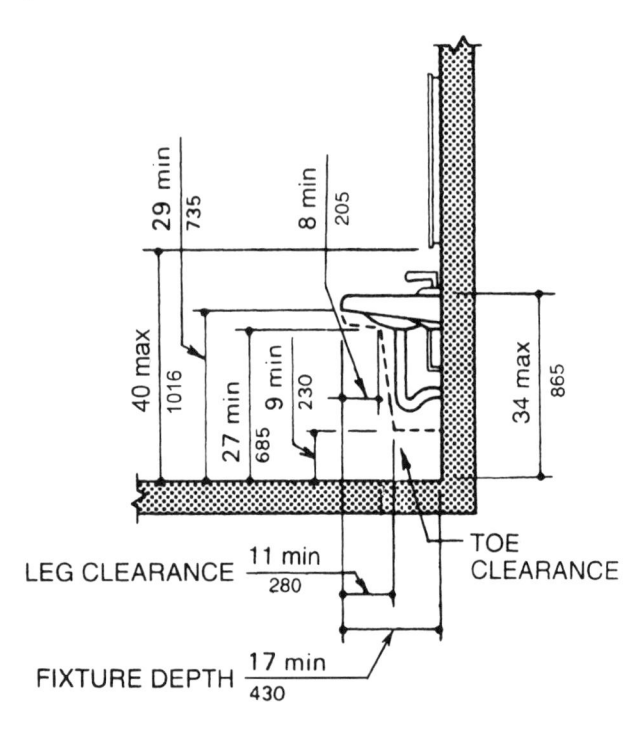

**Figure 6-11 Leg Clearances**

Source: CABO/ANSI A117.1-1992. Reprinted with permission.

*Note:* Dashed line indicates dimensional clearance of optional, under-fixture enclosure.

***603.2.2 Overlap*** Clear floor or ground spaces, clearances at fixtures, and wheelchair turning spaces shall be permitted to overlap.

***603.2.3 Doors*** Doors shall not swing into the clear floor or ground space or clearance for any fixture.

EXCEPTION: Where the room is for individual use and a clear floor or ground space complying with 305.3 is provided within the room beyond the arc of the door swing.[2]

**603.3 Mirrors** Mirrors shall be mounted with the bottom edge of the reflecting surface 40 in. (1015 mm) maximum above the floor or ground. (See Figure 6-11—B4.20.3.1.)

**603.4 Coat hooks and shelves** Coat hooks provided within toilet rooms shall accommodate a forward reach or side reach complying with 308.[2] Where provided, a fold-down shelf shall be 40 in. (1015 mm) minimum and 48 in. (1220 mm) maximum above the floor or ground.

*Toilet and bathing rooms note:*

*When a door opens into a bathroom, sufficient maneuvering space is provided within the room for a person using a wheelchair to enter, close the door, use the fixtures, reopen the door and exit without undue difficulty.*

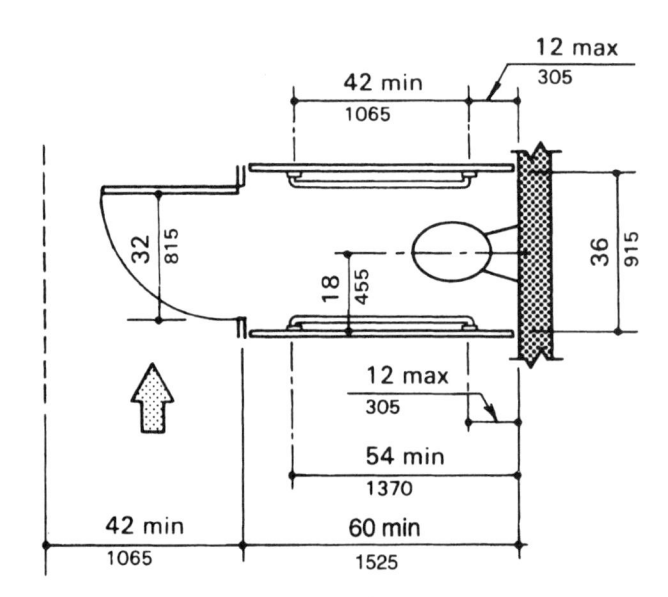

**Figure 6-12 Ambulatory Accessible Stall**

Source: CABO/ANSI A117.1-1992. Reprinted with permission.

# Chapter 6 — Plumbing for Physically Challenged Individuals

*The wheelchair maneuvering space overlaps the required clear floor space at fixtures and extends under the lavatory 19 in. (480 mm) maximum because knee space is provided. However, because toe or knee space is not available at the toilet, the wheelchair maneuvering space is clear of the toilet. Design and location of floor drains should not impede the use of plumbing fixtures.*

*Medical cabinets or other methods for storing medical and personal care items are very useful to physically challenged people. Shelves, drawers, and floor-mounted cabinets should be within the reach ranges of a physically challenged person.*

*If mirrors are to be used by both ambulatory people and wheelchair users, then they should be 74 in. (1880 mm) high minimum at their top-most edge and 40 in (1015 mm) maximum at their lowest edge. A single full-length mirror accommodates all people, including children.*

## 604 Water Closets and Toilet Compartments

**604.1 General** Accessible water closets and toilet compartments shall comply with 604.

**604.2 Location** The water closet shall be positioned with a wall or partition to the rear and to one side. The centerline of the water closet shall be 16 in. (405 mm) minimum to 18 in. (455 mm) maximum from the side wall or partition, except that the water closet shall be centered in the ambulatory accessible compartment specified in 604.8.2. (See Figure 6-12—B4.18.4.}

**604.3 Clearance**

***604.3.1 Size*** Clearance around the water closet shall be 60 in. (1220 mm) minimum, measured perpendicular from the side wall, and 56 in. (1420 mm) minimum, measured perpendicular from the rear wall. No other fixtures or obstructions shall be within the water closet clearance. (See Figure 6-13—B4.18.3.1.)

***604.3.2 Overlap*** The clearance around the water closet shall be permitted to overlap the fixture, associated grab bars, tissue dispensers, accessible routes and clear floor or ground space or clearances at other fixtures and the wheelchair turning space. Clear floor space shall comply with Figure 6-14 (B4.17.2).

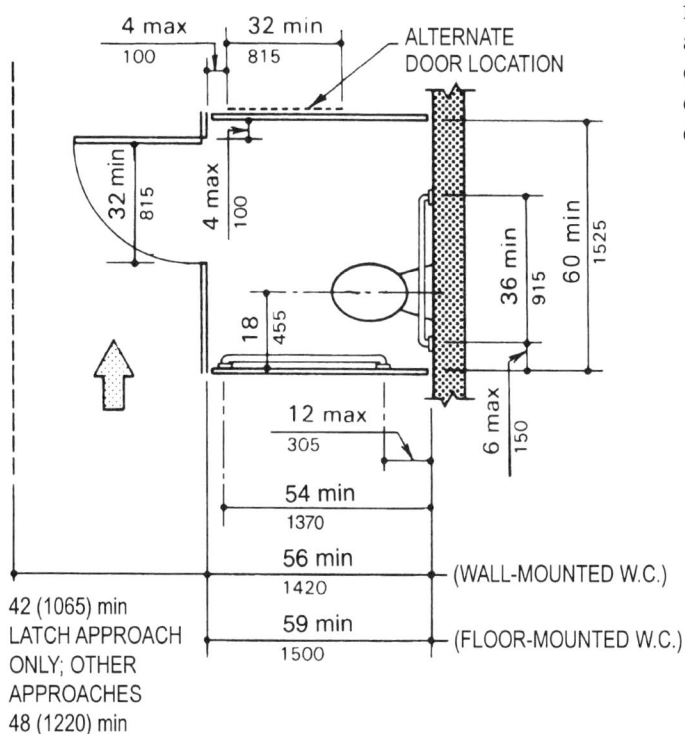

**Figure 6-13 Wheelchair Accessible Toilet Stalls — Door Swing Out**

*Source: CABO/ANSI A117.1-1992. Reprinted with permission.*

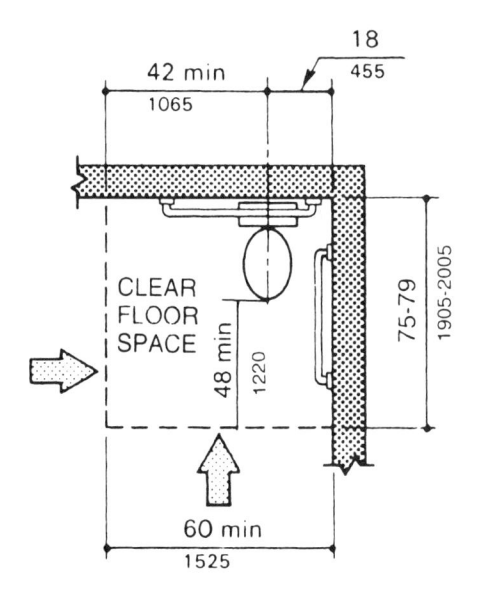

**Figure 6-14 Clear Floor Space at Water Closets**

*Source: CABO/ANSI A117.1-1992. Reprinted with permission.*

**604.4 Height** The top of water closet seats shall be 17 in. (430 mm) minimum and 19 in. (485 mm) maximum above the floor or ground. Seats shall not return automatically to a lifted position. (See Figure 6-15—B4.17.3.)

**604.5 Grab bars** Grab bars for water closets shall comply with 609. Grab bars shall be provided on the rear wall and on the side wall closest to the water closet.

*604.5.1 Side wall* Side wall grab bar shall be 42 in. (1065 mm) long minimum, 12 in. (305 mm) maximum from the rear wall and extending 54 in. (1370 mm) minimum from the rear wall. (See Figure 6-15—B4.17.3.)

*604.5.2 Rear wall* The rear wall grab bar shall be 24 in. (610 mm) long minimum, centered on the water closet. Where space permits, the bar shall be 36 in. (915 mm) long minimum, with the additional length provided on the transfer side of the water closet. (See Figure 6-16—B4.17.4.)

**604.6 Flush controls** Flush controls shall be hand operated or automatic. Hand-operated flush controls shall comply with 309.[2]

**604.7 Dispensers** Toilet paper dispensers shall comply with 309.4 and shall be 7 in. (180 mm) minimum and 9 in. (230 mm) maximum in front of the water closet.[2] The outlet of the dispenser shall be 15 in. (380 mm) minimum and 48 in. (1220 mm) maximum above the floor or ground. There shall be a clearance of 1½ in. (38 mm) minimum below and 12 in. (305 mm) minimum above the grab bar. Dispensers shall not be of a type that control delivery, or that do not allow continuous paper flow.

**604.8 Toilet compartments** Accessible toilet compartments shall comply with 604.8.1 through 604.8.5. Compartments containing more than one plumbing fixture shall comply with 603. Water closets in accessible toilet compartments shall comply with 604.1 through 604.7.

*604.8.1 Wheelchair accessible compartments*

*604.8.1.1 Size* Wheelchair accessible compartments shall be 60 in. (1525 mm) wide minimum measured perpendicular to the side wall, and 56 in. (1420 mm) deep minimum for wall-hung water closets and 59 in. (1500 mm) deep minimum for floor-mounted water closets, measured perpendicular to the rear wall. (See Figure 6-13—B4.18.3.1.)

*604.8.1.2 Doors* Compartment doors shall not swing into the minimum required compartment area. (See Figure 6-17—B4.18.3.2.)

*604.8.1.3 Approach* Compartment arrangements shall be permitted for left-hand or right-hand approach to the water closet.

*604.8.1.4 Toe clearance* In wheelchair-accessible compartments, the front partition and at least one side partition shall provide a toe clearance complying with 306.2 and extending 6 in. (150

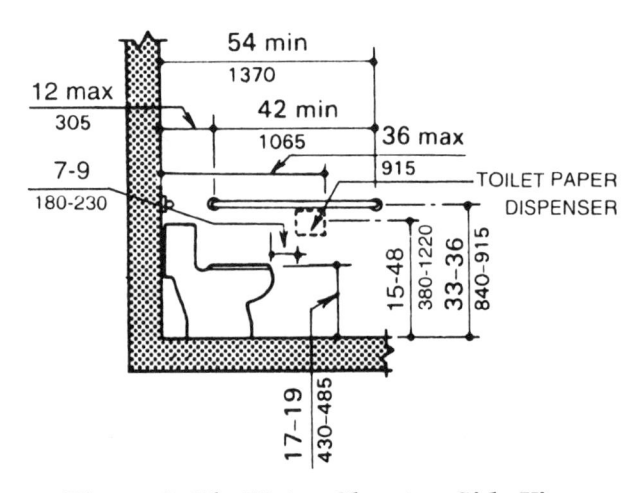

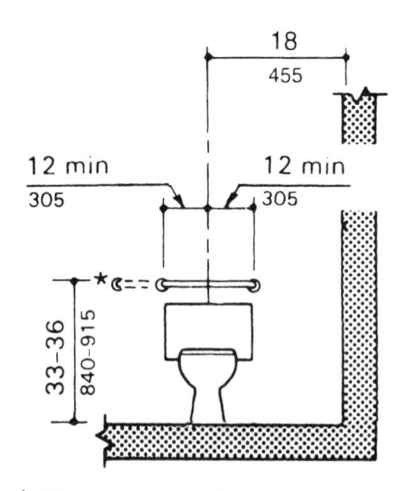

**Figure 6-15 Water Closet — Side View**

*Source:* CABO/ANSI A117.1-1992. Reprinted with permission.

**Figure 6-16 Water Closet — Front View**

*Source:* CABO/ANSI A117.1-1992. Reprinted with permission.

# Chapter 6 — Plumbing for Physically Challenged Individuals

mm) deep beyond the compartment-side face of the partition, exclusive of partition support members.[2] Toe clearance at the front of the partition is not required in a compartment greater than 62 in. (1575 mm) deep with a wall-hung water closet or 65 in. (1650 mm) deep with a floor-mounted water closet. Toe clearance at the side partition is not required in a compartment greater than 66 in. (1675 mm) wide.

***604.8.2 Ambulatory accessible compartments*** Ambulatory-accessible compartments shall be 60 in. (1525 mm) deep minimum and 36 in. (915 mm) wide. Compartment doors shall not swing into the minimum required compartment area. (See Figure 6-12—B4.18.4.)

***604.8.3 Doors*** Toilet compartment doors shall comply with 404, except that if the approach is to the latch side of the compartment door, the clearance between the door side of the compartment and any obstruction shall be 42 in. (1065 mm) minimum. The door shall be hinged 4 in. (100 mm) maximum from the adjacent wall or partition farthest from the water closet. The door shall be self-closing. A door pull complying with 404.2.7 shall be placed on both sides of the door near the latch.[2]

***604.8.4 Grab bars*** Grab bars shall comply with 609.

*604.8.4.1 Wheelchair-accessible compartments* A side-wall grab bar complying with 604.5.1 shall be provided on the wall closest to the water closet, and a rear-wall grab bar complying with 604.5.2 shall be provided. (See Figure 6-13—B4.18.3.1.)

*604.8.4.2 Ambulatory-accessible compartments* A side-wall grab bar complying with 604.5.1 shall be provided on both sides of the compartment. (See Figure 6-12—B4.18.4.)

***604.8.5 Coat hooks and shelves*** Coat hooks provided within toilet compartments shall be 48 in. (1220 mm) maximum above the floor or ground. Where provided, a fold-down shelf shall be 40 in. (1015 mm) minimum and 48 in. (1220 mm) maximum above the floor or ground.

*Water closets and toilet compartments note:*

*The centerline requirement for water closets has been adjusted to allow a range of 16 to 18 in. (407 to 457 mm) from the centerline of the fixture to the side wall, eliminating the fixed 18 in. (457 mm) dimensional requirement. This change has been successfully argued on the merits of allowing some flexibility by code enforcement officials in the field. The greater or lesser accessibility of a water closet installed 16 in. (407 mm) from the side wall to the centerline of the toilet versus a water closet installed 18 in. (457 mm) from the side wall has yet to be answered to a majority of either committee.*

*The toilet seat height of 17 to 19 in. (432 to 483 mm) in public areas is intended to minimize the difference between the seat and the standard wheelchair seat height to aid the transfer process, without elevating the toilet seat to the point that stability problems are created.*

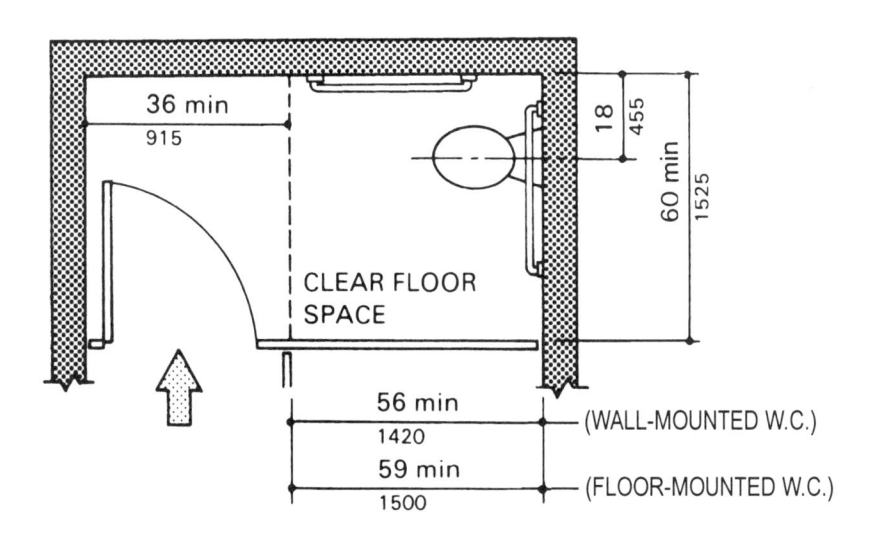

**Figure 6-17 Wheelchair Accessible Toilet Stalls — Door Swing In**

Source: CABO/ANSI A117.1-1992. Reprinted with permission.

# We're fluent in three languages.

If the plumbing equipment you specify hasn't achieved CSA Certification or NSF Certification, and the manufacturer hasn't achieved ISO Registration, you might want to go to school on a company that's done its homework.

Our philosophy has always been to set industry standards in product quality and innovation. Just recently T&S achieved the NSF Certification for both our above and below the deck faucets. All with minimal adjustments to our existing line.

Why did we do this? Because we know the time will come when NSF Certification will be mandatory. So we want specification engineers like you to be totally confident in specing T&S today without concern of retrofit later.

What's more, its important for you to note that there's a major distinction between NSF Certified and tested to NSF standards. T&S achieved the NSF Certification after our products were subjected to rigorous testing by the world's leading environmental and public health testing organization. We are also subject to unannounced inspections which means our quality and performance will remain consistent.

So, if you're interested in learning more about plumbing products of ingenuity, absolute integrity and highest quality, talk to T&S. Rest assured we will speak your language.

*Spec T&S and forget it.*

©T&S Brass and Bronze Works, Inc. • 2 Saddleback Cove • Travelers Rest, South Carolina 29690 • (864)834-4102   (800)476-4103   Fax (800)868-0084

# Chapter 6 — Plumbing for Physically Challenged Individuals

*The 60 in. (1525 mm) wide wheelchair accessible compartment is preferred and should be designed. In the design of alterations to existing structures, it may not be possible to create the preferred compartment by combining two existing compartments, or physical conditions may not permit the full 60 in. (1525 mm) width. In these cases, the authority having jurisdiction may permit a narrower compartment. In no case should a width of less than 48 in. (1220 mm) be used.*

*The needs of a semiambulatory user are best served by a narrower, 36 in. max. (915 mm max.), compartment which premises use of grab bars on either or both sides of the compartment.*

*ADAAG note:*

*ADAAG has an exception to the height requirement of water closets and grab bars for water closets located in a toilet room for a single occupant, accessed only through a private office and not for common or public use. Where six (6) or more compartments are provided in a toilet room, one (1) must be a 60 in. (1525 mm), wheelchair-accessible compartment and one (1) must be a 36 in. (915 mm), ambulatory compartment.*

*The flush valve handles should not exceed 44 in. (1118 mm) above the floor. The handles in standard accessible stalls must be at the wide side of the stall (ADAAG section 4.16.5). This means, depending on how the stall is configured, the handle must be on either the right or left side of the flush valve. This does not apply to tank type units, although several manufacturers have now come up with a right-hand operator.*

## 605 Urinals

**605.1 General** Accessible urinals shall comply with 605.

**605.2 Height** Urinals shall be of the stall type or shall be of the wall-hung type with the rim at 17 in. (430 mm) maximum above the floor or ground.

**605.3 Clear floor or ground space** A clear floor or ground space complying with 305 positioned for forward approach shall be provided.[2]

**605.4 Flush controls** Flush controls shall be hand operated or automatic. Hand-operated flush controls shall comply with 309.[2]

*Urinal note:*

*It should be understood that the referenced urinal is not intended to be used by a wheelchair occupant for the normal urination process. It is intended for the drainage of bladder bags, a function normally performed in a water closet compartment, if available. Where an accessible urinal is required, it can serve as a child's urinal. Urinals must be provided with an elongated rim (ADAAG section 4.18.2). Although ADAAG does not define what constitutes an elongated urinal, the Department of Justice deferred to ANSI, which defines these fixtures as having a lip that protrudes a minimum of 14 in. (356 mm) from the wall. Flush valve handles should not exceed 48 in. (1220 mm) above the floor.*

## 606 Lavatories and Sinks

**606.1 General** Accessible lavatories and sinks shall comply with 606.

**606.2 Clear floor or ground space** A clear floor or ground space complying with 305.3, positioned for forward approach, shall be provided. Knee and toe clearance complying with 306 shall be provided.[2] (See Figure 6-18—B4.20.3.2.)

EXCEPTIONS:

1. A parallel approach shall be permitted to a kitchen sink in a space where a cook-top or conventional range is not provided.

2. The dip of the overflow shall not be considered in determining knee and toe clearances.

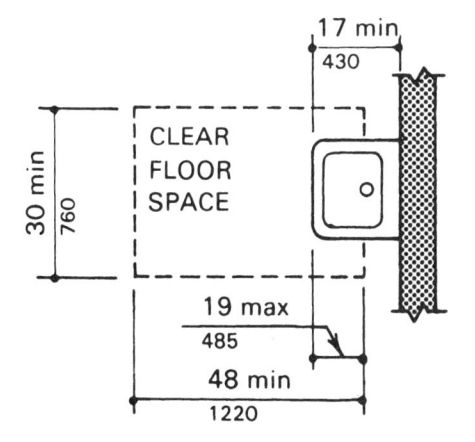

**Figure 6-18 Clear Floor Space at Lavatories and Sinks**

*Source: CABO/ANSI A117.1-1992. Reprinted with permission.*

**606.3 Height and clearances** The front of lavatories and sinks shall be 34 in. (865 mm) maximum above the floor or ground, measured to the higher of the fixture rim or counter surface.

**606.4 Faucets** Faucets shall comply with 309.[2] Hand-operated, self-closing faucets shall remain open for 10 seconds minimum.

**606.5 Bowl depth** Sinks shall be 6½ in. (165 mm) deep maximum. Multiple-compartment sinks shall have at least one compartment complying with this requirement.

**606.6 Exposed pipes and surfaces** Water supply and drain pipes under lavatories and sinks shall be insulated or otherwise configured to protect against contact. (See Figure 6-11—B4.20.3.1.) There shall be no sharp or abrasive surfaces under lavatories and sinks.

*Lavatories and sinks note:*

*Conventional, slab type lavatories are available to meet the dimensional requirements of A117.1, since the dip of the overflow can be ignored.*

*Built-in lavatories in countertops should be placed as close as possible to the front edge of the countertop to minimize the reach to the faucet. Single-lever faucets are preferred, but where aesthetics or fear of vandalism precludes their use, conventional quarter-turn handles are a good choice. Avoid faucets that require finger dexterity for grasping or twisting.*

*Both hot and cold water pipes, as well as drain pipes that are in the vicinity of the designated clear floor space under the fixture must be concealed or insulated to protect wheelchair users who have no functioning sensory nerves. Insulation is not required on pipes beyond possible contact.*

## 607 Bathtubs

**607.1 General** Accessible bathtubs shall comply with 607.

**607.2 Clearance** Clearance in front of bathtubs shall extend the length of the bathtub and shall be 30 in. (760 mm) wide minimum. A lavatory complying with 606 shall be permitted at the foot end of the clearance. (See Figure 6-19—B4.21.2.) Where a permanent seat is provided at the head end of the bathtub, the clearance shall extend a minimum of 15 in. (380 mm) beyond the wall at the head end of the bathtub.

**607.3 Seat** A permanent seat at the head end of the bathtub or a removable in-tub seat shall be provided. Seats shall comply with 610.

**607.4 Grab bars** Grab bars shall comply with 607.4 and 609.

*607.4.1 Bathtubs with permanent seats* For bathtubs with permanent seats, grab bars complying with 607.4.1.1 and 607.4.1.2 shall be provided.

*607.4.1.1 Back wall* Two grab bars shall be provided on the back wall, one complying with 609.4 and the other 9 in. (230 mm) above the rim of the bathtub. Each grab bar shall be 15 in. (380 mm) maximum from the head-end wall and 12 in. (305 mm) maximum from the foot-end wall.

*607.4.1.2 Foot-end wall* A grab bar 24 in. (610 mm) long minimum shall be provided on the foot-end wall at the front edge of the bathtub.

*607.4.2 Bathtubs without permanent seats* For bathtubs without permanent seats; grab bars complying with 607.4.2.1 through 607.4.2.3 shall be provided.

*607.4.2.1 Back wall* Two grab bars shall be provided on the back wall, one complying with 609.4 and the other 9 in. (230 mm) above the rim of the bathtub. Each grab bar shall be 24 in. (610 mm) long minimum and shall be 24 in. (610 mm) maximum from the head-end wall and 12 in. (305 mm) maximum from the foot-end wall.

*607.4.2.2 Foot-end wall* A grab bar 24 in. (610 mm) long minimum shall be provided on the foot-end wall at the front edge of the bathtub.

*607.4.2.3 Head-end wall* A grab bar 12 in. (305 mm) long minimum shall be provided on the head-end wall at the front edge of the bathtub.

**607.5 Controls** Controls, other than drain stoppers, shall be on an end wall. Controls shall be between the bathtub rim and grab bar, and between the open side of the bathtub and the midpoint of the width of the bathtub. Controls shall comply with 309.4.[2] (See Figure 6-20—B4.21.4.)

**607.6 Shower unit** A shower spray unit shall be provided, with a hose 59 in. (1500 mm) long minimum, that can be used as a fixed shower head and as a hand-held shower. If an adjustable-height shower head on a vertical bar is used, the bar shall not obstruct the use of grab bars.

## Chapter 6 — Plumbing for Physically Challenged Individuals

**607.7 Bathtub enclosures** Bathtub enclosures shall not obstruct controls or transfer from wheelchairs onto bathtub seats or into bathtubs. Bathtub enclosures shall not have tracks on the rim of the bathtub.

*Bathtub note:*

*A fixed seat at the head of the tub adds safety and convenience for transfer purposes, as does the 17 to 19 in. (432 to 483 mm) rim height. The rim height that is more in line with the tub seat does not require the use of a deeper tub; it is better to use a tub with a deeper apron or use a tile filler.*

*Due to the probable lack of maneuverability of the user, it is recommended that the plumbing engineer specify a temperature and/or pressure-balanced, water-blending valve with temperature-limit stops.*

### 608 Shower Compartments

**608.1 General** Accessible shower compartments shall comply with 608.

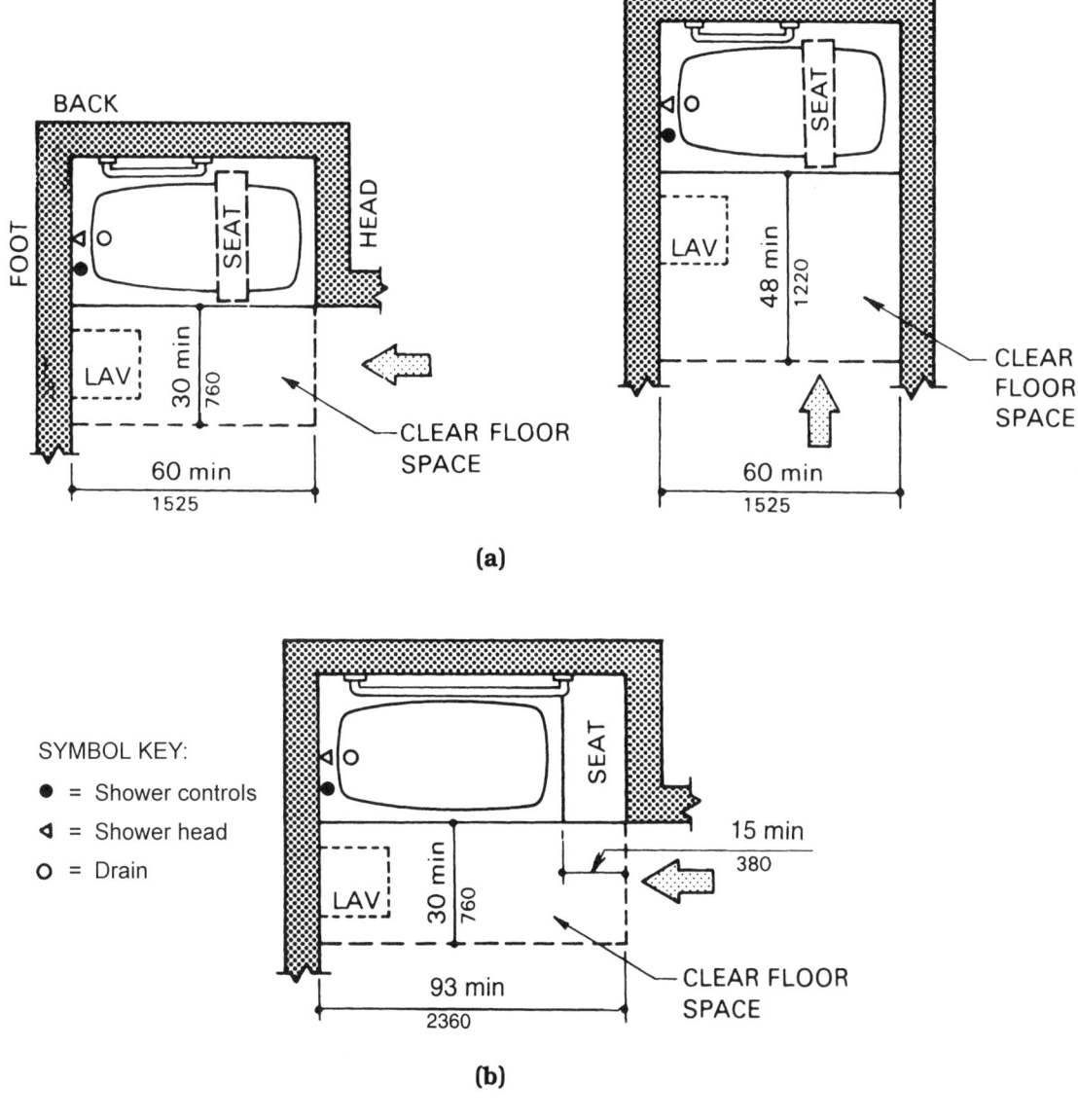

**Figure 6-19 Clear Floor Space at Bathtubs
(a) With Seat in Tub  (b) With Seat at Head of Tub**

*Source:* CABO/ANSI A117.1-1992. Reprinted with permission.

**608.2 Size and clearances**

***608.2.1 Transfer type shower compartments***
Transfer type shower compartments shall be 36 in. (915 mm) wide by 36 in. (915 mm) deep inside finished dimension, measured at the centerpoint of opposing sides, and shall have a minimum 36 in. (915 mm) wide entry on the face of the shower compartment. The clearance in front of the compartment shall be 48 in. (1220 mm) long minimum measured from the control wall and 36 in. (915 mm) wide minimum. (See Figure 6-21—B4.22.2.1.)

***608.2.2 Standard roll-in type shower compartments*** Roll-in type shower compartments shall be 30 in. (760 mm) wide minimum by 60 in.

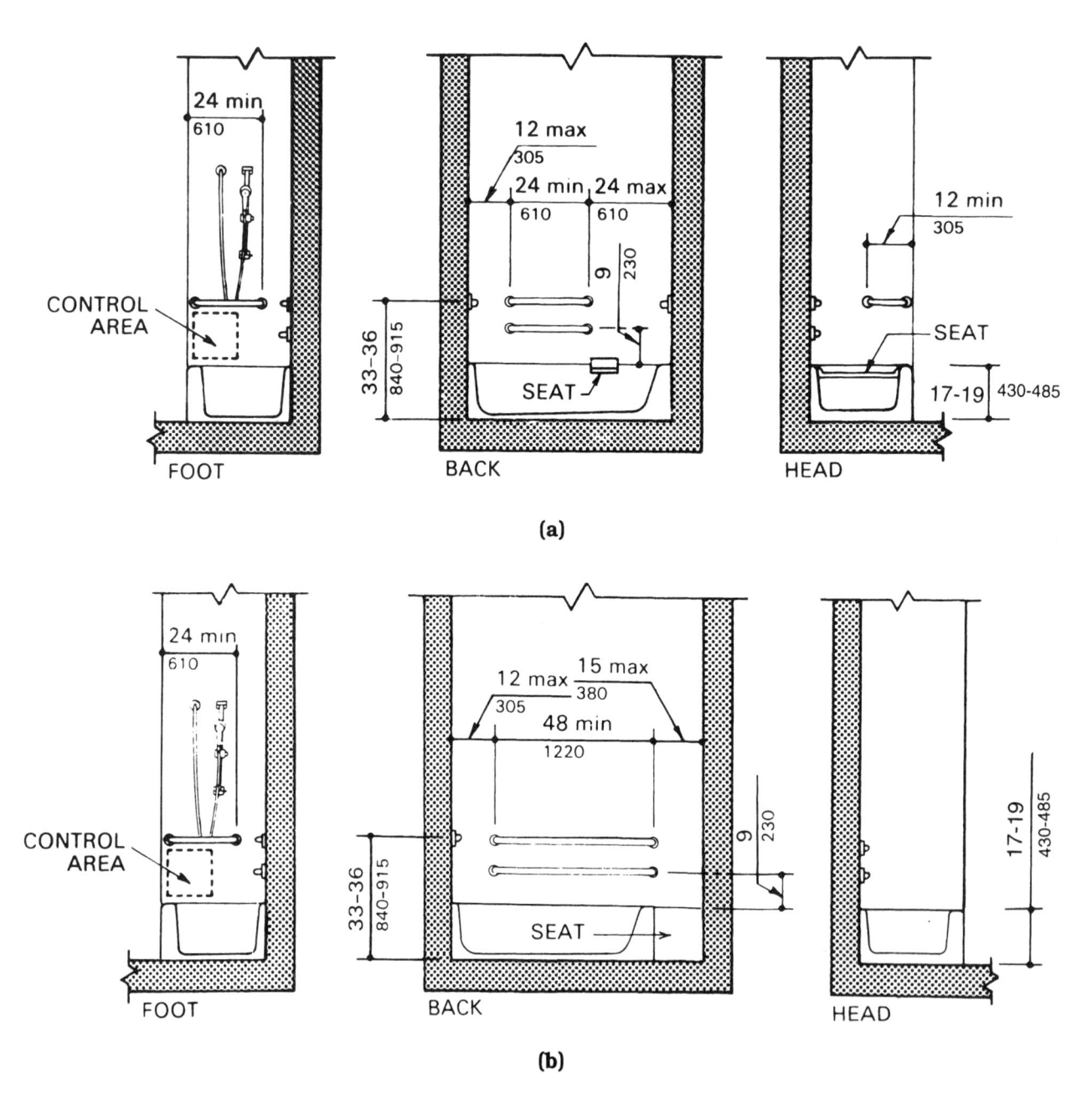

**Figure 6-20 Bathtub Accessories
(a) Without Permanent Seat in Tub  (b) With Permanent Seat at Head of Tub**

*Source:* CABO/ANSI A117.1-1992. Reprinted with permission.

# Chapter 6 — Plumbing for Physically Challenged Individuals

(1525 mm) deep minimum, clear inside dimension, measured at the centerpoint of opposing sides and shall have a minimum 60 in. (1220 mm) wide entry on the face of the shower. A 30 in. (760 mm) wide minimum by 60 in. (1525 mm) long minimum clearance shall be provided adjacent to the open face of the shower compartment. A lavatory complying with 606 shall be permitted at the end of the clear space, opposite the shower-compartment side where shower controls are positioned. (See Figure 6-22—B4.22.2.2.)

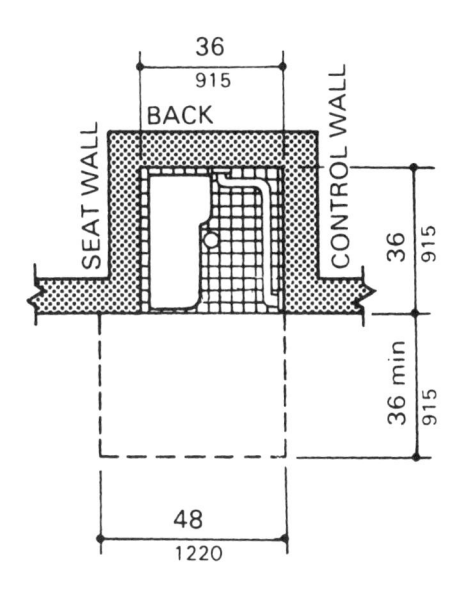

**Figure 6-21 Transfer Type Shower Stall**

*Source:* CABO/ANSI A117.1-1992. Reprinted with permission.

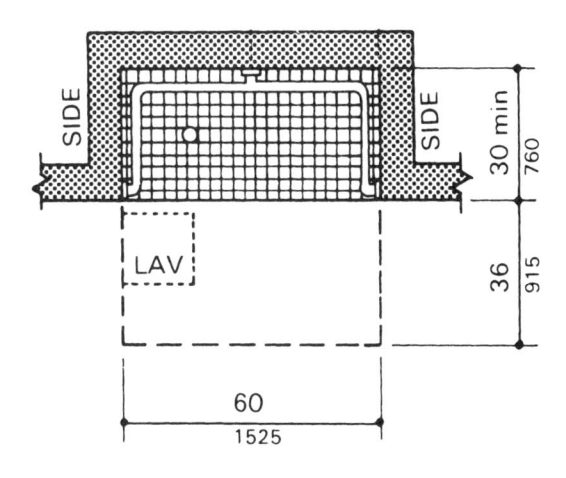

**Figure 6-22 Roll-in Type Shower Stall**

*Source:* CABO/ANSI A117.1-1992. Reprinted with permission.

***608.2.3 Alternate roll-in type shower compartments*** Alternate roll-in shower compartments shall be 36 in. (915 mm) wide and 60 in. (1220 mm) deep minimum. A 36 in. (915 mm) wide minimum entry shall be provided at one end of the long side of the compartment. The shower unit and controls shall be mounted on the end wall farthest from the compartment entry.

**608.3 Grab bars** Grab bars shall comply with 608.3 and 609 and shall be provided.

***608.3.1 Transfer type showers*** Grab bars shall be provided across the control wall and on the back wall to a point 18 in. (455 mm) from the control wall. (See Figure 6-23A—B4.22.4A.)

***608.3.2 Roll-in type showers*** Grab bars shall be provided on the three walls of the shower. (See Figure 6-23B—B4.22.4B.) Grab bars shall be 6 in. (150 mm) maximum from the adjacent wall.

EXCEPTIONS:

1. Where a seat is provided in a roll-in shower, grab bars shall not extend over the seat at the control wall and shall not be behind the seat.

2. In alternate roll-in type showers, grab bars shall not be required on the side wall opposite the control wall and shall not be behind the seat.

**608.4 Seats** An attachable or integral seat shall be provided in transfer type shower compartments. Seats shall comply with 610.

**608.5 Controls** Shower or bathtub/shower facilities shall deliver water that is thermal-shock protected to 120°F (49°C) maximum. Faucets and controls shall comply with 309.4.[2] Controls in roll-in showers shall be above the grab bar but no higher than 48 in. (1220 mm) above the shower floor. (See Figure 6-23B—B4.22.4B.) In transfer type shower compartments, controls, faucets, and the shower unit shall be on the side wall opposite the seat 38 in. (965 mm) minimum and 48 in. (1220 mm) maximum above the shower floor. (See Figure 6-23A—B4.22.4A.)

**608.6 Shower unit** A shower spray unit shall be provided, with a hose 59 in. (1500 mm) long minimum, that can be used as a fixed shower head and as a hand-held shower. In transfer type showers, the controls and shower unit shall be

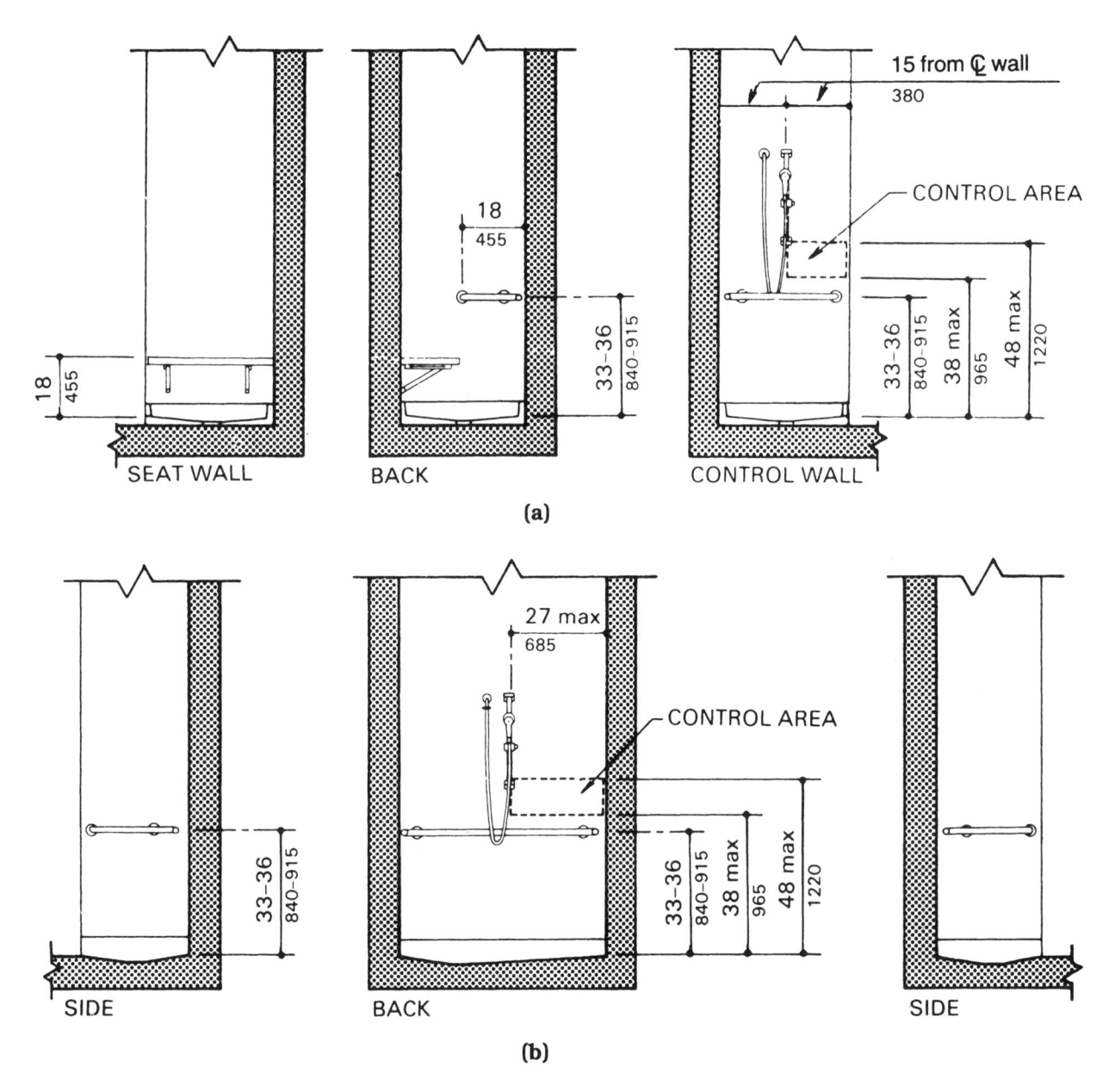

**Figure 6-23 Grab Bars at Shower Stalls**
**(a) 36 × 36-in (915 × 915-mm) Stall  (b) 30 × 60-in (760 × 1525-mm) Stall**

Source: CABO/ANSI A117.1-1992. Reprinted with permission.

Note, Figure 6-23b: Shower head and control area may be on back wall (as shown) or on either side wall.

on the control wall within 15 in. (380 mm), left or right, of the centerline of the seat. In roll-in type showers, shower spray units mounted on the back wall shall be 27 in. (685 mm) maximum from the side wall. If an adjustable-height shower head mounted on a vertical bar is used, the bar shall not obstruct the use of grab bars.

**608.7 Thresholds** Shower compartment thresholds shall be ½ in. (13 mm) high maximum and shall comply with 303.[2]

**608.8 Shower enclosures** Shower compartment enclosures for shower compartments shall not obstruct controls or obstruct transfer from wheelchairs onto shower seats.

# Chapter 6 — Plumbing for Physically Challenged Individuals

*Shower compartments note:*

*The recommended shower compartments are for independent use by an individual. Compartments between the two recommended sizes do not effectively serve physically challenged persons who wish to use a shower without assistance.*

*Transfer type shower compartments that are 36 in. by 36 in. (915 mm by 915 mm) provide additional safety to people who have difficulty maintaining balance because all grab bars and walls are within easy reach. Seated people use the walls of these showers for back support.*

*The shower compartment with inside finish dimensions of 36 in. by 36 in. (915 mm by 915 mm) has been designated a transfer type compartment to indicate that wheelchair users can transfer from their chair to the required seat. These dimensions will allow a person of average size to reach and operate the controls without difficulty, while providing reasonable knee space for larger users. A transfer type shower is also intended to serve persons without disabilities so a folding seat would provide more space for a standing person. Temperature may be limited to 105 to 110°F (40.5 to 43°C), depending on local code requirements.*

## 609 Grab Bars

**609.1 General** Grab bars in accessible toilet or bathing facilities shall comply with 609.

**609.2 Size** Grab bars shall have a circular cross section with a diameter of 1¼ in. (32 mm) minimum and 2 in. (51 mm) maximum, or shall provide equivalent graspability complying with 505.7.1.[2]

**609.2.1 Noncircular cross sections** Grab bars with other shapes shall be permitted, provided they have a perimeter dimension of 4 in. (100 mm) minimum and 4.8 in. (160 mm) maximum and edges having a ⅛ in. (3.2 mm) minimum radius.

**609.3 Spacing** The space between the wall and the grab bar shall be 1½ in. (38 mm). The space between the grab bar and objects below and at the ends shall be 1½ in. (38 mm) minimum. The space between the grab bar and projecting objects above shall be 15 in. (355 mm) minimum. (See Figure 6-24—B4.24.2.1.)

EXCEPTION: The space between the grab bars and shower controls, shower fittings, and other grab bars above shall be 1½ in. (38 mm) minimum.

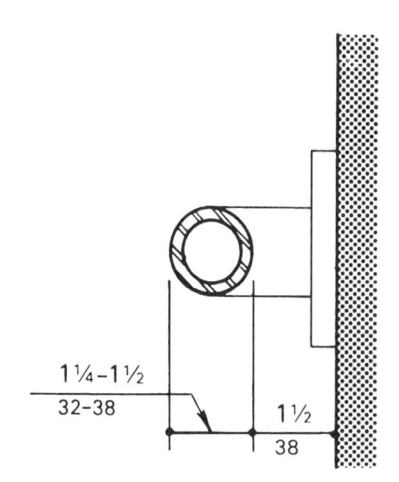

**Figure 6-24 Size and Spacing of Grab Bars**

*Source:* CABO/ANSI A117.1-1992. Reprinted with permission.

**609.4 Position of grab bars** Grab bars shall be mounted in a horizontal position, 33 in. (840 mm) minimum and 36 in. (915 mm) maximum above the floor.

EXCEPTION: Height of grab bars on the back wall of a bathtub shall comply with 607.4.1.1 and 607.4.2.1.

**609.5 Surface hazards** Grab bars and any wall or other surfaces adjacent to grab bars shall be free of sharp or abrasive elements. Edges shall have a radius of ⅛ in. (3 mm) minimum.

**609.6 Fittings** Grab bars shall not rotate within their fittings.

**609.7 Installation** Grab bars shall be installed in any manner that provides a gripping surface at the locations specified in this standard and that does not obstruct the clear floor space.

**609.8 Structural strength** Allowable stresses in bending, shear and tension shall not be exceeded for materials used where a vertical or horizontal force of 250 lb (113.5 kg) is applied at any point on the grab bar, fastener mounting device or supporting structure.

*Grab bars note:*

*Many physically challenged people rely heavily upon grab bars to maintain balance and prevent serious falls. Many people brace their forearms between supports and walls to give them more*

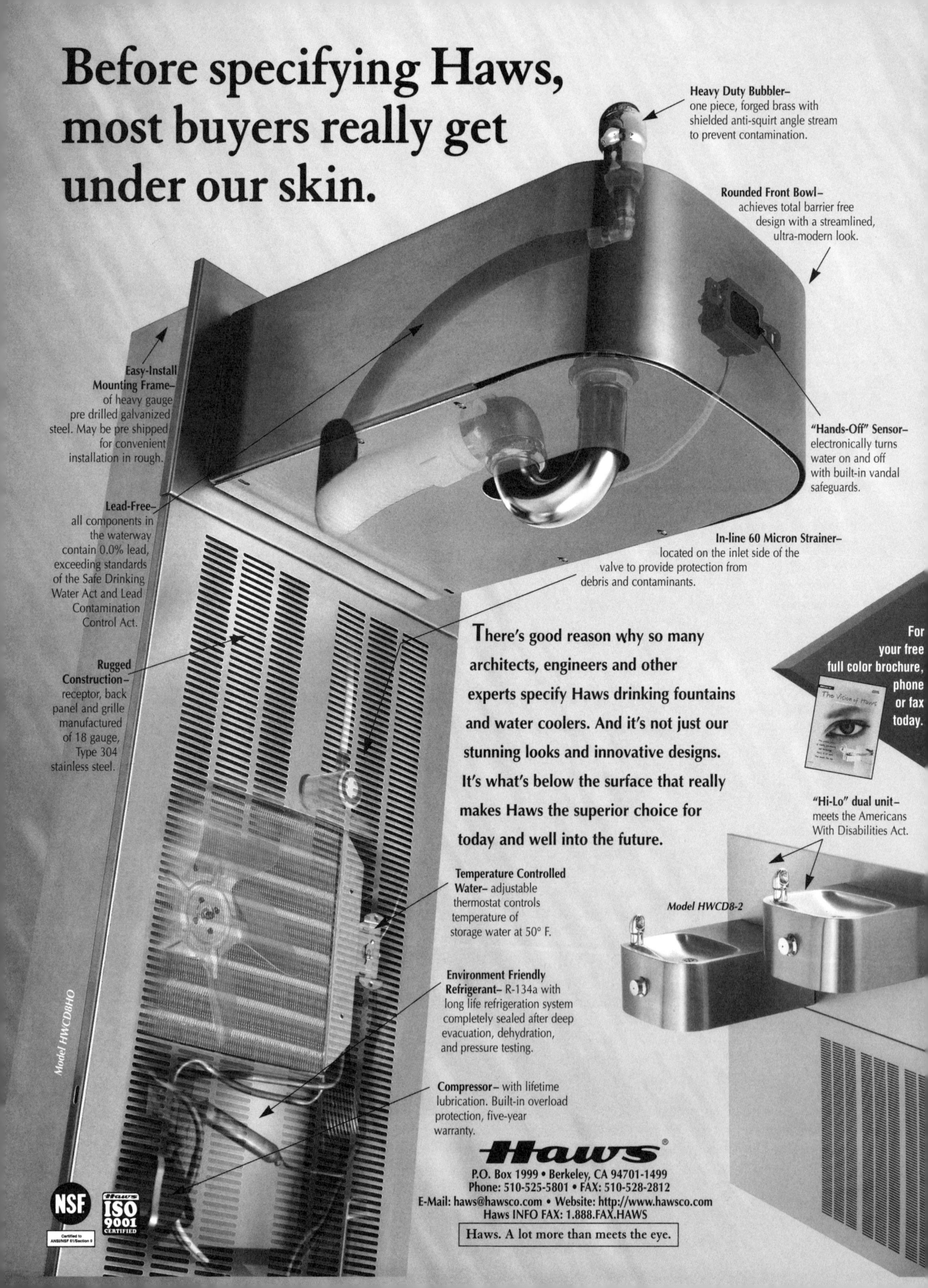

# Before specifying Haws, most buyers really get under our skin.

**Heavy Duty Bubbler–** one piece, forged brass with shielded anti-squirt angle stream to prevent contamination.

**Rounded Front Bowl–** achieves total barrier free design with a streamlined, ultra-modern look.

**Easy-Install Mounting Frame–** of heavy gauge pre drilled galvanized steel. May be pre shipped for convenient installation in rough.

**"Hands-Off" Sensor–** electronically turns water on and off with built-in vandal safeguards.

**Lead-Free–** all components in the waterway contain 0.0% lead, exceeding standards of the Safe Drinking Water Act and Lead Contamination Control Act.

**In-line 60 Micron Strainer–** located on the inlet side of the valve to provide protection from debris and contaminants.

**Rugged Construction–** receptor, back panel and grille manufactured of 18 gauge, Type 304 stainless steel.

There's good reason why so many architects, engineers and other experts specify Haws drinking fountains and water coolers. And it's not just our stunning looks and innovative designs. It's what's below the surface that really makes Haws the superior choice for today and well into the future.

For your free full color brochure, phone or fax today.

**"Hi-Lo" dual unit–** meets the Americans With Disabilities Act.

**Temperature Controlled Water–** adjustable thermostat controls temperature of storage water at 50° F.

*Model HWCD8-2*

**Environment Friendly Refrigerant–** R-134a with long life refrigeration system completely sealed after deep evacuation, dehydration, and pressure testing.

**Compressor–** with lifetime lubrication. Built-in overload protection, five-year warranty.

*Model HWCD8HO*

## Haws®

P.O. Box 1999 • Berkeley, CA 94701-1499
Phone: 510-525-5801 • FAX: 510-528-2812
E-Mail: haws@hawsco.com • Website: http://www.hawsco.com
Haws INFO FAX: 1.888.FAX.HAWS

Haws. A lot more than meets the eye.

NSF | Haws ISO 9001 CERTIFIED
Certified to ANSI/NSF 61/Section 9

# Chapter 6 — Plumbing for Physically Challenged Individuals

*leverage and stability in maintaining balance or for lifting. The grab bars clearance of 1½ in. (38 mm) required in this standard is a safety clearance to prevent injuries from arms slipping through the opening. This clearance also provides a minimum space for gripping.*

*Grab bars that are wall mounted do not affect the measurement of required clear floor space where the space below the grab bar is clear and does not present a knee space encroachment.*

## 610 Seats

**610.1 General** Seats in accessible bathtubs and shower compartments shall comply with 610.

**610.2 Bathtub seats** A removable in-tub seat shall be 15 in. (380 mm) minimum and 16 in. (405 mm) deep maximum, and shall be capable of secure placement. A permanent seat shall be 15 in. (380 mm) deep minimum and be positioned at the head end of the bathtub. The top of the seat shall be 17 in. (430 mm) minimum and 19 in. (485 mm) maximum above the bathroom floor.

**610.3 Shower compartment seats** Where a seat is provided in a roll-in shower compartment, it shall be a folding type and shall be on the wall adjacent to the controls. Seats shall be L-shaped or rectangular. The top of the seat shall be 17 in. (430 mm) minimum and 19 in. (485 mm) maximum above the bathroom floor. In a transfer type shower, the seat shall extend from the back wall to a point within 3 in. (75 mm) of the compartment entry. In a roll-in type shower, the seat shall extend from the control wall to a point within 3 in. (75 mm) of the minimum required seat wall width.

*610.3.1 Rectangular seats* The rear edge of a rectangular seat shall be 2½ in. (64 mm) maximum from the seat wall, and the front edge 15 in. (380 mm) minimum and 16 in. (405 mm) maximum from the seat wall. In a transfer type shower, the side edge of a rectangular seat shall be 1½ in. (38 mm) maximum. In a roll-in type shower, the side edge of a rectangular seat shall be 1½ in. (38 mm) maximum from the control wall.

*610.3.2 L-shaped seats* The rear edge of an L-shaped seat shall be 2½ in. (64 mm) maximum from the seat wall, and the front edge 15 in. (380 mm) minimum and 16 in. (405 mm) maximum from the seat wall. The rear edge of the "L" portion of the seat shall be 1½ in. (38 mm) maximum from the wall and the front edge shall be 14 in. (355 mm) minimum and 15 in. (380 mm) maximum from the wall. The end of the "L" shall be 22 in. (560 mm) minimum and 23 in. (585 mm) maximum from the main seat wall. (See Figure 6-25—B4.22.3.)

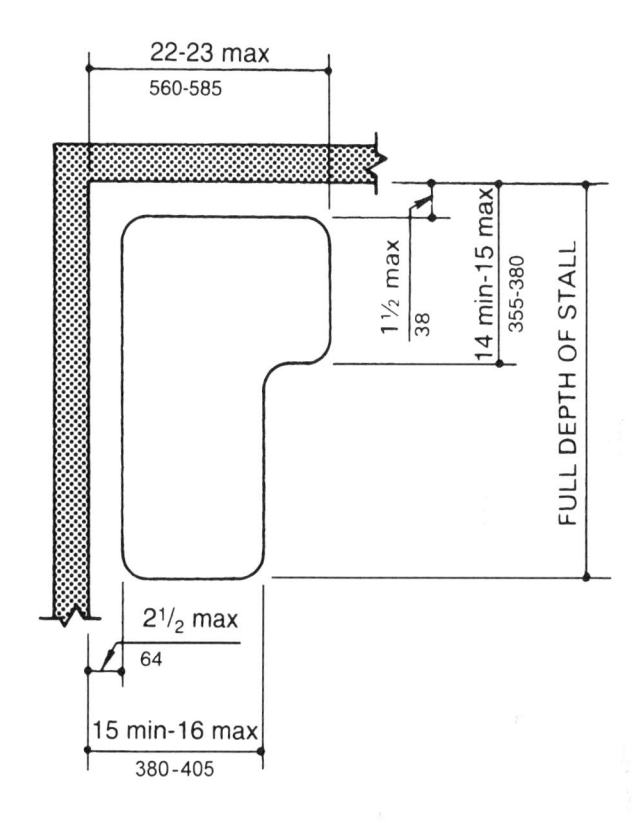

**Figure 6-25  Shower Seat Design**

*Source:* CABO/ANSI A117.1-1992. Reprinted with permission.

**610.4 Structural strength** Allowable stresses in bending, shear and tension shall not be exceeded for materials used where a vertical or horizontal force of 250 lb (113.5 kg) is applied at any point on the seat, fastener mounting device or supporting structure.

*Seats note:*

*The seat in a shower is required to be nearly the full depth of the compartment; it should be as close to the front edge of the seat wall as possible to minimize the distance between the seat and the wheelchair so as to facilitate a transfer. The seat wall must be free of grab bars to allow a person to slide onto the seat and a portion of the adjacent back wall must be without a grab bar so the person's back can be placed against the walls for support.*

## 611 Laundry Equipment

**611.1 General**  Accessible washing machines and clothes dryers shall comply with 611.

**611.2 Clear floor or ground space**  A clear floor or ground space complying with 305 positioned for parallel approach shall be provided. The clear floor or ground space shall be centered on the appliance.[2]

**611.3 Operable parts**  Operable parts, including doors, lint screens, detergent and bleach compartments, shall comply with 309.[2]

**611.4 Height**  Top-loading machines shall have the door to the laundry compartment 34 in. (865 mm) maximum above the floor or ground. Front-loading machines shall have the bottom of the opening to the laundry compartment 15 in. (380 mm) minimum and 34 in. (865 mm) maximum above the floor or ground.

## REFERENCES

1. ADAAG Review Federal Advisory Committee. September 30, 1996. Recommendations for a new ADAAG.

2. Council of American Building Officials (CABO)/International Code Council, Inc. 1998 [1992]. CABO/ANSI A117.1, Accessible and usable buildings and facilities. Falls Church, VA.

# 7 Energy Conservation in Plumbing Systems

## INTRODUCTION

Until the early 1970s and prior to the 1973-1974 OPEC oil embargo, many people (including plumbing engineers) considered energy to be inexhaustible and expendable, and these assumptions seemed verified by the low cost of most forms of energy. The problems fostered by these attitudes are manifested in the design and operation of schools, hospitals, and commercial and industrial buildings, which, by the standards of today, utilize excessive amounts of costly energy.

Plumbing systems being designed now and those to be designed in the future can and will utilize many new techniques and options that can lead to maximum energy efficiency. Two building energy conservation methods are being advanced in plumbing design efforts today.

The first method involves implementation of specific end-use restrictions provided the owner in the design of the plumbing system, e.g., controlling hot water temperatures and water pressures and providing spray-type faucets and shower heads with flow restrictors. Ease of initial implementation is the primary advantage of such an approach. Unfortunately, the end-use restriction method has numerous drawbacks. Key among these is that if, for whatever reason, a system is inefficient, it will waste energy every time it is used. End-use restrictions tend to ignore the significant energy savings that can be realized by designing plumbing systems that operate as efficiently as possible, taking into account the total building as a unique complex where many elements interrelate.

The second energy conservation approach is sometimes called the "total energy management" concept, or TEM. This technique involves the plumbing engineer's understanding of how the building consumes energy. When this is understood, energy conservation design practices become integrated into the more efficient operation of the building, using the least amount of energy to get the job done while still meeting the specific needs of the occupants.

This chapter is intended not only to provide the plumbing engineer with design techniques for specific end-use energy conservation savings, but also to assist the engineer (working in conjunction with engineers involved in electrical and heating, ventilating, and air conditioning systems) in the selection of equipment and systems to achieve total energy management. The recommendations set forth in this chapter should not be construed as to violate existing or new energy or building codes as they may relate to energy consumption and the health and safety of the building's occupants.

## SAVING ENERGY IN PLUMBING SYSTEMS

The plumbing designer can make significant contributions toward the conservation of energy in many areas. Energy reductions can be in the way of specific end-use restrictions or total energy management and are detailed below.

### Reducing Domestic-Water Temperatures

Domestic hot water often consumes between 2 and 4% of the total energy used in large office buildings, retail stores, and smaller buildings. Residential (natural gas) water heaters can account for about 8% of the total energy consumption, and in an all-electric home, the water heater is the second largest energy user, at about 23% of all consumption.

Many domestic-water heating systems are designed to deliver 140°F (60°C) water to the points of consumption on the premise that kitchen and janitorial uses require this elevated temperature, though water for human contact is normally consumed at 105°F (40.6°C). Where the blending of hot water with cold water is designed into a domestic-water heating system (see the ASPE *Data Book* chapter, "Domestic Hot Water Systems"), the reduction in domestic-water temperature, in itself, does not result in a decrease in energy input. Many energy codes and standards for new buildings require domestic hot water systems to be set at 110°F (43.3°C) under the impression that this will automatically reduce energy in direct proportion to the reduced temperature differential (ΔT). For fossil fuel systems, care must be taken to maintain the stored water at a temperature higher than the dew point of the flue gas to prevent condensation. Blending 140°F (60°C) and 125°F (51.7°C) water with cold water, at say 50°F (10°C), to achieve the desired usable temperature actually reduces the amount of hot water required but not the energy required to heat the water–whereas with 100°F (43.3°C) water there is no blending, often requiring the entire water consumed at 105°F (40.6°C) to be heated. In other words, the proportion of hot water increases at lower temperatures, and can affect the hot water storage capacity of the storage heater. If the building is provided with a kitchen dishwasher requiring 180°F (78°C) water temperature, then the booster heater will need to be sized carefully to adequately function at the increased ΔT due to the reduction of the building domestic water system temperature.

The temperature after mixing two or more volumes or flows of water is calculated using the following equation:

### Equation 7-1

$$t_m = \frac{Q_1 \times t_1 + Q_2 \times t_2}{Q_1 + Q_2}$$

*where:*

- $t_m$ = Temperature of mixture
- $t_1$ = Temperature of flow $Q_1$
- $t_2$ = Temperature of flow $Q_2$
- $Q_1$ = Cold water, gpm (L/s)
- $Q_2$ = Hot water, gpm (L/s)

*Example 7-1*

What is the temperature of 45 gpm (2.84 L/s) of 155°F (68.5°C) water mixed with 55 gpm (3.47 L/s) of 75°F (23.9°C) water?

$$\frac{45 \times 155 + 55 \times 75}{45 + 55} = 100°F$$

*in SI units:*

$$\left(\frac{2.84 \times 68.5 + 3.47 \times 23.9}{2.84 + 3.47} = 43°C\right)$$

The ratio (%) of hot water required to be mixed with cold water to provide a mixed water requirement is determined using the following equation:

### Equation 7-2

$$\text{Ratio HW} = \frac{t_m - t_1}{t_2 - t_1}$$

*Example 7-2*

(A) How much hot water is required to provide 80 gph (0.084 L/s) of 110°F (43°C) mixed water with 155°F (68.5°C) hot water and 75°F (23.9°C) cold water?

$$\frac{110 - 75}{155 - 75} = .44 \text{ or } 44\% \text{ hot water}$$

80 gph × 0.44 = 35 gph of 155°F hot water
(0.084 L/s × 0.44 = 0.037 L/s of 68.5°C hot water)

(B) How much hot water is required to provide 80 gph (0.084 L/s) of 110°F (43°C) mixed water with 125°F (51.5°C) hot water and 75°F (23.9°C) cold water?

$$\frac{110 - 75}{125 - 75} = .70 \text{ or } 70\% \text{ hot water}$$

80 gph × 0.70 = 56 gph of 125°F hot water
(0.084 L/s × 0.70 = 0.059 L/s of 51.5°C hot water)

As shown, the reduction in domestic-water temperature, in itself, does not necessarily result in a reduction in energy input related to the water consumed.

## Reduced Water Flow Rates

In 1992, the enactment of the Energy Efficiency Act set maximum water usage for specific fixtures, e.g., 1.6 gal (6 L) per flush for water closets. Reduced flow rates result in less pumping, smaller pipe sizes, and less heat loss from piping; consequently, energy conservation is achieved

# Chapter 7 — Energy Conservation In Plumbing Systems Design

by reducing water flow to fixtures. Fixture flow rates vary with the supply fitting design and water pressure. Manufacturers' test results show that flows can be quite high at lavatories and showers—the prime candidates for fixture flow reduction. Fixture flow rates can be reduced by providing automatic flow control fittings. On lavatories, types of faucet and spout dictate the location of these fittings. In showers, the types of head and arm are the determining factors in flow device location. Reduced flow rates of 1 gpm (0.063 L/s) or less are usually applicable for lavatories and of 3 gpm (0.0189 L/s) or less for showers.

Figure 7-1 provides a simple way to translate faucet flow rate in plumbing fixtures to annual consumption and is useful in determining the most energy-efficient design flow rate. By varying the percent of hot water flow rate at the fixture, annual energy consumption can be predicted.

*Example 7-3*

Faucet use at 3.25 gal (12.3 L) of 150°F (66°C) hot water per day with a 100% faucet flow rate equates to an annual energy use of 800 × 10$^6$ Btu (844 × 10$^6$ kJ) per year. A 67% flow rate reduces energy use to 475 × 10$^6$ Btu (507 × 10$^6$ kJ) per year, and a 33% flow rate reduces energy use to 225 × 10$^6$ Btu (237.4 × 10$^6$ kJ) per year or a 62% reduction from full faucet flow rate.

Figure 7-1 can be used as a design tool to predict energy consumption, which correlates to cost and payback calculations for these kinds of energy conservation devices. Manufacturers of flow-control devices describe in greater detail their design and installation requirements. The installation of these water-conserving devices has resulted in the saving of millions of gallons of water per year throughout the country. The reduction in water demand translates into water the local utility company does not have to pump, the purification plant does not have to handle and process, and the waste-treatment plant does not have to treat and pump.

## Economic Thermal Insulation Thickness

Economic thickness of insulation is that thickness that produces the lowest sum of the annual cost of energy and the annual cost of insulation. (See the "Insulation" chapter of the ASPE *Data Book* for the proper selection criteria for plumbing

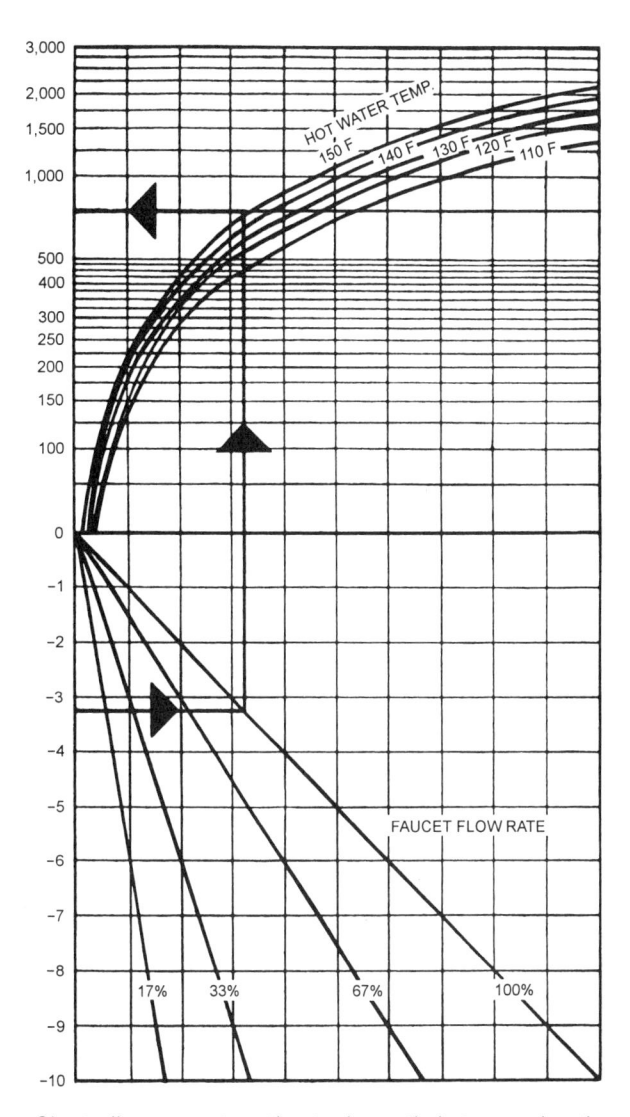

*Chart allows user to estimate domestic hot water heating use in terms of water temperature and faucet flow rate.*

**Figure 7-1 Energy Savings from Reduced Faucet Flow Rates**

*Source*: Cassidy 1982.

insulation materials.) In addition to conserving energy by retarding heat loss, insulation provides such additional benefits as protection against personnel burns, reduction of noise, and the control of condensation. The National Insulation Contractors' Association (NICA) is currently using and promoting a computer program called "Economic Thickness of Insulation" (ETI). This program determines cost-effective insulation thickness for a particular project and allows the designer to factor in the effects of rising utility costs.

### Table 7-1 Energy Savings Chart for Steel Hot Water Pipes and Tanks

| ΔT °F (°C) | ½ (12.7) | ¾ (19.1) | 1 (25.4) | 1¼ (31.8) | 1½ (38.1) | 2 (50.8) | 2½ (63.5) | Hot Water Tanks, Btu/h/ft² (kJ/h/m²) w/ Insulation | w/o Insulation |
|---|---|---|---|---|---|---|---|---|---|
| 40 (4.4) | 14 (48.44) | 17 (58.8) | 21 (72.7) | 26 (90.0) | 29 (100.3) | 35 (121.1) | 42 (145.3) | 6 (68.1) | 57 (647.3) |
| 45 (7.2) | 16 (55.36) | 20 (69.2) | 24 (83.0) | 30 (103.8) | 33 (114.2) | 41 (141.9) | 48 (166.1) | 6 (68.1) | 65 (738.2) |
| 50 (10.0) | 18 (62.28) | 22 (76.1) | 27 (93.4) | 34 (117.6) | 38 (131.5) | 47 (162.6) | 55 (190.3) | 7 (79.5) | 73 (829.1) |
| 55 (12.8) | 20 (69.20) | 25 (86.5) | 31 (107.3) | 38 (131.5) | 42 (145.3) | 52 (179.9) | 62 (214.5) | 7 (79.5) | 83 (942.6) |
| 60 (13.6) | 23 (79.58) | 28 (96.9) | 35 (121.1) | 42 (145.3) | 48 (166.1) | 58 (200.7) | 69 (238.7) | 9 (102.2) | 92 (1044.8) |
| 65 (18.3) | 25 (86.50) | 31 (107.3) | 38 (131.5) | 47 (162.6) | 53 (183.4) | 65 (224.9) | 77 (266.4) | 9 (102.2) | 102 (1158.4) |
| 70 (21.1) | 28 (96.88) | 34 (117.6) | 42 (145.3) | 52 (179.9) | 58 (200.7) | 71 (245.7) | 84 (290.6) | 10 (113.6) | 112 (1272.0) |
| 75 (23.9) | 30 (103.8) | 36 (124.6) | 46 (159.2) | 56 (193.8) | 64 (221.4) | 78 (269.9) | 91 (314.9) | 11 (124.9) | 122 (1385.6) |
| 80 (26.7) | 33 (114.2) | 41 (141.9) | 50 (173.0) | 61 (211.1) | 69 (238.7) | 84 (290.6) | 99 (342.5) | 11 (124.9) | 132 (1499.1) |
| 85 (28.4) | 36 (124.6) | 44 (152.2) | 54 (186.8) | 67 (231.8) | 74 (256.0) | 91 (314.9) | 107 (370.2) | 12 (136.3) | 142 (1612.7) |
| 90 (32.2) | 38 (131.5) | 47 (162.6) | 58 (200.7) | 72 (249.1) | 80 (276.8) | 98 (339.1) | 116 (401.4) | 12 (136.3) | 154 (1749.0) |
| 95 (35.0) | 42 (145.3) | 51 (176.5) | 62 (214.5) | 77 (266.4) | 86 (297.6) | 105 (363.3) | 124 (429.0) | 14 (159.0) | 164 (1862.5) |
| 100 (37.8) | 45 (155.7) | 54 (186.8) | 66 (228.4) | 82 (283.7) | 93 (321.8) | 113 (391.0) | 133 (460.2) | 14 (159.0) | 175 (1987.5) |
| 105 (38) | 47 (162.6) | 58 (200.7) | 72 (249.1) | 87 (301.0) | 98 (339.1) | 120 (415.2) | 141 (487.9) | 15 (170.4) | 187 (2123.8) |
| 110 (43) | 51 (176.5) | 62 (214.5) | 75 (259.5) | 93 (321.8) | 104 (359.8) | 128 (442.9) | 150 (519) | 16 (181.7) | 198 (2248.7) |
| 115 (46) | 54 (186.8) | 65 (224.9) | 80 (276.8) | 98 (339.1) | 110 (380.6) | 135 (467.1) | 159 (550.1) | 16 (181.7) | 210 (2385.0) |
| 120 (49) | 56 (193.8) | 69 (238.7) | 85 (294.1) | 104 (359.8) | 117 (404.8) | 143 (494.8) | 169 (584.7) | 17 (193.1) | 222 (2521.3) |

*Source*: San Diego Gas & Electric Co.

*Notes*: 1. Savings are in Btu/h/linear ft. (kJ/h/linear m), unless otherwise indicated.
2. Figures are based on an assumption of 1 in. (25.4 mm) of insulation.

### Table 7-2 Energy Savings Chart for Copper Hot Water Pipes

| ΔT °F (°C) | ½ (12.7) | ¾ (19.1) | 1 (25.4) | 1¼ (31.8) | 1½ (38.1) | 2 (50.8) | 2½ (63.5) | 3 (76.2) |
|---|---|---|---|---|---|---|---|---|
| 40 (4.4) | 8 (27.68) | 12 (41.5) | 14 (48.4) | 17 (58.8) | 20 (69.2) | 25 (86.5) | 30 (103.8) | 35 (121.1) |
| 45 (7.2) | 10 (34.6) | 13 (45.0) | 16 (55.5) | 20 (69.2) | 23 (79.6) | 29 (100.3) | 35 (121.1) | 40 (138.4) |
| 50 (10.0) | 12 (41.5) | 15 (51.9) | 19 (65.7) | 23 (79.6) | 26 (90.0) | 33 (114.2) | 40 (138.4) | 46 (159.2) |
| 55 (12.8) | 13 (45.0) | 17 (58.8) | 21 (72.7) | 26 (90.0) | 30 (103.8) | 38 (131.5) | 45 (155.7) | 52 (179.9) |
| 60 (13.6) | 15 (51.9) | 20 (69.2) | 24 (83.0) | 29 (100.3) | 34 (117.6) | 42 (145.3) | 51 (176.5) | 58 (200.7) |
| 65 (18.3) | 16 (55.4) | 21 (72.7) | 27 (93.4) | 32 (110.7) | 37 (128.0) | 47 (162.6) | 56 (193.8) | 65 (224.9) |
| 70 (21.1) | 18 (62.3) | 24 (83.0) | 30 (103.8) | 35 (121.1) | 41 (141.9) | 52 (180.0) | 62 (214.5) | 71 (245.7) |
| 75 (23.9) | 20 (69.2) | 26 (90.0) | 33 (114.2) | 39 (134.9) | 44 (152.2) | 56 (193.8) | 67 (231.8) | 76 (263.0) |
| 80 (26.7) | 21 (72.7) | 28 (96.7) | 35 (121.1) | 42 (145.3) | 49 (169.5) | 61 (211.1) | 73 (252.6) | 85 (294.1) |
| 85 (29.4) | 22 (76.1) | 31 (107.3) | 38 (131.5) | 45 (155.7) | 53 (183.4) | 66 (228.4) | 79 (273.3) | 92 (318.3) |
| 90 (32.2) | 24 (83.0) | 33 (114.2) | 41 (141.9) | 49 (169.5) | 57 (197.2) | 71 (245.7) | 85 (294.1) | 99 (342.5) |
| 95 (35.0) | 26 (90.0) | 36 (124.6) | 44 (152.2) | 53 (183.4) | 61 (211.1) | 76 (263.0) | 91 (314.9) | 106 (366.7) |
| 100 (37.8) | 28 (96.7) | 38 (131.5) | 48 (166.1) | 57 (197.2) | 65 (224.9) | 82 (283.7) | 98 (339.1) | 113 (391.0) |
| 105 (38) | 30 (103.8) | 41 (141.9) | 51 (176.5) | 60 (207.6) | 70 (242.2) | 87 (301.0) | 104 (359.8) | 121 (418.7) |
| 110 (43) | 32 (110.7) | 43 (148.8) | 54 (186.8) | 65 (224.9) | 74 (256.0) | 93 (321.8) | 111 (384.1) | 128 (442.9) |
| 115 (46) | 34 (117.6) | 46 (159.2) | 57 (197.2) | 68 (235.3) | 78 (269.9) | 98 (339.1) | 118 (408.3) | 136 (470.6) |
| 120 (49) | 36 (124.6) | 49 (169.5) | 61 (211.1) | 72 (249.1) | 83 (287.2) | 104 (359.8) | 125 (432.5) | 144 (498.2) |

*Source*: San Diego Gas & Electric Co.

*Notes*: 1. Savings are in Btu/h/linear ft (kJ/h/linear m).
2. Figures are based on an assumption of 1 in. (25.4 mm) of insulation.

# Chapter 7 — Energy Conservation In Plumbing Systems Design

Savings in Btu (J) can be determined by the following formula: Determine the hot water circulating temperature ($t_o$) and determine the temperature of the air surrounding the piping system ($t_a$). The temperature difference ($\Delta T$) equals $t_o - t_a$. Table 7-1 (if steel pipe) or 7-2 (if copper pipe) can be used to find the Btu/linear ft (kJ/linear m) savings for the various sizes of pipe. Total system savings can be determined by using the following formula:

### Equation 7-3

$$S = g \times L$$

*where:*

S = Energy savings, Btu/h (kJ/h)

g = Factors taken from Table 7-1 or 7-2 at a particular $\Delta T$, Btu/h/ft (kJ/h/m)

L = System length, ft (m)

## Hot Water System Improvement

The areas of potential energy reduction listed below offer a look at hot water system improvement from both an owner's and a designer's standpoint.

### From the owner's viewpoint

1. Inspect water supply system and repair all leaks, including those at the faucets.
2. Inspect and test hot water controls to determine if they are working properly. If not, regulate, repair, or replace.
3. Inspect insulation on storage tanks and piping and repair or replace as needed.

### From the plumbing engineer's viewpoint

1. Increase the amount of insulation installed on hot water piping and storage tanks.
2. Consider the specification of hot water faucets with spray type action and flow restrictors.
3. Consider specifying spring-activated, self-closing, hot or tempered water faucets.
4. Specify the use of a pressure-reducing valve on groups of fixtures using hot water if the water pressure exceeds 40 psig (276 kPa).
5. Design the hot water system for minimum hot water temperatures in an energy-efficient manner. Boost hot water temperatures locally for kitchens and other areas where needed rather than providing higher-than-necessary temperatures for the entire building.
6. Connect the electric water heater and circulating return water pump to an energy management system limiting the duty cycle to avoid adding these loads to the building during peak electric demands.
7. Consider shutting off both return and forced circulation pump systems when the building is unoccupied.
8. Locate the water heater as close to the point of use as possible.
9. Use waste heat for preheating domestic water.
10. Evaluate point-of-use water heating vs. recirculating hot water vs. heat maintenance tape.

## Standby Losses and Circulating Vs. Noncirculating Systems

An opportunity for the plumbing designer to provide substantial savings in system energy use is the automatic shut down of the water heater and circulating system during building unoccupied hours. One hundred thirteen (113) hours of "off" time per week can be developed in system operation if a building is only occupied 50 hours per week and the system can be brought up to operating temperature 1 hour each day. Assuming a domestic hot water system can be shut off 113 hours per week and the system contains 2000 gal (7570 L) of water, Table 7-3 indicates the effect of stopping circulation. However, if fossil fuel is used, one can expect condensation when the system is brought up to temperature.

### Table 7-3  The Effect of Stopping Circulation

| Operating Temperature, °F (°C) | Piping Insulation Thickness, in. (mm) | Energy Conserved, Btu/yr (kJ/yr) |
|---|---|---|
| 140 (60)   | ½ (12.7)  | $1428 \times 10^6$ ($1506.5 \times 10^6$) |
| 125 (51.5) | ½ (12.7)  | $1153 \times 10^6$ ($1216 \times 10^6$) |
| 110 (43)   | ½ (12.7)  | $824 \times 10^6$ ($869.3 \times 10^6$) |
| 140 (60)   | 1 (25.4)  | $934 \times 10^6$ ($985.4 \times 10^6$) |
| 125 (51.5) | 1 (25.4)  | $714 \times 10^6$ ($753.3 \times 10^6$) |
| 110 (43)   | 1 (25.4)  | $522 \times 10^6$ ($550.7 \times 10^6$) |

It is obvious that the operating temperature and insulation thickness affect the amount of energy that can be saved in this conservation strategy and greatly affect the overall thermal efficiency of the hot water system whether at operating temperature or idle in off times.

Energy savings from using time clocks on water circulating pumps can be calculated as follows:

***Equation 7-4***

Motor kW × off hours × electric rate ($/kWh) = total savings ($)

## Use of Waste Heat to Heat Water

In recent years, the use of waste heat from air conditioning, refrigeration, laundries, and industrial processes for domestic water heating has been a consideration of mechanical engineers involved in the design of such systems.

The common waste heat sources can be itemized as follows:

1. Heat rejected from air conditioning and commercial refrigeration processes.
2. Heat reclaimed from steam condensate.
3. Heat generated by cogeneration plants.
4. Heat pumps and heat reclamation systems.
5. Heat from waste water in operations such as laundries.

Before any decision about the use of waste heat for heating water is made, a life-cycle cost analysis should be made to establish economic justification for the proposed system.

**Heat rejected from air conditioning and commercial refrigeration processes** Rejected heat from air conditioning and refrigeration systems can be reclaimed from the following sources:

1. Systems with air-cooled or evaporative condensers.
2. Systems with water-cooled condensers.

For every Btu/h (kJ/h) of cooling effect produced at 40°F (4.4°C) evaporating and 105°F (38°C) condensing temperature, heat rejection at the condenser will be approximately 1.15 Btu/h (1.21 kJ/h). Systems with air-cooled or evaporative condensers can be provided with a heat exchanger in the compressor hot gas discharge line. (Refer to Figure 7-2.)

Systems with water-cooled condensers can be provided with a heat exchanger in the hot condenser water return line to the cooling tower. (Refer to Figure 7-3.) System efficiency can be improved by providing a storage tank with a tube bundle. (Refer to Figure 7-4.)

The advantage of the system shown in Figure 7-4 is that simultaneous usage of the domestic water and refrigeration systems need not occur for heat recovery. The recovered heat contributes to matching the required storage temperature, with the direct-fired water heater acting as the backup to bring the storage tank to proper design temperature. The backup heater can be a fossil fuel (gas, oil, or electric) heater or a separate steam, medium or high-temperature, water-fired heater. Furthermore, the storage tank shown in Figure 7-4 can be fitted with a tube bundle utilizing steam or hot water.

**Heat reclaimed from steam condensate** Wherever steam is used as a source of heat for space heating, water heating, or process work, generally there is steam condensate. The heat content of the condensate can be made available for heating with the use of heat exchangers. This reclaimed heat can then be used for preheating makeup water. Laundries are a prime example of facilities where heat reclaimed from steam condensate can be put to use in heat recovery. It is essential to select a system with adequate storage to compensate for fluctuations in the condensate flow and corresponding domestic-water flow. However, this will not save energy unless the boiler used to raise the temperature of the returned condensate is more thermally efficient than the primary water heater.

**Cogeneration systems** The heat produced as a byproduct of generating electricity through the use of reciprocating engines or gas turbines can be reclaimed from the engine's cooling systems and exhaust gases by use of waste heat boilers and heat exchangers. This heat is used to provide steam or medium-temperature water. The latter media, through the use of heat exchangers, can heat makeup water or maintain temperatures in hot-water storage systems. To be economically viable, most systems must have a year-round thermal heat load. Reheating makeup water and maintaining temperatures in a domestic hot water system are excellent ways to maintain high overall thermal efficiencies.

**Heat pump and heat reclamation systems**

***Heat pumps*** In today's high-technology buildings, where computer rooms generate heat

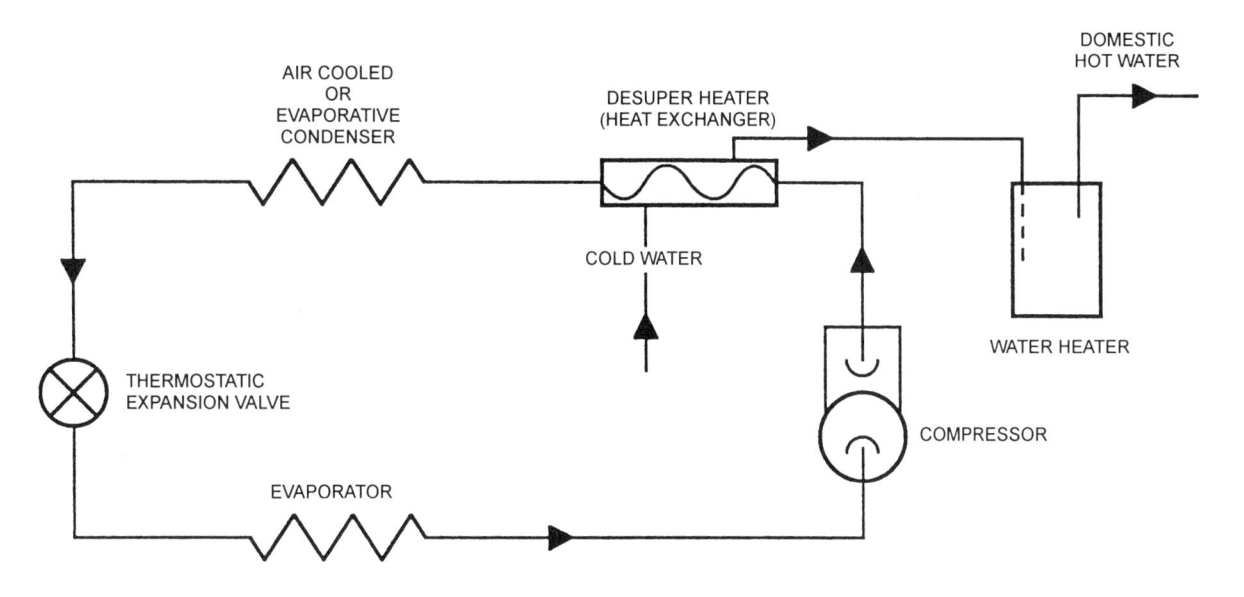

# Chapter 7 — Energy Conservation In Plumbing Systems Design

**Figure 7-2 Refrigeration Waste Heat Recovery**

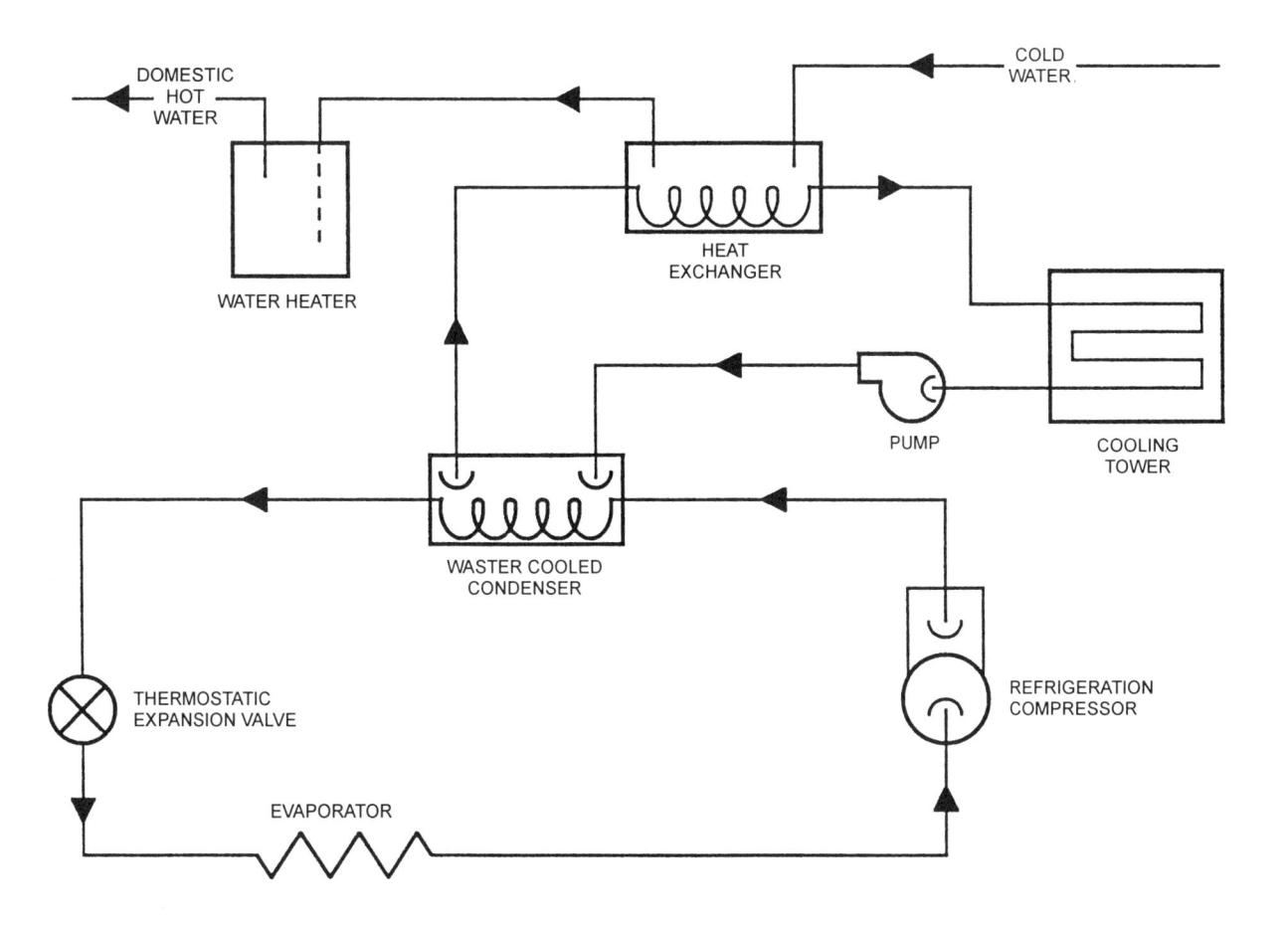

**Figure 7-3 Condenser Water Heat Recovery**

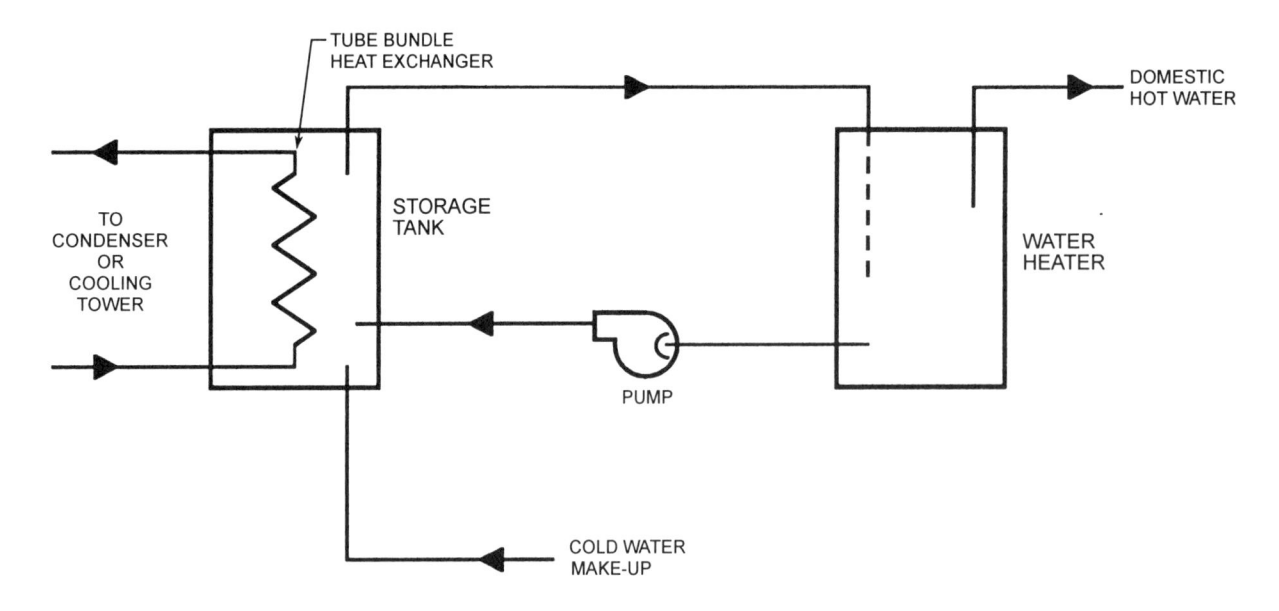

**Figure 7-4  Condenser Water Heat Recovery with Storage Tank**

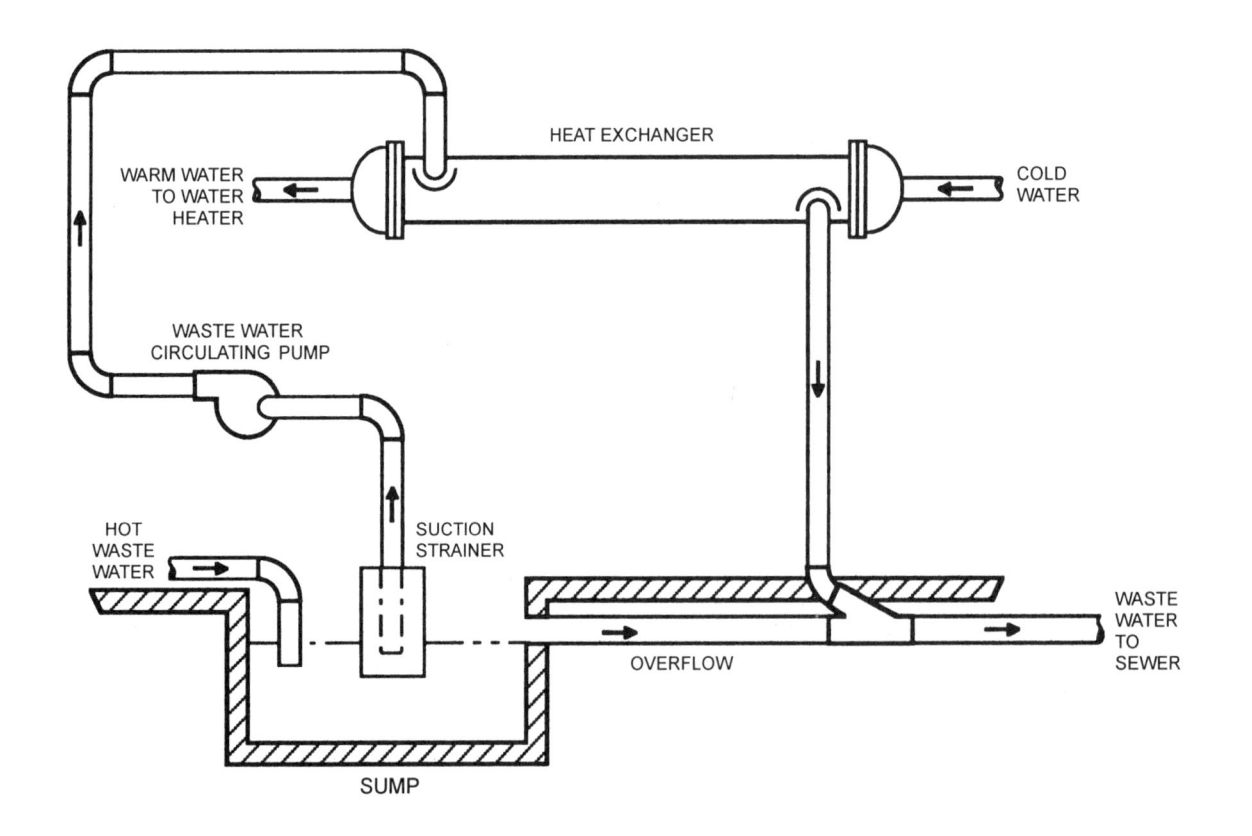

**Figure 7-5  Wastewater Heat Recovery**

# Chapter 7 — Energy Conservation In Plumbing Systems Design

continuously all year, and in industrial plants, where process requirements produce waste heat as a byproduct, the use of heat pumps to transfer this heat to domestic hot water systems can result in energy conservation.

Heat pumps, of either the direct expansion or chilled water type, depending on size, transfer the heat, through the refrigeration process, to water storage tanks. Heat exchangers and tube bundles can be effectively put to work providing domestic hot water using the refrigerant or condenser water as the heating medium.

The mechanics of such a system are to extract heat either directly from a warm environment (such as air or a fluid) or through a heat exchanger or cooling coil circulating the warm air or fluid. The heat thus extracted can be pumped to the domestic water system.

***Heat reclamation*** Many processes, such as those of an industrial laundry, result in hot waste water being discharged. Heat can be reclaimed from such processes using a system such as that shown in Figure 7-5.

## SAVING UTILITY COSTS IN PLUMBING SYSTEMS

### Using Off-Peak Power

Sizing domestic water heating equipment to meet the needs of occupants in the most energy-efficient manner is the responsibility of the plumbing engineer. The use of off-peak power in heating and circulating water does not change the number of British thermal units (Btu) required but does allow the system owner to benefit from the lowest utility charges available. Power companies are anxious for their customers to purchase power during off-peak hours in order to flatten or even out the demand on their generating equipment, which makes their systems operate efficiently. Some utility companies not only offer lower and very attractive rates for electric energy purchased during off and semi-peak periods, but in many instances, have no customer demand charges. It is advisable for the plumbing engineer to obtain electric rate schedules from the utility serving the site and observe the off-peak periods to program the operation of domestic-water heating equipment. In many cases, this technique has great economic benefits. A design that includes heating water by off-peak power, holding it at elevated temperatures, and then blending it to achieve safe temperature levels (depending on particular electric rate schedules) can generally pay for the additional equipment required in a very few years, even after equipment heat losses during periods of standby are considered.

### Using Minimum-Energy-Consuming Equipment

Plumbing engineers, in their quest for design techniques to save energy, need to examine their equipment specifications to make sure only equipment having minimum energy requirements is approved for installations, e.g., water heaters labeled "energy efficient"; variable-speed, multiple pumps; water-saving fixtures; and central water cooler systems for building drinking water requirements.

The requirements for a pumped domestic-water system require that the designer understand different hydraulic pump characteristics to select the correct pump for a specific application. The "standard" pump rarely exists. Each pump application requires analysis. Optimum pump selection requires the examination of the following two performance characteristics: head-flow curves and pump performance curves. Packaged pump systems for multistory buildings are available with two main pumps and a jockey pump that maintains building pressure during low water usage. The jockey pump is equipped with a small motor, which saves energy over the conventional three-pump system. The "Pumps" chapter of the ASPE *Data Book* should be referred to in detail relative to the types of pump available and their selection criteria.

The following factors contribute to the net efficiency of gas-fired water heaters and need to be taken into consideration in their selection: 1) combustion equipment and its adjustment; 2) tank insulation; 3) heat exchanger effectiveness, firing rate, pickup, and demand; and 4) standby stack losses.

### Water Reuse Systems

Many states and municipalities recognize the many benefits of water reuse systems, including the reduction of water use, the reduction of water heating demand, and the significant decrease in wastewater flows that must be treated.

# NONDEPLETABLE AND ALTERNATE ENERGY SOURCES

In many parts of the country, energy code compliance subjects the domestic-water heating system to certain restrictions and is calculated as part of the total energy chargeable to the proposed design. The designer may choose to use a nondepletable energy source (solar or geothermal) or an alternate source (solid waste disposal) for all or part of the system's calculated energy. With the so-called "inexhaustible" fossil fuel supply now expected to be in short supply within the next 50 years, nondepletable energy sources have taken on a new importance.

## Solar Energy

Of all nondepletable energy sources, solar energy seems to offer the greatest promise as a main energy source for heating domestic water. The energy from the sun is, for all practical purposes, inexhaustible and of no cost to the building owner. The collector efficiency and the part of the country the solar system is erected in will dictate the amount of thermal energy created to offset traditional supplies. The *ASPE Solar Energy System Design Handbook* is recommended as a source of information in the use and selection of solar system equipment.

## Geothermal Energy

In states where geothermal energy (heat from the earth) is believed to be available at reasonable depths, the US Department of Energy is supporting various state energy commissions in their funding of geothermal assessment programs. The viability of geothermal energy is dictated by the temperature of the liquid or gas available and the cost of retrieval. Geothermal steam is currently being used in the generation of electricity, while hot water with a minimum temperature of 150°F (66°C) is used in building domestic hot-water systems. Industrial parks that use geothermal energy for space and water-heating needs (and are sometimes called "geothermal parks of commerce") are being developed in this country. Three prime areas of concern must be addressed in the planning and development of geothermal energy: 1) competitive institutional processes, 2) adequate temperature and flow rate, and 3) thermal loads to make the system economically viable.

A geothermal energy system consists of: production and disposal wells; water-to-water heat exchangers, usually of the shell-and-tube type (two required–one can be cleaned of deposits while the other is in use); and an insulated piping, circulating pump, and control system. The plumbing engineer should consult with the state energy office (Department of Gas and Oil Resources or the Geothermal Resources Council) for resource information to apply this high-capital, low-operating-cost, alternate energy source.

## Solid Waste Disposal

Solid waste collection and disposal systems produce gases during decomposition. These gases may be collected and recovered then used to produce heat by burning the collected methane. Also, they can be used for leachate evaporation systems for landfill closures, utilizing the methane as the fuel source.

Various solid waste incineration systems are in use and others are being constructed that meet stringent pollution-control rules and regulations and provide large volumes of steam and/or hot water for domestic hot-water systems. The use of this alternate energy source is like the use of a geothermal energy source in that the use of hot water should be in reasonably close proximity to the resource. Typical applications include industrial plants with large volumes of burnable materials —trash, paper, scrap wood, plastics, etc. The system consists of the waste disposal plant with conveyer, loading system, boiler, ash disposal equipment, heat exchanger, insulated piping, circulating pump, and controls.

Geohermal and solid waste disposal energy can have cascaded uses, e.g., 500 gpm (31.5 L/s) at 180°F (82°C) water can be used for industrial process heat, flowing at 140°F (60°C) to space heating systems, followed by 115°F (46°C) water flowing into a domestic hot-water system, and finally at 95°F (35°C) being used in groundwater heat pump systems. Cost and energy savings are thereby provided.

# PERFORMANCE EFFICIENCY OF HEATING AND HOT WATER STORAGE EQUIPMENT

The performance efficiency of equipment specified by the plumbing engineer should match that recommended in the "Domestic Water Heating

# Chapter 7 — Energy Conservation In Plumbing Systems Design

Systems" chapter of the ASPE *Data Book*. The recovery efficiency and standby losses of water heating equipment should comply with the latest codes and regulations for their manufacture, e.g., American National Standards Institute ANSI C72.1, ANSI Z22.10.3, latest editions. State energy codes also mandate the use of energy-efficient equipment and should be checked by the plumbing engineer prior to the preparation of specifications.

## ENERGY CODE COMPLIANCE

The plumbing engineer is subject to compliance with a variety of codes in the design of potable water, domestic hot-water and chilled-water systems, including energy codes. Local building department and state governing authorities dictate which code compliance is necessary for a particular building requirement and should be consulted prior to the beginning of design.

## GLOSSARY

***British thermal unit (Btu)*** A heat unit equal to the amount of heat required to raise 1 pound of water 1 degree Fahrenheit.

***Coefficient of performance (COP)*** The ratio of the rate of heat removal to the rate of energy input, in consistent units, generally relating to a refrigeration system under designated operating conditions.

***Condenser*** A heat exchanger that removes heat from a vapor changing it to its liquid state.

***Delta T (ΔT)*** Temperature differential.

***Domestic-water heating*** Supply of hot water for domestic or commercial purposes other than comfort heating.

***Domestic-water heating demand*** The maximum design rate of energy withdrawal from a domestic-water heating system in a specified period of time.

***Efficiency, thermal (overall system)*** The ratio of useful energy at the point of ultimate use to the energy input.

***Energy*** The force required for doing work.

***Energy, nondepletable*** Energy derived from incoming solar radiation and phenomena resulting therefrom, including wind, waves, and tides, and lake or pond thermal differences, and energy derived from the internal heat of the earth (geothermal)–including nocturnal thermal exchanges.

***Energy, recovered*** A byproduct of energy used in a primary system that would otherwise be wasted from an energy utilization system.

***Heat, latent*** The quantity of heat required to effect a change in state.

***Heat, sensible*** Heat that results in a temperature change but not a change in state.

***Life-cycle cost*** The cost of the equipment over its entire life, including operating and maintenance costs.

***Makeup*** Water supplied to a system to replace that lost by blowdown, leakage, evaporation, etc.

***Solar energy source*** Source of chemical, thermal, or electrical energy derived from the conversion of incident solar radiation.

***System*** An arrangement of components (including controls, accessories, interconnecting means, and terminal elements) by which energy is transformed to perform a specific function.

***Terminal element*** The means by which the transformed energy from a system is ultimately delivered.

## REFERENCES

1. Cassidy, Victor M. 1982. Energy saving and the plumbing system. *Specifying Engineering* (February).
2. San Diego Gas & Electric Company. *Commercial energy conservation manual.*

# 8 Corrosion

## INTRODUCTION

Corrosion is the degradation of a material by its environment. In the case of metals, corrosion is an electrochemical reaction between a metal and its environment. For iron piping, the iron reacts with oxygen to form iron oxide, or rust, which is the basic constituent of the magnetic iron ore (hematite) from which the iron was refined. The many processes necessary to produce iron or steel pipe–from refining through rolling, stamping, and fabricating to finished product–all impart large amounts of energy to the iron. The iron in a finished pipe is in a highly energized state and reacts readily with oxygen in the environment to form rust. Corrosion results from a flow of direct current through an electrolyte (soil or water) from one location on the metal surface to another location on the metal surface. The current flow is caused by a voltage difference between the two locations.

This chapter covers the fundamentals of corrosion as they relate to a building's utility systems, essentially dealing with piping materials for the conveyance of fluids, both liquid and gas. These pipes are installed either under or above ground, thus making the external environment of the pipe earth or air, respectively. The internal environment is the fluid conveyed inside the pipe. There are many environmental conditions that may affect the performance of any given piping material.

## FUNDAMENTAL CORROSION CELL

### Basic Relations

Corrosion is, in effect, similar to a dry cell. In order for corrosion to occur, there must be four elements, namely: electrolyte, anode, cathode, and a return circuit. The electrolyte is an ionized material, such as earth or water, capable of conducting an electric current.

Figure 8-1 shows the actual corrosion cell. Figure 8-2 (practical case) shows the current flows associated with corrosion:

1. Current flows through electrolyte from the anode to the cathode. It returns to the anode through the return circuit.

2. Corrosion occurs wherever current leaves the metal and enters the electrolyte. The point where current leaves is called the anode. Corrosion, therefore, occurs at the anode.

3. Current is picked up at the cathode. No corrosion occurs here, as the cathode is protected against corrosion (this is the basis of cathodic protection). Polarization (hydrogen film buildup) occurs at the cathode.

4. The flow of the current is caused by a potential (voltage) difference between the anode and the cathode.

### Electrochemical Equivalents

Dissimilar metals, when coupled together in a suitable environment, will corrode according to Faraday's law; that is, it will require 26.8 ampere-hours (A-h), or 96,500 coulombs (C), to remove 1 gram-equivalent of the metal. At this rate of attack, the amount of metal that is removed by a current of 1 A flowing for 1 year is shown in Table 8-1.

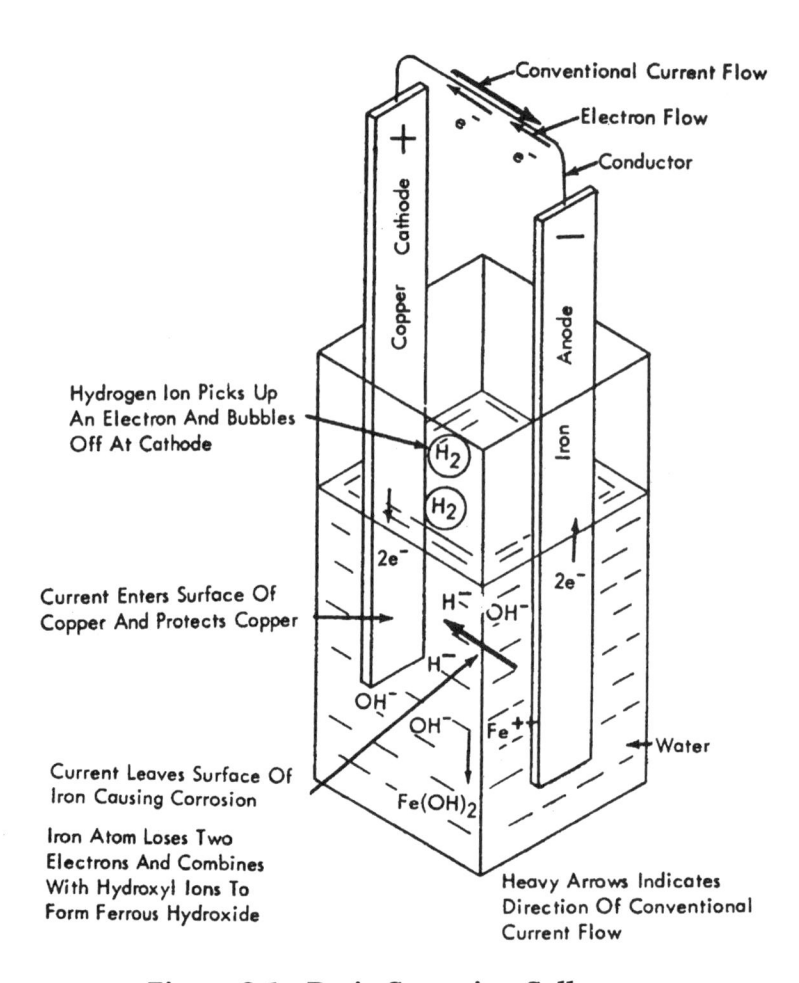

Figure 8-1 Basic Corrosion Cell

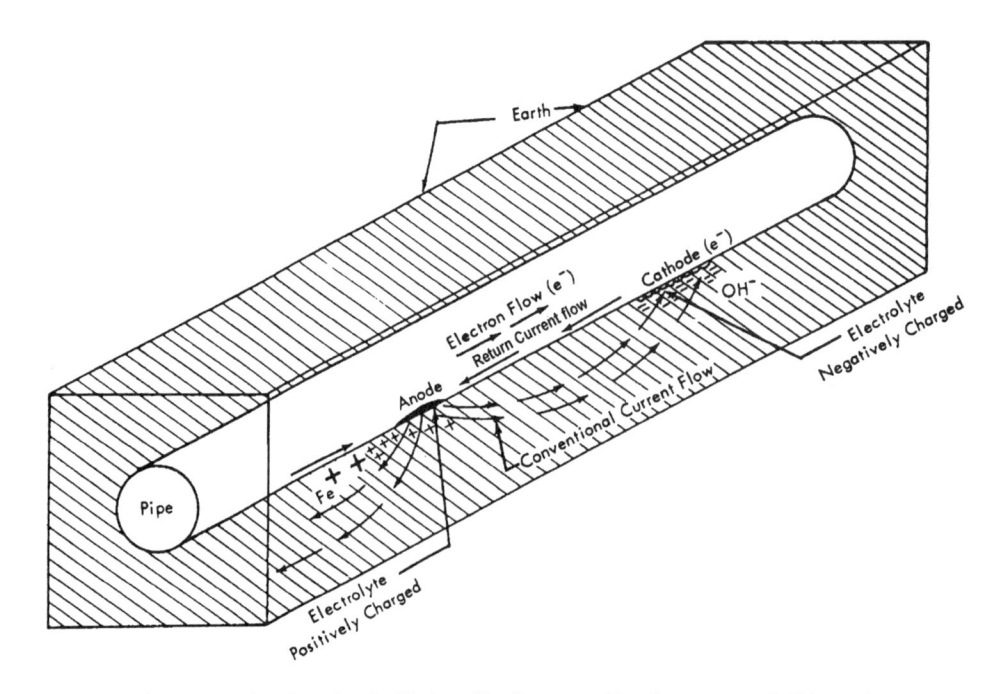

Figure 8-2 Basic Cell Applied to an Underground Structure

# Chapter 8 — Corrosion

**Table 8-1  Electrochemical Metal Losses of Some Common Metals**

| Metal | Loss, lb/A-yr (kg/C) |
|---|---|
| Iron ($Fe^{2+}$) | 20.1 (72.4) |
| Aluminum ($Al^{3+}$) | 6.5 (23.4) |
| Lead ($Pb^{2+}$) | 74.5 (268.3) |
| Copper ($Cu^{2+}$) | 45.0 (162.0) |
| Zinc ($Zn^{2+}$) | 23.6 (85.0) |
| Magnesium ($Mg^{2+}$) | 8.8 (31.7) |
| Nickel ($Ni^{2+}$) | 21.1 (76.0) |
| Tin ($Sn^{+}$) | 42.0 (151.2) |
| Silver ($Ag^{+}$) | 77.6 (279.4) |
| Carbon ($C^{4+}$) | 2.2 (7.9) |

## COMMON FORMS OF CORROSION

Corrosion occurs in a number of common forms as follows:

***Uniform attack***   (Figure 8-3) Uniform attack is characterized by a general dissolving of the metal wall. The material and its corrosion products are readily dissolved in the corrosive media.

***Pitting corrosion***   (Figure 8-4) Pitting corrosion is usually the result of the localized breakdown of a protective film or layer of corrosion products. Anodic areas form at the breaks in the film and cathodic areas form at the unbroken portion of the film. The result is localized, concentrated corrosion, which forms deep pits.

***Galvanic corrosion***   (Figure 8-5) Galvanic corrosion occurs when two dissimilar metals are in contact with an electrolyte. The example shown is iron and copper in a salt solution, the iron being the anode corroding toward the copper cathode.

***Concentration cell attack***   (Figure 8-6) Concen-tration cell attack is caused by differences in the concentration of a solution, such as differences in oxygen concentration or metal-ion concentration. These can occur in crevices, as shown in the example, or under mounds of contamination on the metal surface. The area of low oxygen or metal-ion concentration becomes anodic to areas of higher concentration.

***Crevice corrosion***   A form of concentration cell attack (see separate listing).

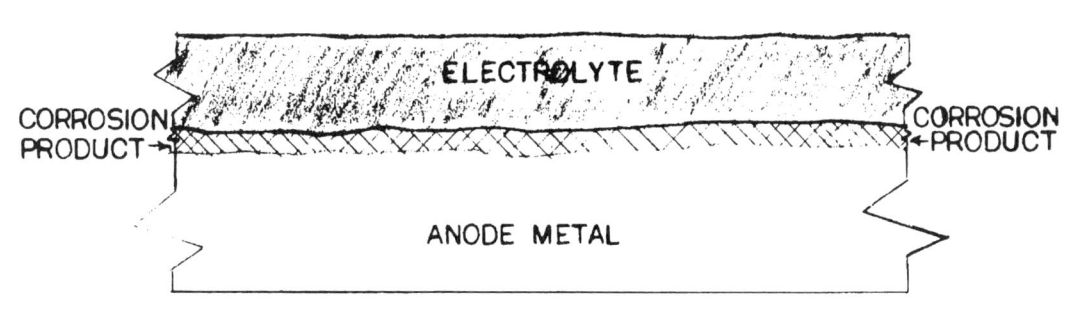

**Figure 8-3  Uniform Attack**

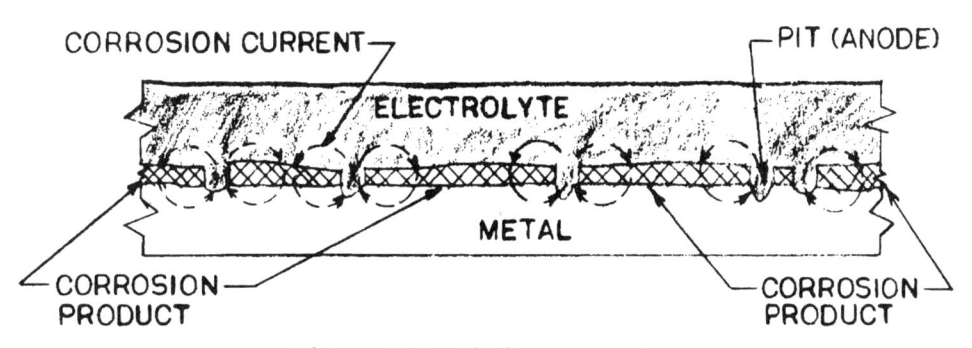

**Figure 8-4  Pitting Corrosion**

***Impingement attack*** (Figure 8-7) Impingement attack is the result of turbulent fluid, at high velocity, breaking through protective or corrosion films on a metal surface. There usually is a definite direction to the corrosion formed.

***Stress corrosion cracking*** (Figure 8-8) Stress corrosion cracking results from placing highly stressed parts in corrosive environments. Corrosion causes concentration of the stress, which eventually exceeds the yield strength of the material, and cracking occurs.

***Selective attack*** (Figure 8-9) Selective attack is the corrosive destruction of one element of an alloy. Examples are dezincification of brass and graphitization of cast iron.

***Stray current*** (Figure 8-10) Stray current corrosion is caused by the effects of a direct current source such as a cathodic protection rectifier. Protective current may be picked up on a pipeline or structure that is not part of the protected system. This current follows to the other structure and at some point leaves the other structure and travels through the electrolyte (soil or water) back to the protected structure. This causes severe corrosion at the point of current discharge.

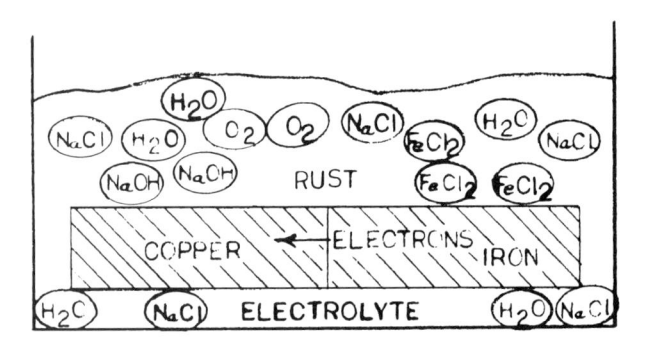

***Corrosion by differential environmental conditions*** (Figure 8-11) Examples of differential environmental cells are shown in Figure 8-11. It should be noted that variations in moisture content, availability of oxygen, change in soil resistivity, or variations of all three may occur in some cases. As in all corrosion phenomena, changes or variations in the environment are a contributing factor.

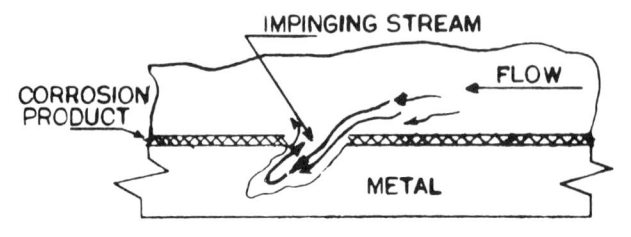

**Figure 8-7  Impingement Attack**

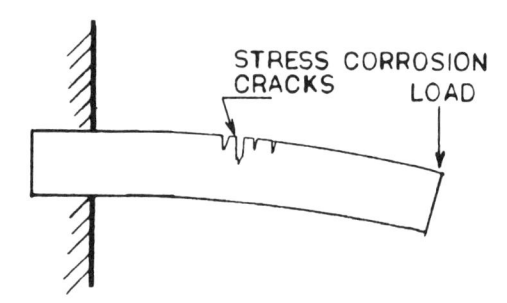

**Figure 8-8  Stress Corrosion**

**Figure 8-5  Galvanic Corrosion**

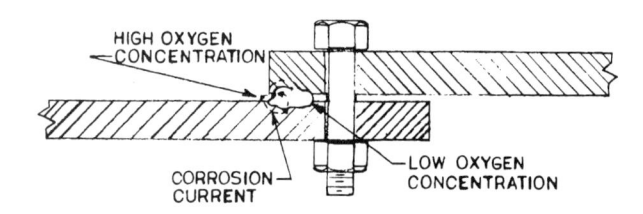

**Figure 8-6  Concentration Cells**

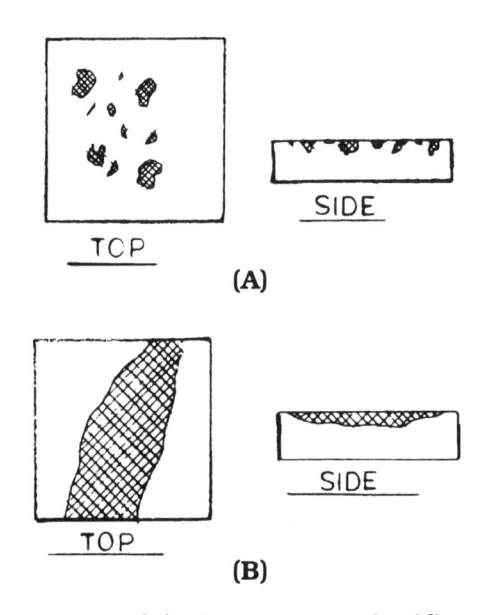

**Figure 8-9  (A) Plug-Type Dezincification (B) Layer-Type Dezincification**

Chapter 8 — Corrosion

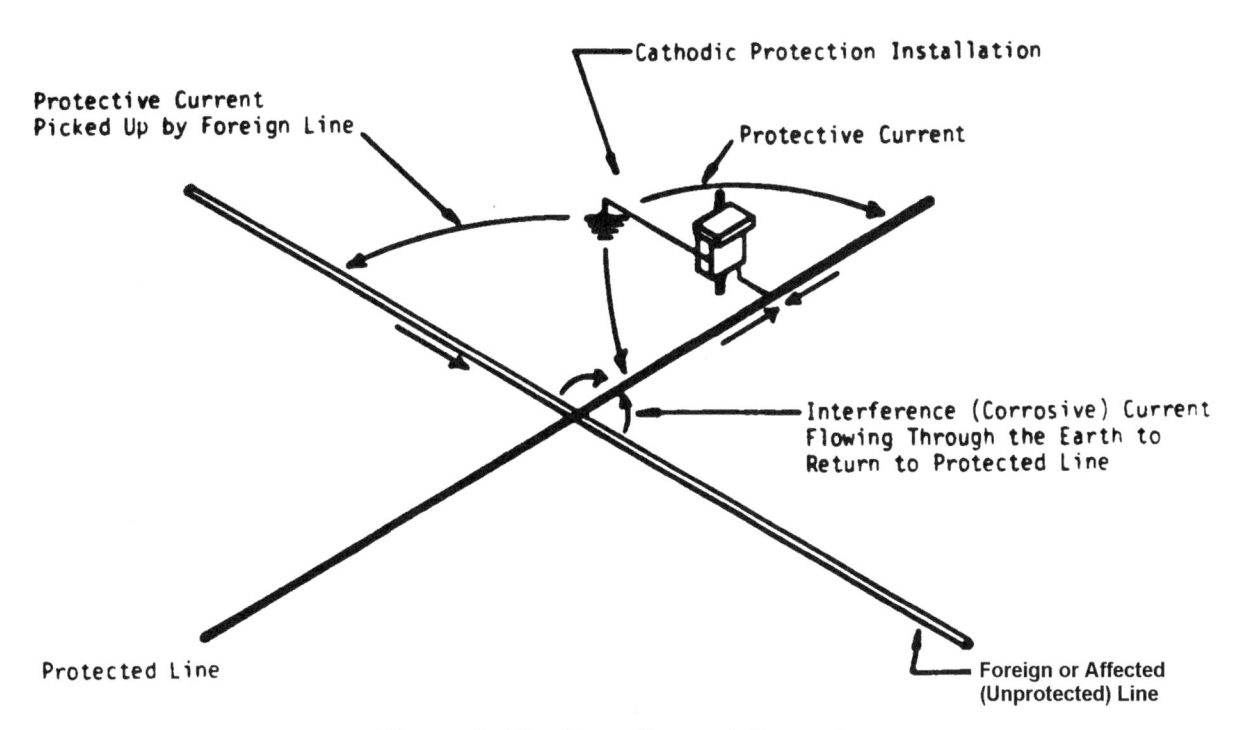

Figure 8-10  Stray Current Corrosion

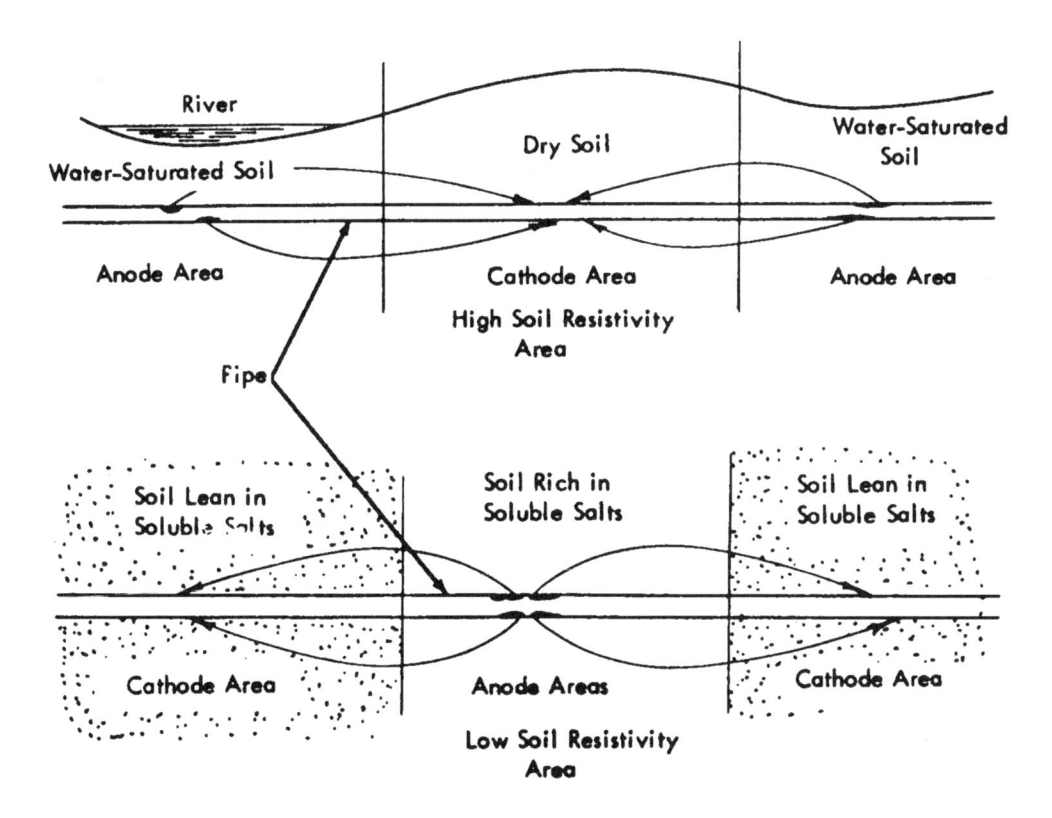

Figure 8-11  Corrosion by Differential Environmental Conditions

## THE GALVANIC SERIES

The galvanic series of metals, listed in Table 8-2, is useful in predicting the effects of coupling various metals. Metals that are far apart in the series have a greater potential for galvanic corrosion than do metals in the same group or metals close to each other in the series. Metals listed above other metals in the series are generally anodic (corrode) to metals listed below them. The relative area of the metals in the couple must be considered along with the polarization characteristic of each metal.

### Table 8-2  Galvanic Series of Metals

**Corroded end (anodic)**
Magnesium
Magnesium alloys
Zinc
Aluminum 100
Cadmium
Aluminum 2017
Steel or iron
Cast iron
Chromium-iron (active)
Ni-resist irons
18-8 SS (active)
18-8-3 SS (active)
Lead-tin solders
Lead
Tin
Nickel (active)
Inconel (active)
Hastelloy C (active)
Brasses
Copper
Bronzes
Copper-nickel alloys
Monel
Silver solder
Nickel (passive)
Inconel (passive)
Chromium-iron (passive)
Titanium
18-8 SS (passive)
18-8-3 SS (passive)
Hastelloy C (passive)
Silver
Graphite
Gold
Platinum
**Protected end (cathode)**

## ELECTROMOTIVE FORCE SERIES

An "electromotive force" is defined as a force that tends to cause a movement of electrical current through a conductor. Table 8-3, known as the "electromotive force series," lists the metals in their electromotive force order and defines their potential with respect to a saturated copper-copper sulfite half-cell. This list is arranged according to their standard electrode potentials, with positive potentials (greater than 1.0) for elements that are cathodic to a standard hydrogen electrode and negative potentials (less than 1.0) for elements that are anodic to a standard hydrogen electrode. In most cases, any metal in this series will displace the more positive metal from a solution and thus corrode to protect the more positive metal. There are exceptions to this rule because of the effect of ion concentrations in a solution and because of different environments found in practice. This exception usually applies to metals close together in the series, which may suffer reversals of potential. Metals far apart in the series will behave as expected, the more negative will corrode to the more positive. In an electrochemical reaction, the atoms of an element are changed to ions. If an atom loses one or more electrons (e$^-$), it becomes an ion that is positively charged and is called a cation (example: Fe$^{2+}$). An atom that takes on one or more electrons also becomes an ion, but it is negatively charged and is called an anion (example: OH$^-$). The charges coincide with the valence of the elements.

The arrangement of a list of metals and alloys according to their relative potentials in a given environment is a galvanic series. By definition, a different series could be developed for each environment.

## FACTORS AFFECTING THE RATE OF CORROSION

### General

The rate of corrosion is directly proportional to the amount of current leaving the anode surface. This current is related to both the potential (voltage) between the anode and cathode and the circuit resistance. Voltage, resistance, and current are governed by Ohm's Law:

***Equation 8-1***

$$I = \frac{E}{R}$$

# Chapter 8 — Corrosion

*where:*

I = Current (A or mA)
E = Voltage (V or mV)
R = Resistance (Ω)

Essentially, Ohm's Law states that current is directly proportional to the voltage and inversely proportional to the resistance.

### Table 8-3  Electromotive Force Series

| | Potential of Metals |
|---|---|
| Magnesium (galvomag alloy)[a] | 1.75 |
| Magnesium (H-l alloy)[a] | 1.55 |
| Zinc | 1.10 |
| Aluminum | 1.01 |
| Cast iron | 0.68 |
| Carbon steel | 0.68 |
| Stainless steel type 430 (17% Cr)[b] | 0.64 |
| Ni-resist cast iron (20% Ni) | 0.61 |
| Stainless steel type 304 (18% Cr, 8% Ni)[b] | 0.60 |
| Stainless steel type 410 (13% Cr)[b] | 0.59 |
| Ni-resist cast iron (30% Ni) | 0.56 |
| Ni-resist cast iron (20% Ni+Cu) | 0.53 |
| Naval rolled brass | 0.47 |
| Yellow brass | 0.43 |
| Copper | 0.43 |
| Red brass | 0.40 |
| Bronze | 0.38 |
| Admiralty brass | 0.36 |
| 90:10 Cu·Ni+ (0.8% Fe) | 0.35 |
| 70:30 Cu·Ni+ (0.06% Fe) | 0.34 |
| 70:30 Cu·Ni+ (0.47% Fe) | 0.32 |
| Stainless steel type 430 (17% Cr)[b] | 0.29 |
| Nickel | 0.27 |
| Stainless steel type 316 (18% Cr, 12% Ni, 3% Mo)[b] | 0.25 |
| Inconel | 0.24 |
| Stainless steel type 410 (13% Cr)[b] | 0.22 |
| Titanium (commercial) | 0.22 |
| Silver | 0.20 |
| Titanium (high purity) | 0.20 |
| Stainless steel type 304 (18% Cr, 8% Ni)[b] | 0.15 |
| Hastelloy C | 0.15 |
| Monel | 0.15 |
| Stainless steel type 316 (18% Cr, 12% Ni, 3% Mo)[b] | 0.12 |

*Note:* Based on potential measurements in sea water, velocity of flow 13 ft/s (3.96 m/s), temperature 77°F (25°C).

[a] Based on data provided by the Dow Chemical Co.

[b] The stainless steels, as a class, exhibited erratic potentials depending on the incidence of pitting and corrosion in the crevices formed around the specimen supports. The values listed represent the extremes observed and, due to their erratic nature, should not be considered as establishing an invariable potential relation among the alloys that are covered.

## Effect of the Metal Itself

For a given current flow, the rate of corrosion of a metal depends on Faraday's Law.

### Equation 8-2

$$w = KIt$$

*where:*

w = Weight loss
K = Electrochemical equivalent
I = Current
t = Time

For practical purposes, the weight loss is usually expressed in pounds per ampere year (kilo-grams per coulomb). Loss rates for some common metals are given in Table 8-4.

### Table 8-4  Corrosion Rates for Common Metals

| Metal | Loss Rate, lb/A-yr (kg/C) |
|---|---|
| Iron or steel | 20 (6.1) |
| Lead | 74 (22.5) |
| Copper | 45 162.0) |
| Zinc | 23 (7.0) |
| Aluminum | 6.5 (23.4) |
| Carbon | 2.2 (7.9) |

This indicates that if 1 ampere is discharged from a steel pipeline over a period of 1 year, 20 pounds (6.1 kilograms) of steel will be lost.

Corrosion of metals in aqueous solutions is also influenced by the following factors: acidity, oxygen content, film formation, temperature, velocity, and homogeneity of the metal and the electrolyte. These factors are discussed below, since they are factors that can be measured or detected by suitable instruments.

## Acidity

The acidity of a solution represents the concentration of hydrogen ions or the pH. In general, low pH (acid) solutions are more corrosive than neutral (7.0 pH) or high pH (alkaline) solutions. Iron or steel, for example, suffers accelerated corrosion in solutions where the pH is 4.5 or less. Exceptions to this rule are amphoteric materials such as aluminum or lead, which corrode more rapidly in alkaline solutions.

## Oxygen Content

The oxygen content of aqueous solutions causes corrosion by reacting with hydrogen at the metal surface to depolarize the cathode, resulting in the exposure of additional metal. Iron or steel corrodes at a rate proportional to the oxygen content. Most natural waters originating from rivers, lakes, or streams are saturated with oxygen. Reduction of oxygen is a part of the corrosion process in most of the corrosion found in practice. The possibility of corrosion being influenced by atmospheric oxygen should not be overlooked in design work.

## Film Formation

Corrosion and its progress are often controlled by the corrosion products formed on the metal surface. The ability of these films to protect metal depends on how they form when the metal is originally exposed to the environment. Thin, hard, dense, tightly adherent films afford protection, whereas thick, porous, loose films allow corrosion to proceed without providing any protection. As an example, the iron oxide film that usually forms on iron pipe in contact with water is porous and easily washed away to expose more metal to corrosion. The effective use of corrosion inhibitors in many cases depends on the type of film it forms on the surface to be protected.

## Temperature

The effect of temperature on corrosion is complex because of its influence on other corrosion factors. Temperature can determine oxygen solubility, content of dissolved gases, and nature of protective film formation, thereby resulting in variations in the corrosion rate. Generally, in aqueous solutions, higher temperatures increase corrosion rates. In domestic hot water systems, for example, corrosion rates double for each 10°F (6°C) rise above 140°F (60°C) water temperature. Temperature can also reverse potentials, such as in the case of zinc-coated iron at approximately 160°F (71.1°C) water temperature, when the zinc coating can become cathodic to the iron surface, accelerating the corrosion of iron.

## Velocity

Velocity of the solution in many cases controls the rate of corrosion. Increasing velocity usually increases corrosion rates. The more rapid movement of the solution causes corrosion chemicals, including oxygen, to be brought into contact with the metal surface at an increased rate. Corrosion products or protective films are carried away from the surface at a faster rate.

Another important effect of high velocity is that turbulence can result in local differential oxygen cells or metal-ion concentration cells causing severe local attack. High velocities also tend to remove protective films causing rapid corrosion of the metal surfaces.

## Homogeneity

The homogeneity of the metal and of the electrolyte is extremely important to corrosion rates. In general, nonhomogeneous metals or electrolytes cause local attack or pitting, which occurs at concentrated areas and is, therefore, more serious than the general overall corrosion of a material. Examples include: concentration cells, galvanic cells, microstructural differences, and differences in temperature and velocity.

# CORROSION CONTROL

Corrosion control is the regulation, control, or prevention of a corrosion reaction for a specific goal. This may be accomplished through any one or a combination of the following factors:

1. Materials selection.
2. Design to reduce corrosion.
3. Passivation.
4. Coating.
5. Cathodic protection.
6. Inhibitors (water treatment).

## Materials Selection

Corrosion resistance, along with other important properties, must be considered in selecting a material for any given environment. When a material is to be specified, the following steps should be used:

1. Determine the application requirements.
2. Evaluate possible material choices that meet the requirements.
3. Specify the most economical method.

# Chapter 8 — Corrosion

Factors to be considered include:

1. Material cost.
2. Corrosion resistance data.
3. Ability to be formed or joined by welding or soldering.
4. Fabricating characteristics (bending, stamping, cutting, etc.).
5. Mechanical properties (tensile and yield strength, impact resistance, hardness, ductility, etc.).
6. Availability of material.
7. Electrical or thermal properties.
8. Compatibility with other materials in system.
9. Specific properties, such as nuclear radiation absorption, low or high-temperature properties.

Initial cost is an important consideration, but the life cost as applied to the system, as a whole, is more important. For example, if an inexpensive part must be periodically replaced, the cost of downtime and labor to install it may make the inexpensive part the most expensive part when all factors are considered.

## Design to Reduce Corrosion

Corrosion can be eliminated or substantially reduced by incorporating some basic design suggestions in the system design. The following five design suggestions can minimize corrosive attack:

1. *Provide dielectric insulation between dissimilar metals*, when dissimilar metals such as copper and steel are connected together, e.g., at a water heater. In a pipeline, for example, dielectric insulation should be installed to prevent contact of the two metals. Without such insulation the metal higher in the galvanic series (steel) will suffer accelerated corrosion because of the galvanic cell between copper and steel. When designing systems requiring dissimilar metals, the need for dielectric insulation should be investigated.
2. *Avoid surface damage or marking.* Areas on surfaces that have been damaged or marked can initiate corrosion. These areas usually become anodic to the adjacent untouched areas and can lead to failures. The design, therefore, should consider this when there is a need for machining or fabrication so that unnecessary damage does not occur.
3. *Do not use excessive welding or soldering heat.* Areas that are heated excessively during welding or soldering can result in changes to the metals' microstructure. Large grain growth can result in accelerated corrosion. The grain growth changes the physical properties of the metal and results in nonhomogeneity of the metal wall. Designs can minimize this effect by using heavier wall thicknesses in areas to be welded.
4. *Crevices should be avoided.* Concentration cells usually form in crevices and can cause premature failures. Regardless of the amount of force applied in bolting two plates together, it is not possible to prevent gradual penetration of liquid into the crevice between the plates. This forms concentration cells where the fluid in the crevices is depleted and forms anodic areas. The most practical way of avoiding crevices is to design welded connections in place of mechanical fasteners.
5. *Other design suggestions*: Corrosion can be minimized if heat or chemicals near metal walls are avoided. Condensation of moisture from the air on cold metal surfaces can cause extensive corrosion if not prevented. The cold metal surface should be thermally insulated if possible. Any beams, angles, etc. should be installed so they drain easily and cannot collect moisture, or drain holes must be provided.

## Passivation

Passivation is the accelerated formation of a protective coating on metal pipe (primarily stainless steel) by contact with a chemical specifically developed for this purpose.

## Coating

Materials exposed to the atmosphere that do not have the ability to form natural protective coatings, such as nickel and aluminum, are best protected by the application of artificial protective coatings. The coating is applied to keep the corroding material from the surface at all times.

One of the most important considerations in coating application is surface preparation. The surface must be properly cleaned, free of scale, rust, grease, and dirt to allow the coating to bond properly to the surface. The best coating in the world will give unsatisfactory results if the surface is poorly prepared. The surface may require pickling, sandblasting, scratch brushing, or flame cleaning to properly prepare it for application of a coating.

The actual coating that is applied depends on the application and may be either a metallic (such as galvanizing) or a nonmetallic, organic (such as vinyl or epoxy) coating. The coating may actually be a coating system, such as primer, intermediate coat (to bond primer and top coat), and finish or top coat. Coating manufacturers' literature should be consulted regarding coating performance, surface preparatory application, handling of coated surfaces, etc.

For atmospheric exposure, coatings alone are relied on to provide protection in many applications. Coatings by themselves, however, are not considered adequate for corrosion control of buried or submerged structures because there is no such thing as a perfect coating. All coatings have inherent holes or holidays. Often the coating is damaged during installation or adjacent construction. Concentrated corrosion at coating breaks often causes failures sooner on coated structures than on bare ones. In stray current areas, severe damage occurs frequently on coated pipe because of the high density of discharge current at coating faults.

The most important function of coating is in its relation to cathodic protection. Cathodic protection current requirements, and hence operating costs, are proportional to the amount of bare surface exposed to soil. When structures are coated, it is necessary only to protect the small areas of coating faults. Careful applications of coating and careful handling of coated structures lead to maximum coating effectiveness, thus minimizing protective current requirements and costs. Also, lower current usage generally means less chance of stray current effects on other structures.

## Cathodic Protection

Cathodic protection is an effective tool to control corrosion of metallic structures, such as water lines and tanks, buried or immersed in a continuous electrolyte by making the metal structure the cathode and applying direct current from an anode source. By making the entire structure the cathode, all anode areas from the local corrosion cells are eliminated and DC current is prevented from leaving the structure, thereby stopping further corrosion.

The most common sacrificial anode is made of magnesium. Magnesium has the highest natural potential of the metals listed in the electromotive series and, therefore, the greatest current-producing capacity of the series. Zinc anodes are sometimes used in very low-resistivity soils where current-producing capacity such as that of magnesium is not required.

The two proven methods of applying cathodic protection are with (1) galvanic anodes and (2) impressed current systems. The basic difference between the two types of protection is as follows: The galvanic anode system depends on the voltage difference generated between the anode material and the structure material to cause a flow of DC current to the structure. The impressed current system utilizes an AC/DC rectifier to provide current to relatively inert anodes and can be adjusted to provide the necessary voltage to drive the required current to the structure surfaces. Choice of the proper system depends on a number of factors. Each has its advantages, as are discussed below.

**Galvanic anodes** Galvanic anodes are used most advantageously on coated structures in low soil resistivity where current requirements are low. Some advantages and disadvantages of galvanic anodes are as follows:

*Advantages:*

1. Relatively low installation cost.
2. Do not require external power source.
3. Low maintenance requirements.
4. Usually do not cause adverse effects on foreign structures.
5. Can be installed with pipe, minimizing right-of-way cost.

*Disadvantages:*

1. Driving voltage is low (approximately 0.15 V).
2. Current output is limited by soil resistivity.
3. Not applicable for large current requirements.

The galvanic anode system of an active metal anode, such as magnesium or zinc, is placed in the electrolyte (soil or water) near the structure and connected to it with a wire. This is illustrated in Figures 8-12 and 8-13. Cathodic

Chapter 8 — Corrosion 145

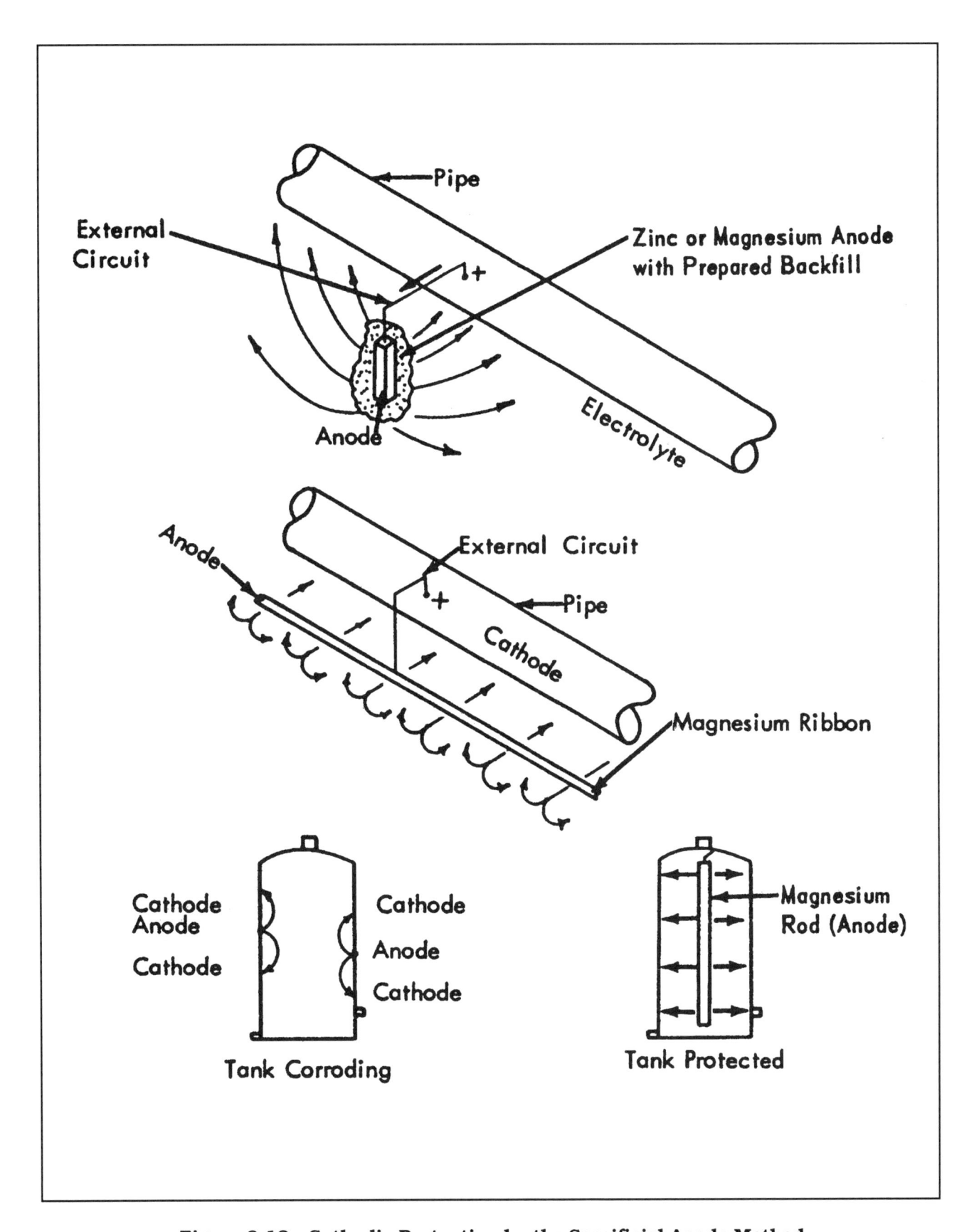

Figure 8-12  Cathodic Protection by the Sacrificial Anode Method

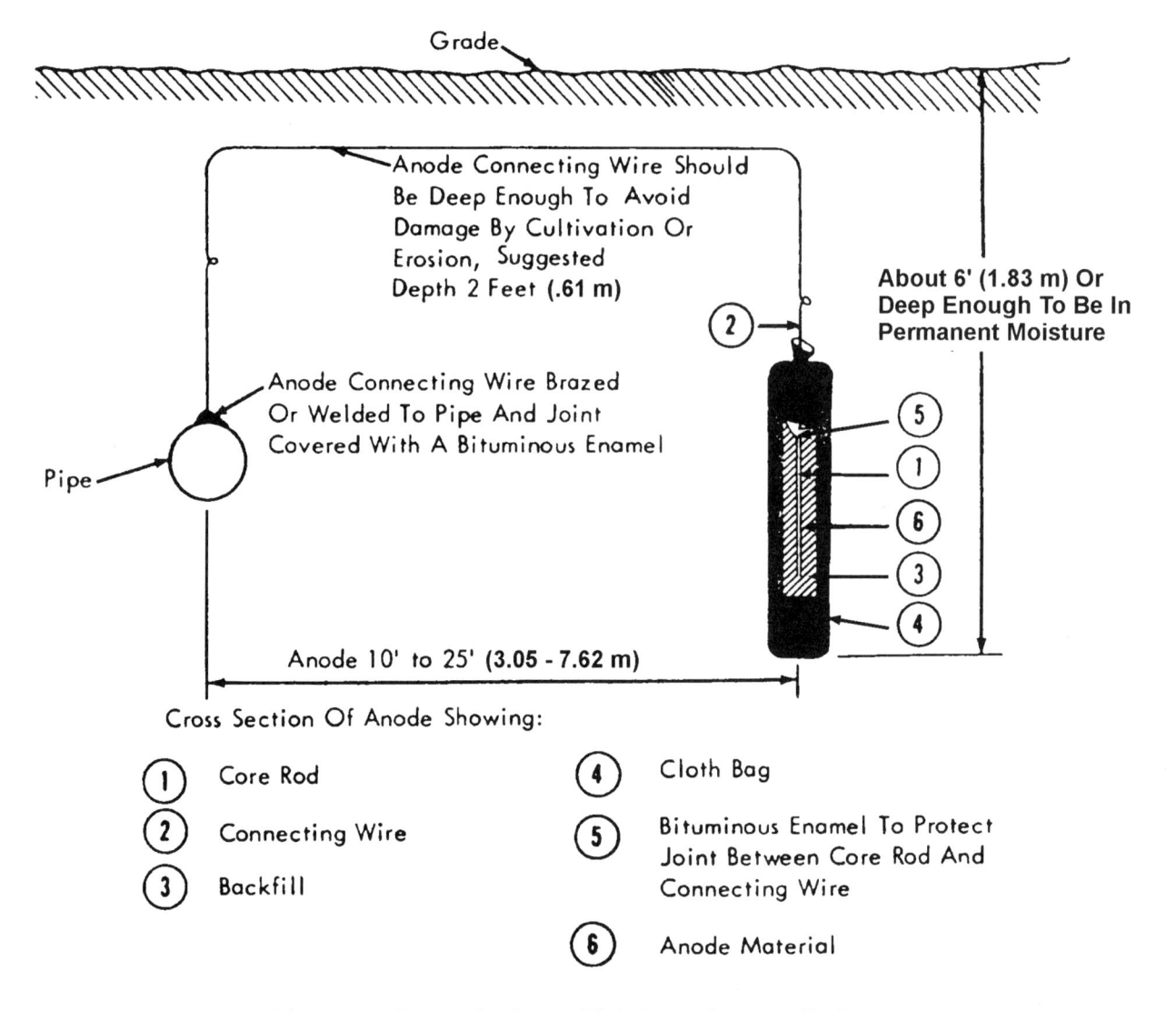

**Figure 8-13  Typical Sacrificial Anode Installation**

protection is achieved by current flow due to the potential difference between the anode (metal) and the cathode (structure). A corrosion cell or battery is created and current flows from the corroding anode material through the soil to the cathode or protected structure. Hence the galvanic anode is deliberately caused to waste itself to prevent corrosion of the protected structure. Because the galvanic anode system relies on the difference in voltage between two metals, which in most cases is limited to 1.0 V or less, the current generated by the anodes is usually low (approximately 0.1 to 0.5 A per anode).

Galvanic anode systems are usually used for structures having small current requirements, such as well-coated, small-diameter pipes; water heaters; sewage lift stations; some offshore structures; and structures in congested areas where currents must be kept low to avoid detrimental effects on other structures. Galvanic anodes may be installed in banks at specific locations. They are, however, usually distributed around protected structures because of their limited current output.

As an example, considering a pipe-to-soil potential of 0.85 V as protection for a steel pipeline, the driving potential of zinc anodes is 0.25 V and for magnesium is 500 A-h/lb (1795 C/kg). The actual life of anodes of a given weight at a known current output can be calculated using the following formulas:

# Chapter 8 — Corrosion

### Equation 8-3

$$LM = \frac{57.08 \times w}{i}$$

### Equation 8-4

$$Lz = \frac{38.2 \times w}{i}$$

*where:*

- $LM$ = Life of magnesium anode (yr)
- $Lz$ = Life of zinc anode (yr)
- $w$ = Weight of anode, lb (kg)
- $i$ = Output of anode (mA)

The controlling factor for current output of zinc and magnesium anodes is soil resistivity. When soil resistivity is known or determined, then the current output of variously sized anodes for either magnesium or zinc can be estimated as follows:

### Equation 8-5

$$iM = \frac{150,000 \times f}{p}$$

### Equation 8-6

$$iZ = \frac{150,000 \times f \times 0.27}{p}$$

*where:*

- $iM$ = Current output of magnesium (mA)
- $iZ$ = Current output of zinc (mA)
- $p$ = Soil resistivity (Ω-cm)
- $f$ = Anode size factor

Cost of galvanic cathodic protection generally favors the use of zinc anodes over magnesium at soil resistances below 1500 Ω-cm and the use of magnesium at soil resistances over 1500 Ω-cm.

**Impressed current** The impressed current system, illustrated in Figure 8-14, differs substantially from the galvanic anode system in that it is externally powered, usually by an AC-DC rectifier, which allows great freedom in adjustment of current output. Current requirements of several hundred amperes can be handled by impressed current systems. The impressed current system usually consists of graphite or high-silicon iron anodes connected to an AC-DC rectifier, which, in turn, is wired to the structure being protected. Current output is determined by adjustment of the rectifier voltage to provide current as required. The system is not limited by potential difference between metals, and voltage can be adjusted to provide adequate driving force to emit the necessary current. Impressed current systems are used for structures having large current requirements, such as bare pipe; tank farms; large-diameter, cross-country pipe lines; cast-iron water lines; and many offshore facilities.

Impressed current cathodic protection has the following advantages and disadvantages:

*Advantages:*

1. Large current output.
2. Voltage adjustment over a wide range.
3. Can be used with a high soil resistivity environment.
4. Can protect uncoated structures.
5. Can be used to protect larger structures.

*Disadvantages:*

1. Higher installation and maintenance cost.
2. Power costs.
3. Can cause adverse effects (stray current) with foreign structures.

When designing impressed current cathodic protection systems, the type and condition of the structure must be determined. Obtaining knowledge of the presence or lack of coating, size of structure, electrical continuity, and location is a necessary first step. Next, the availability of power and ease of installing the ground bed are required. After all of the above are satisfactorily done, it is generally necessary to perform a current requirement test utilizing a portable DC generator or storage batteries. This defines an apparent DC current requirement to protect the structure. Tests to determine any adverse effects should also be conducted on foreign structures at this time. Any current drained to foreign structures should be added to the current requirements. After the total current requirement is known, the ground bed is designed so that the circuit resistance is relatively low. Actual ground-bed design is dependent on soil resistivity. A number of empirical formulas are available to determine the number of parallel anodes required for a certain circuit resistance.

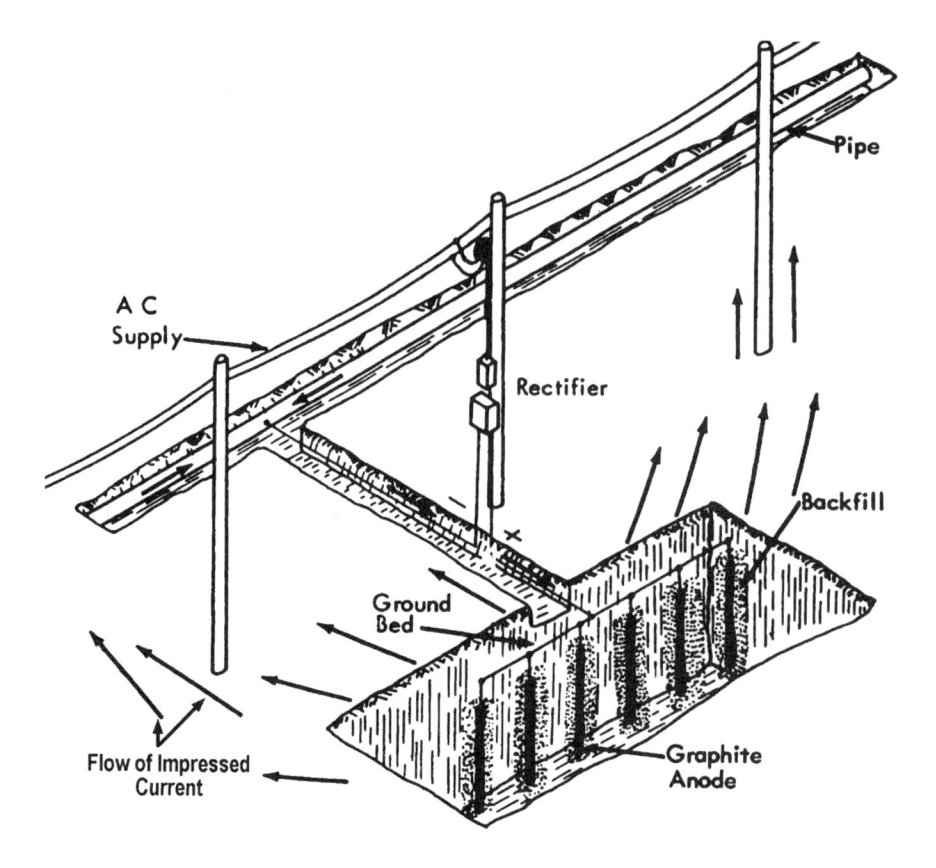

**Figure 8-14  Cathodic Protection by the Impressed Current Method**

**Cathodic protection criteria** Criteria for determining adequate cathodic protection have been established by The National Association of Corrosion Engineers (NACE). These criteria are based on measuring structure-to-electrode potentials with a copper-sulfate reference electrode. The criteria are listed for various metals, such as steel, cast iron, aluminum, and copper, and may be found in NACE Standard RP-01.

Cathodic protection serves its purpose best, and is by far the most economical, when it is properly coordinated with the other methods of corrosion control, especially coating. In general, the least expensive, easiest to maintain, and most practical system is to apply a good-quality coating to a new structure and then use cathodic protection to eliminate corrosion at the inevitable breaks in the coating. The reason for this is that it takes much more current and anodes to protect bare metal than it does to protect coated metal. The amount of protective current required is proportional to the area of metal exposed to the electrolyte.

In addition to using coatings, it is necessary to assure continuity of the structures to provide protection of the whole structure. This also prevents undesirable accelerated stray current corrosion to the parts of the structure that are not electrically continuous. Therefore, all noncontinuous joints, such as mechanical, push-on, or screwed joints in pipelines, must be bonded. All tanks in a tank farm or piles on a wharf must be bonded together to ensure electrical continuity.

Other important components used in effective cathodic protection systems are dielectric insulation and test stations. Dielectric insulation is sometimes used to isolate underground protected structures from above-ground structures to reduce the amount of cathodic protection current required. Care must be taken to avoid short-circuiting (bypassing) the insulation, or protection can be destroyed. Test stations are wires attached to the underground structure (pipeline or tank) to provide electrical contact for the purpose of determining protection effectiveness. Test stations are also used to make bonds

# Chapter 8 — Corrosion

or connections between structures when required to mitigate stray-current effects.

**Costs of cathodic protection** Corrosion of underground, ferrous metal structures can be economically controlled by cathodic protection. Cathodic protection costs are added to the initial investment since they are a capital expense. To be economically sound, the spending of the funds must yield a fair return over the expected life of the facility.

To protect a new facility requires an initial increase of perhaps 10% in capital investment. Payout time is usually 10 to 15 years; thereafter, appreciable savings accrue due to this investment, which prevents or reduces the frequency of leaks. Effective corrosion control through the application of cathodic protection reduces the leak frequency for a structure to the minimum with minimum cost.

Cathodic protection systems must be properly maintained. Rectifier outputs must be checked monthly. Changes or additions to the protected structure must be considered to see if changes or additions to the cathodic protection system are required. Annual inspections by a corrosion engineer are required to ensure that all malfunctions are corrected and cathodic protection continues unhampered.

### Inhibitors (Water Treatment)

Plant utility services such as boiler feed water, condensate, refrigerants, and cooling water require the addition of inhibitors or water treatment. Boiler feed water must be treated to maintain proper pH control, dissolved solid levels, and oxygen content. Condensate requires treatment to control corrosion by oxygen and carbon dioxide. Brine refrigerants and cooling water in closed-loop circulating systems require proper inhibitors to prevent corrosion.

Water treatment may consist of a simple adjustment of water hardness to produce naturally forming carbonate films. This carbonate film, if properly adjusted, will form to a controlled thickness just sufficient to prevent corrosion by keeping water from contacting the metal surface. In cooling water, where hardness control is not practical, inhibitors or film-forming compounds may be required.

Sodium silicate and sodium hexametaphosphate are examples of film-forming additives in potable water treatment. A tight, thin, continuous film of silica (water glass) or phosphate adheres to the metal surface, preventing pipe contact with the water. (Phosphate additives to potable water are limited or prohibited in some jurisdictions.)

In closed-loop cooling systems, and systems involving heat-exchange surfaces, it may not be possible to use film-forming treatment because of detrimental effects on heat transfer. In these cases, inhibitors are used; these control corrosion by increasing polarization of anodic or cathodic surfaces and are called "anodic" or "cathodic inhibitors," respectively. The anodic or cathodic surfaces are covered, preventing completion of the corrosion cell by elimination of either the anode or cathode.

When water treatment or inhibitors are used, a testing program must be established to ensure that proper additive levels are maintained. In some cases, continuous monitoring is required. Also, environmental considerations in local areas must be determined before additives are used or before any treated water is discharged to the sanitary sewer or storm drainage system.

## GLOSSARY

***Active*** The state in which a metal is in the process of corroding.

***Active potential*** The capability of a metal corroding based on a transfer of electrical current.

***Aeration cell*** An oxygen concentration cell–an electrolytic cell resulting from differences in the quantity of dissolved oxygen at two points.

***Amphoteric corrosion*** Corrosion usually caused by a chemical reaction resulting from a concentration of alkaline products formed by the electrochemical process. Amphoteric materials are those materials that are subject to attack from both acidic and alkaline environments. Aluminum and lead, commonly used in construction, are subject to amphoteric corrosion in highly alkaline environments. The use of cathodic protection in highly alkaline environments, therefore, intensifies the formation of alkaline byproducts.

***Anaerobic*** Free of air or uncombined oxygen.

***Anion*** A negatively charged ion of an electrolyte that migrates toward the anode under the influence of a potential gradient.

***Anode*** Negative in relation to the electrochemical process. The electrode at which oxidation or corrosion occurs.

***Anodic protection*** An appreciable reduction in corrosion by making a metal an anode and maintaining this highly polarized condition with very little current flow.

***Cathode*** Positive in relation to the electrochemical process. The electrode where reduction (and practically no corrosion) occurs.

***Cathodic corrosion*** An unusual condition in which corrosion is accelerated at the cathode because cathodic reaction creates an alkaline condition corrosive to certain metals, such as aluminum, zinc, and lead.

***Cathodic protection*** Reduction or elimination of corrosion by making the metal a cathode by means of an impressed DC current or attachment to a sacrificial anode.

***Cathodic*** The electrolyte of an electrolytic cell adjacent to the cathode.

***Cation*** A positively charged ion of an electrolyte that migrates toward the cathode under the influence of a potential gradient.

***Caustic embrittlement*** Weakening of a metal resulting from contact with an alkaline solution.

***Cavitation*** Formation and sudden collapse of vapor bubbles in a liquid, usually resulting from local low pressures, such as on the trailing edge of an impeller. This condition develops momentary high local pressure which can mechanically destroy a portion of the surface on which the bubbles collapse.

***Cavitation-corrosion*** Corrosion damage resulting from cavitation and corrosion: metal corrodes, pressure develops from collapse of the cavity and removes the corrosion product, exposing bare metal to repeated corrosion.

***Cell*** A circuit consisting of an anode and a cathode in electrical contact in a solid or liquid electrolyte.

***Concentration cell*** A cell involving an electrolyte and two identical electrodes, with the potential resulting from differences in the chemistry of the environments adjacent to the two electrodes.

***Concentration polarization*** That portion of the polarization of an electrolytic cell produced by concentration changes resulting from passage of electric current through the electrolyte.

***Contact corrosion*** Corrosion of a metal at an area where contact is made with a (usually nonmetallic) material.

***Corrosion*** Degradation of a metal by chemical or electrochemical reaction with its environment.

***Corrosion fatigue*** Reduction of fatigue durability by a corrosive environment.

***Corrosion fatigue limit*** The maximum repeated stress endured by a metal without failure in a stated number of stress applications under defined conditions of corrosion and stressing.

***Corrosion mitigation*** The reduction of metal loss or damage through use of protective methods and devices.

***Corrosion prevention*** The halting or elimination of metal damage through use of corrosion-resisting materials, protective methods, and protective devices.

***Corrosion potential*** The potential that a corroding metal exhibits under specific conditions of concentration, time, temperature, aeration, velocity, etc.

***Couple*** A cell developed in an electrolyte resulting from electrical contact between two dissimilar metals.

***Cracking*** Separation in a brittle manner along a single or branched path.

***Crevice corrosion*** Localized corrosion resulting from the formation of a concentration cell in a crack formed between a metal and a nonmetal, or between two metal surfaces.

***Deactivation*** The process of prior removal of the active corrosion constituents, usually oxygen, from a corrosive liquid by controlled corrosion of expendable metal or by other chemical means.

***Dealloying*** The selective leaching or corrosion of a specific constituent from an alloy.

# Chapter 8 — Corrosion

***Decomposition potential (or voltage)*** The practical minimum potential difference necessary to decompose the electrolyte of a cell at a continuous rate.

***Depolarization*** The elimination or reduction of polarization by physical or chemical means; depolarization results in increased corrosion.

***Deposit attack (deposition corrosion)*** Pitting corrosion resulting from accumulations on a metal surface that cause concentration cells.

***Differential aeration cell*** An oxygen concentration cell resulting from a potential difference caused by different amounts of oxygen dissolved at two locations.

***Drainage*** Conduction of current (positive electricity) from an underground metallic structure by means of a metallic conductor.

***Electrode*** A metal in contact with an electrolyte that serves as a site where an electrical current enters the metal or leaves the metal to enter the solution.

***Electrolyte*** An ionic conductor (usually in aqueous solution).

***Electromotive force series (e.m.f. series)*** A list of elements arranged according to their standard electrode potentials, the sign being positive for elements having potentials that are cathodic to hydrogen and negative for elements having potentials that are anodic to hydrogen. (This convention of sign, historically and currently used in European literature, has been adopted by the Electrochemical Society and the National Bureau of Standards; it is employed in this publication. The opposite convention of G. N. Lewis has been adopted by the American Chemical Society.)

***Electronegative potential*** A potential corresponding in sign to those of the active or anodic members of the e.m.f. series. Because of the existing confusion of sign in the literature, it is suggested that "anodic potential" be used whenever "electronegative potential" is implied. (See "electromotive force series.")

***Electropositive potential*** A potential corresponding in sign to potentials of the noble or cathodic members of the e.m.f. series. It is suggested that "cathodic potential" be used whenever "electropositive potential" is implied. (See "electromotive force series.")

***Forced drainage*** Drainage applied to underground metallic structures by means of an applied e.m.f. or sacrificial anode.

***Galvanic cell*** A cell consisting of two dissimilar conductors in contact with an electrolyte, or two singular conductors in contact with dissimilar electrolytes. More generally, a galvanic cell converts energy liberated by a spontaneous chemical reaction directly into electrical energy.

***Galvanic corrosion*** Corrosion that is increased because of the current caused by a galvanic cell (sometimes called "couple action").

***Galvanic series*** A list of metals arranged according to their relative corrosion potential in some specific environment; sea water is often used.

***General corrosion*** Corrosion in a uniform manner.

***Graphitization (graphitic corrosion)*** Corrosion of gray cast iron in which the metallic constituents are converted to corrosion products, leaving the graphite flakes intact. Graphitization is also used in a metallurgical sense to mean the decomposition of iron carbide to form iron and graphite.

***Hydrogen embrittlement*** Hydrogen embrittlement causes a weakening of the metal by the entrance of hydrogen into the metal through, for example, pickling or cathodic polarization.

***Hydrogen overvoltage*** A higher than expected difference in potential associated with the liberation of hydrogen gas.

***Impingement attack*** Localized erosion-corrosion caused by turbulence or impinging flow at certain points.

***Inhibitor*** A substance that, when added in small amounts to water, acid, or other liquids, sharply reduces corrosion.

***Ion*** An electrically charged atom or group of atoms known as "radicals."

***Natural drainage*** Drainage from an underground metallic structure to a more negative structure, such as the negative bus of a trolley substation.

***Noble potential*** A potential substantially cathodic compared to the standard hydrogen potential.

***Open-circuit potential*** The measured potential of a cell during which no significant current flows in the external circuit.

***Overvoltage*** The difference between the potential of an electrode at which a reaction is actively taking place and another electrode at equilibrium for the same reaction.

***Oxidation*** Loss of electrons, as when a metal goes from the metallic state to the corroded state. Thus, when a metal reacts with oxygen, sulfur, etc., to form a compound as oxide, sulfide, etc., it is oxidized.

***Oxygen concentration cell*** A galvanic cell caused by a difference in oxygen concentration at two points on a metal surface.

***Passive*** The state of a metal when its behavior is much more noble (resists corrosion) than its position in the e.m.f. series would predict. This is a surface phenomenon.

***pH*** A measure of the acidity or alkalinity of a solution (from 0 to 14). A value of seven (7) is neutral; low numbers (0-6) are acidic, large numbers (8-14) are alkaline.

***Pitting*** Localized light corrosion resulting in deep penetration at a small number of points.

***Polarization*** The shift in electrode potential resulting from the effects of current flow, measured with respect to the "zero-flow" (reversible) potential, i.e., the counter-e.m.f. caused by the products formed or concentration changes in the electrode.

***Protective potential*** A term sometimes used in cathodic protection to define the minimum potential required to suppress corrosion. For steel in sea water, this is claimed to be about 0.85 V as measured against a saturated calomel cell.

***Remote electrode (remote earth)*** Remote earth is any location away from the structure at which the potential gradient of the structure to earth is constant. The potential of a structure-to-earth will change rapidly near the structure and if remote earth is reached, there will be little or no variation in the voltage.

***Resistivity*** The specific opposition of a material. Measured in ohms ($\Omega$) to the flow of electricity.

***Rusting*** Corrosion of iron or an iron-base alloy to form a reddish-brown product that is primarily hydrated ferric oxide.

***Stray current corrosion*** Corrosion that is caused by stray currents from some external source.

***Stress corrosion/stress-accelerated corrosion*** Corrosion that is accelerated by stress.

***Stress corrosion cracking*** Cracking that results from stress corrosion.

***Tuberculation*** Localized corrosion at scattered locations resulting in knob-like mounds.

***Under-film corrosion*** Corrosion that occurs under lacquers and similar organic films in the form of randomly distributed hairlines (most common) or spots.

***Weld decay*** Corrosion, notably at specific zones away from a weld.

# REFERENCES

1. Bosich, Joseph F. 1970. *Corrosion prevention for practicing engineers.*

2. Claes and Fitzgerald. 1975-1976. *Fundamentals of underground corrosion control.* Plant Engineering Technical Publishing. Plant Engineering. New York: McGraw-Hill.

3. Fontana, Mars G., and Norbert D. Greene. 1967. *Corrosion engineering.* New York: McGraw-Hill.

4. Kullen, Howard P. Corrosion. *Power.* December 1956: 74-106.

5. Laque, F. L., and H. R. Copson. 1965. *Corrosion and resistance of metals and alloys.* 2nd ed. New York: Reinhold Publishing.

6. National Association of Corrosion Engineers. 1971. *NACE basic corrosion course.* Houston: National Association of Corrosion Engineers.

7. Peabody, A. W. 1967. *Control of pipeline corrosion.* Houston: National Association of Corrosion Engineers.

8. Shreir, L. L. 1963. Corrosion control. Vol. 2 of *Corrosion.* New York: John Wiley and Sons.

9. Speller, Frank N. 1963. *Corrosion causes and prevention.* New York: McGraw-Hill.

10. Uhlig, Herbert H. 1940. *Corrosion handbook.* New York: John Wiley and Sons.

# AMERICA'S FINEST ACID WASTE AND HIGH PURITY PIPING SYSTEMS

## HIGH PURITY PRODUCTS

### Whiteline Polypropylene High Purity Systems

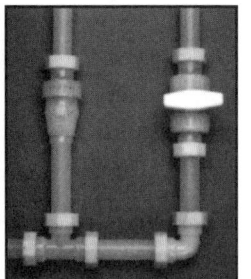

For an economical high purity system there's no reason to look any further than our value-packed Whiteline system.

Whiteline pipe and fittings are designed for efficient transportation of de-ionized and distilled water, as well as many organic and inorganic chemicals.

The Whiteline system is manufactured from pure, unpigmented, virgin type I homopolymer polypropylene using no plasticizers or pigments. The end product is a piping system with a smooth, tight, nonporous surface for less flow resistance that retards fungi, bacteria and other biological impurities.

### PVDF High Purity Systems

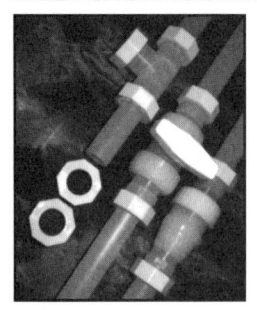

In critical high-purity liquid transportation situations—when you absolutely, positively can't tolerate contamination — there's only one material that gets the job done — KYNAR™ brand of polyvinylidene fluoride (PVDF).

PVDF contains no additives, is non-leaching and has extra smooth walls. It is extremely resistant to chemicals and ultraviolet radiation, has a wide temperature tolerance, is impact resistant, can withstand direct sunlight—or be buried—and still retain its physical properties for a minimum of 20 years.

*Choice of sizes and joining methods*
For maximum installation flexibility, both Whiteline and PVDF systems are available with socket fusion or easy-to-install Rionite mechanical joining methods. Both systems are available in sizes 1/2" through 4".

### High Purity Laboratory Faucets

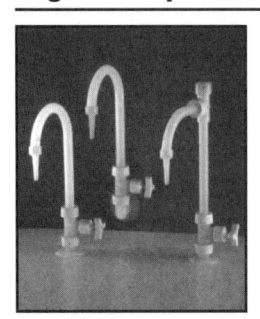

No matter your laboratory faucet needs — chemical, cosmetic, pharmaceutical, educational or medical—Orion has the faucet you need.

Faucets molded from polypropylene or PVDF provide choice and flexibility. Models are available with integral vacuum breakers and in right or left hand configurations.

## ACID WASTE PRODUCTS

### Polypropylene Pipe & Fittings

Blueline drainage systems are molded from flame retardant polypropylene. Blueline systems offer excellent resistance to the most common organic and mineral acids, salts, strong and weak alkalis and most organic chemicals.

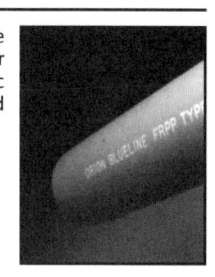

### Superblue™ PVDF Pipe & Fittings

Superblue™ pipe and fittings are molded from KYNAR™ brand of PVDF—available only from Orion. PVDF is the best choice for tough applications because it offers unequaled chemical resistance to a wide variety of weak bases and salts, aliphatic, aromatic and chlorinated solvents. It is also highly resistant to strong oxidants, bases and halogens.

*Choice of sizes and joining methods*
For maximum flexibility, Orion drainage systems are offered in a wide variety of sizes and joining methods. Sizes through 12" are available. Choose mechanical joint, Rionfuse® electrofusion, socket fusion, slip fit or butt fusion joining.

### Double Containment Systems

Our double containment products assure effluent containment in case of a rupture or leak. Available in Polypropylene and PVDF configurations. All fittings are prefabricated. Systems are available in sizes through 12". Exclusive Rionflex™ and Rionlock™ restraining systems available.

### Monitoring Systems

Orion is a leader in providing industry with safe, innovative and economical neutralizing and monitoring devices. These systems help protect the environment by collecting and analyzing effluent and treating and recording pH levels.

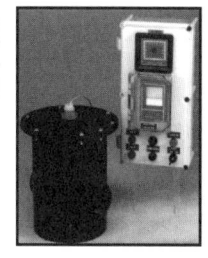

### Neutralization Tanks

Neutralization tanks relieve, dilute and neutralize harmful corrosive chemical wastes. Over 70 models are available.

---

**ORION**

Orion Fittings, Inc.
P.O. Box 17-1580
Kansas City, KS 66117

PHONE (913) 342-1653
FAX (800) 777-1653
www.orionfittings.com

ORION products are proudly made in the U.S.A.

KYNAR® is a registered trademark of Elf Atochem North America, Inc.

# The New Aqua-Saver From Eljer®
## *Performance Under Pressure*

*Eljer introduces the most comprehensive line of pressure-assisted toilets in the industry! We provide the products you need for those typical and not so typical installations.*

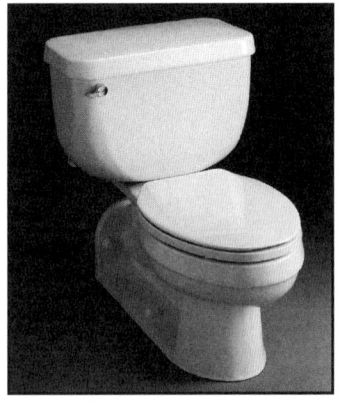

**091-7015** Aqua-Saver

**091-7025** Aqua-Saver

**091-7875** Hylando

**091-7925** Walford

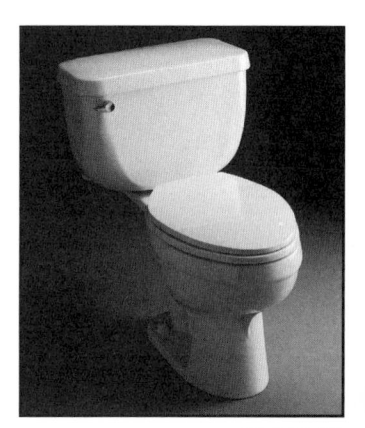

**091-7045** Aqua-Saver

*The improved PF/2 pressure unit is so compact, it fits into a smaller tank for a more streamlined appearance.*

*Old System*

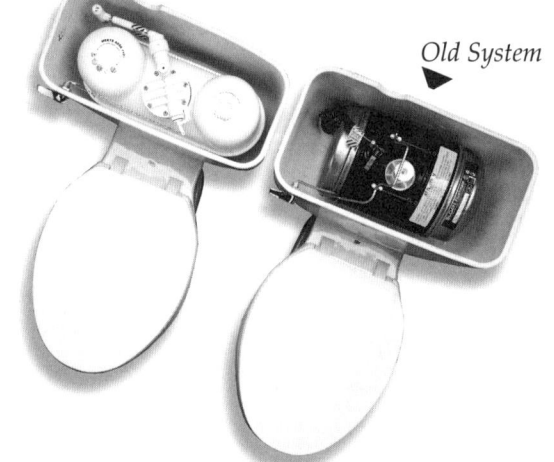

**You can count on Eljer for the features and selection of models that are second to none. The New Aqua-Saver:**

- Requires 60% less force to actuate than our previous system

- Is factory tested to ensure flawless performance

- Redesigned system provides a quieter flush

- Boasts a large 11-1/2" x 15" water surface

PF/2™ ENERGIZED FLUSH

*Award Winning Technology*

14801 Quorum Drive • Dallas, Texas 75240-7584 • 1-800-435-5372 • Visit our website at www.eljer.com

# 9 Seismic Protection of Plumbing Equipment

## INTRODUCTION

Every structure is designed for vertical, or gravity, loads. In the case of pipes, gravity loads include the weight of the pipe and its contents, and the direction of the loading is downward. Seismic loads are the horizontal forces exerted on a structure during an earthquake. Earthquake forces can be in any direction. The ordinary supports designed for gravity loads generally take care of the vertical loads during an earthquake. Therefore, the primary emphasis in seismic design is on lateral, or horizontal, forces.

Study of seismic risk maps, Figures 9-1 and 9-2, indicates that the potential for damaging earthquake motion is far more pervasive than is commonly known. Complete seismic design requirements, including construction of nonstructural elements (piping, ductwork, conduit, etc.), are in effect in only a small fraction of the areas that could be rated as having a high or moderate risk. Seismic design requirements for nonstructural elements, except for heavy cladding panels, are seldom enforced even in California, which is considered the innovator in state building code requirements related to seismic movement. However, the nonstructural damage resulting from recent small earthquakes and the large US and Japanese shocks shows that the major advancements in building structural design, by themselves, may not have produced an acceptable level of overall seismic protection. Now that—at least for modern structures designed and built in accordance with current seismic codes—the potential for collapse or other direct, life-endangering structural behavior is quite small, attention has shifted to nonstructural life safety hazards, continued functionality, and economic issues. The cost of an interruption in a building's ability to function, which could cause a loss of rent, disruption of normal business affairs, or curtailment of production, is coming more into focus.

The costs of seismic protection of plumbing components and equipment range from small—such as those to anchor small tanks—to a considerable percentage of installation costs—such as those for complete pipe bracing systems. Beyond protection of life, the purpose or cost-benefit relationship of seismic protection must be clearly understood before the appropriate response to the risk can be made. The design professional responsible for any given element or system in a building is in the best position to provide that response. Seldom, however, can rational seismic protection be supplied solely by a single discipline. Building systems are interdependent in both design and function, and good seismic protection, like good overall building design, is best provided by employing a cooperative, interdisciplinary approach.

This chapter is intended to provide a basic understanding of the mechanisms of seismic damage and the particular vulnerabilities of plumbing systems and equipment. It is desirable that the professional sufficiently understand the problem in order to select the appropriate seismic protection in any situation, based on a ranking of the damage susceptibility and a knowledge of the scope of mitigation techniques.

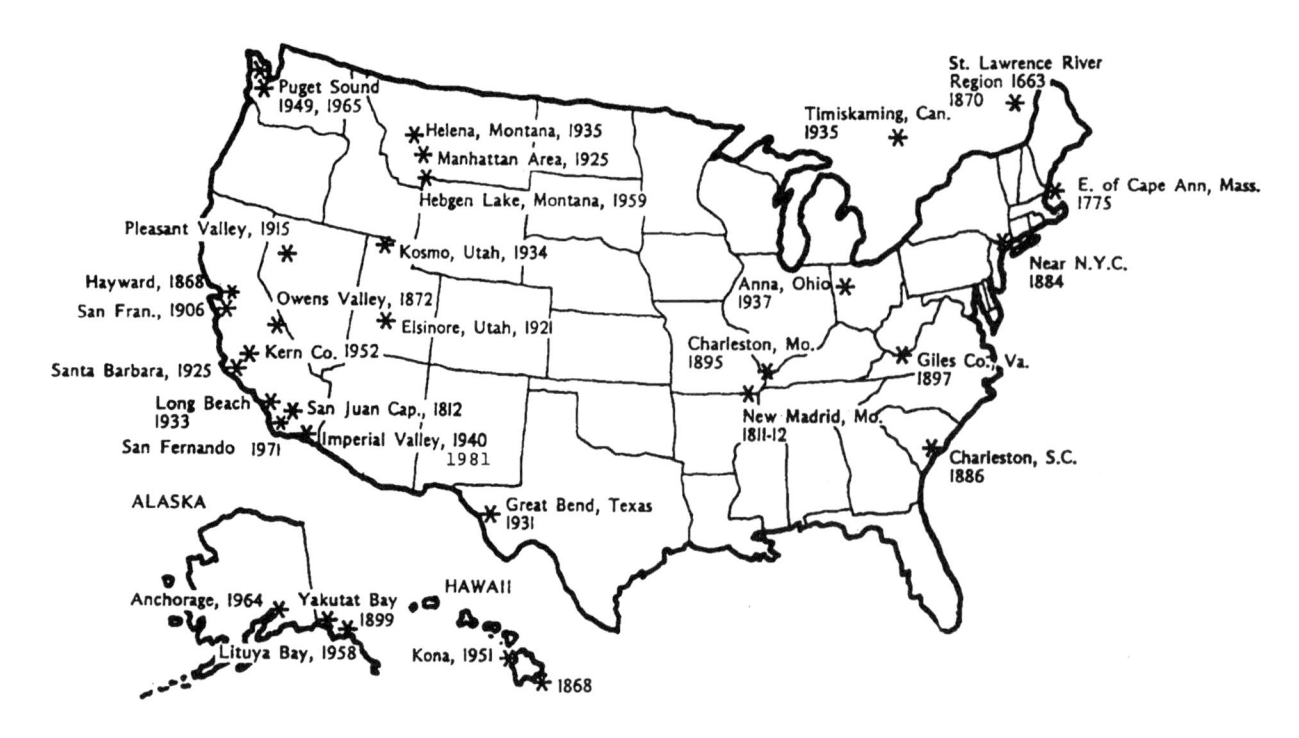

Figure 9-1 Significant Earthquakes in the United States

The seismic protection techniques currently in use for buildings are described in general. Although specific seismic protection details for some situations are discussed, it is suggested that structural design assistance be obtained from a professional of that discipline. Care should be taken in the design of seismic control systems. Proper design may require assistance from an engineer experienced in these systems. In all cases, the current local building code requirements for seismic movement should be consulted and used as minimum standards.

The detailed analysis and design techniques used for nuclear power plants and other heavy industrial applications, while similar in nature to those discussed here, are considered inappropriate for most buildings and are beyond the scope of this chapter. References are given throughout the text for additional study in specific areas of interest.

## CAUSES AND EFFECTS OF EARTHQUAKES

### Plate Tectonics and Faults

All seismic activity on the earth's surface, including earthquakes and volcanoes, are now understood to be caused by the relative movement of pieces of the earth's crust. Ten of the largest pieces, called plates, and their prevailing motions, are shown in Figure 9-3. The edges of these plates make up the world's primary fault systems, along which 90% of all earthquakes occur. The balance of earthquakes occur on countless additional, smaller faults that lie within plate boundaries. The causes and exact mechanisms of these intraplate earthquakes, which affect much of the middle and eastern United States, are not well understood.

The relative movement at plate boundaries is often a sliding action, such as occurs along the San Andreas Fault along the west coast of North America. The plates can also converge, when one plate slides beneath another, or diverge, when molten rock from below rises to fill the voids that gradually form. Although overall plate movement is extremely slow, properly measured only in a geologic time frame, the local relative movement directly at the fault plane can occur either gradually (creep) or suddenly, when tremendous energy is released into the surrounding mass.

The most common mechanism used to describe earthquakes is the "elastic rebound theory," wherein a length of fault that is locked

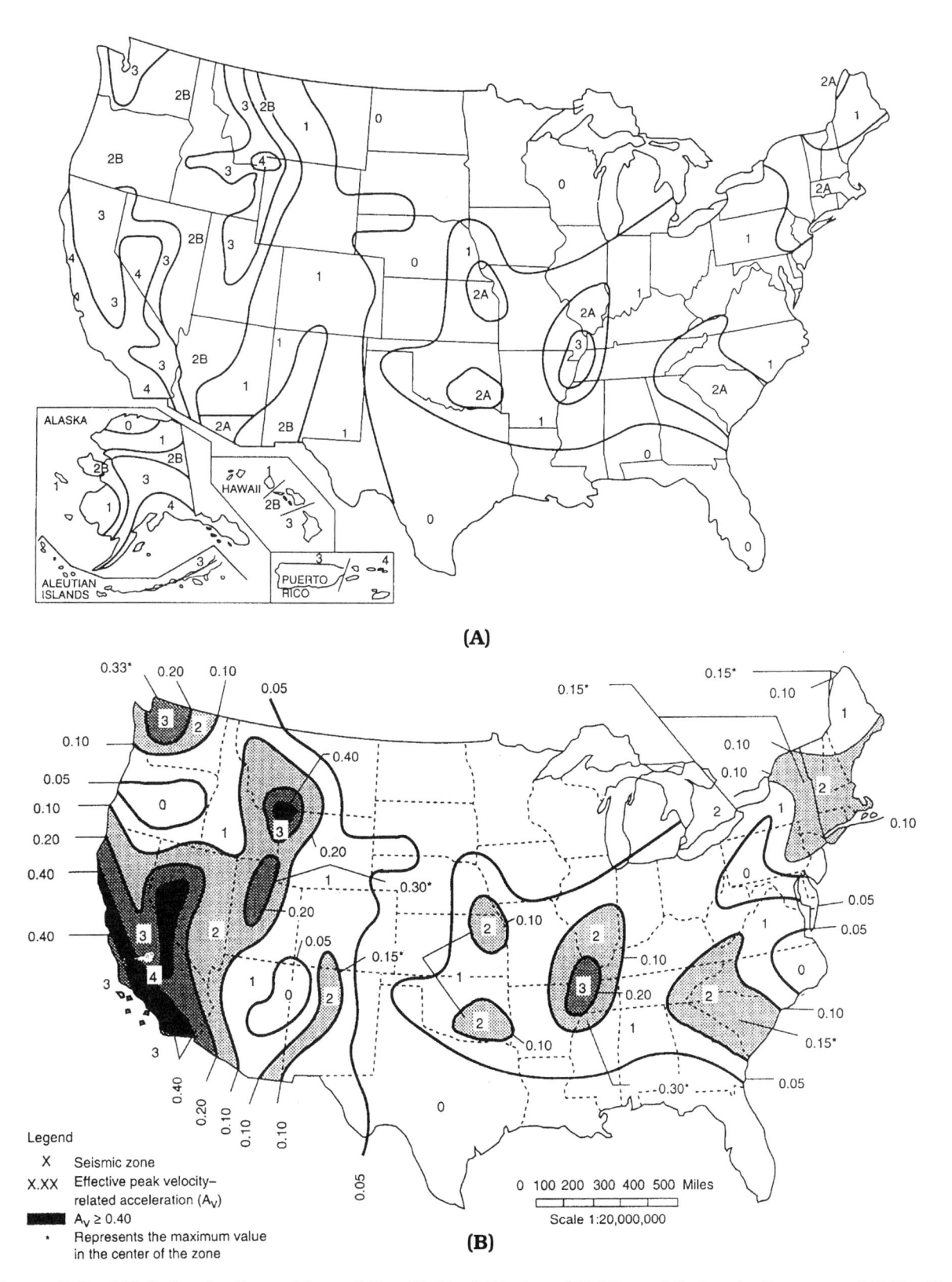

# Chapter 9 — Seismic Protection of Plumbing Equipment

**Figure 9-2** (A) Seismic Zone Map of the United States; (B) Map of Seismic Zones and Effective, Peak-Velocity-Related Acceleration ($A_v$) for Contiguous 48 States.

*Note*: Linear interpolation between contours is acceptable.

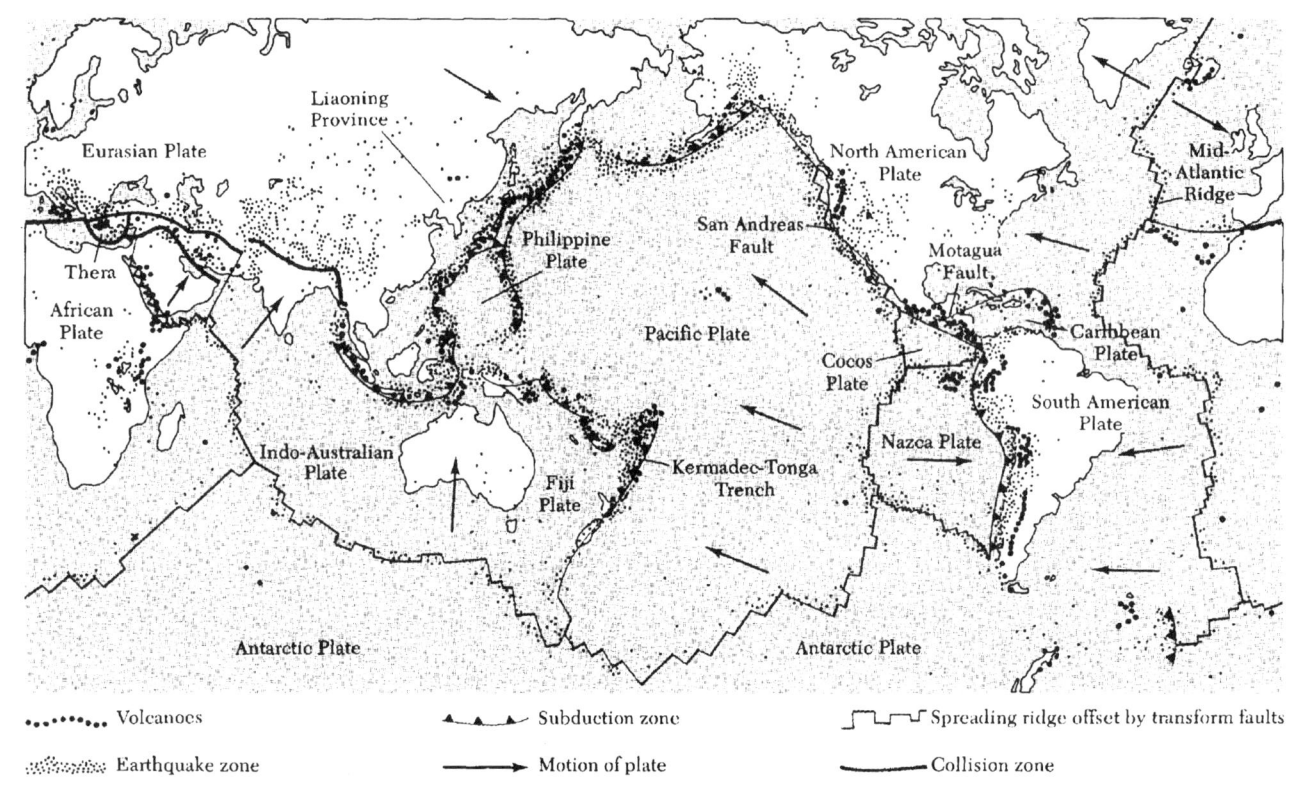

**Figure 9-3  World Map Showing Relation Between the Major Tectonic Plates and Recent Earthquakes and Volcanoes.**

*Note*: Earthquake epicenters are denoted by small dots, volcanoes by large dots.

together by friction is strained to its capacity by the continuing plate movement and both sides spring back to their original positions. See Figure 9-4. Waves in a variety of patterns emanate from this fault movement and spread in every direction. These waves change throughout the duration of the earthquake, add to one another, and result in extremely complicated wave motions and vibrations. At any site away from the fault, the three-dimensional movement of the surface, which is caused by combinations of direct, reflected, and refracted waves, is known simply as "ground shaking." Energy content or intensity of the ground shaking decreases with distance from the causative fault, although because certain structures can be tuned into the motion, this is not always apparent. The horizontal, vertical, and rotational forces on structures are unpredictable in direction, strength, and duration. The structural load is proportional to the intensity of shaking and to the weight of the supported elements.

By combining knowledge of known fault locations with historical and instrumented ground motion records, seismologists can construct maps showing zones of varying expected ground motion. Figure 9-2 shows such maps, which were used to develop design criteria zoning for a national seismic code.

### Damage from Earthquakes

Four separate phenomena created by earthquakes can cause damage:

1. Surface fault slip (ground rupture).
2. Wave action in water created by seismic movement (called tsunamis in open bodies of water, seiches in closed bodies of water).
3. Ground shaking.
4. Ground failure, such as a sudden change to liquid characteristics in certain sands caused by increased pore water pressure called "liquefaction" and "landslides."

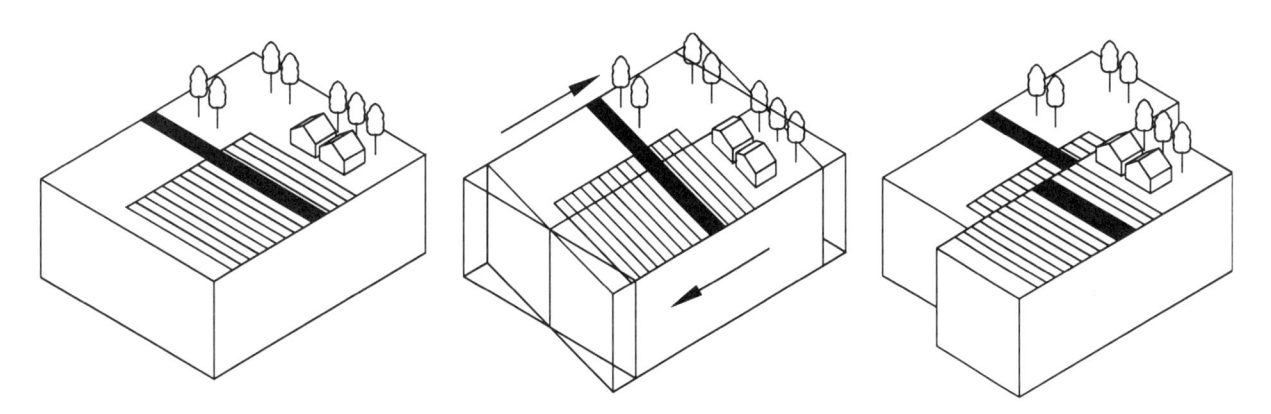

**Figure 9-4 Elastic Rebound Theory of Earthquake Movement**

According to the Elastic Rebound Theory, a fault is incapable of movement until strain has built up in the rocks on either side. As this strain accumulates, the earth's crust gradually shifts (at a rate of about 2 in. a year along the San Andreas Fault). Rocks become distorted but hold their original positions. When the accumulated stress finally overcomes the resistance of the rocks, the earth snaps back into an unrestrained position. The "fling" of the rocks past each other creates the shock waves we know as earthquakes.

It is accepted that buildings and their contents are not designed to withstand ground rupture caused by seismic events. Protection from this is obtained by avoiding potentially dangerous sites. Underground piping can be severely damaged by either fault rupture or ground failure, and frequently pipe lines must cross areas with these potential problems. Seismic design provisions for underground systems in these cases consist of special provisions for the considerable distortion expected in the ground or redundant systems and valving such that local damage can be accepted without serious consequences.

## EARTHQUAKE MEASUREMENT AND SEISMIC DESIGN

### Ground Shaking and Dynamic Response

The primary thrust of seismic design, as it relates to buildings, is to protect against the effects of ground shaking. Although recently there has been concern that surface waves may damage structures by pure distortion, virtually all design is done assuming the entire ground surface beneath a structure moves as a unit, producing a shaking or random motion whose unidirectional components can be studied mathematically and whose effects on structures can be analyzed using structural dynamics and modeling. The movement of the ground mass under a building during an earthquake is measured and recorded using the normal parameters of motion, displacement, velocity, and acceleration. Two orthogonal plan components and one vertical component are used to completely describe the motion. The effect of each orthogonal plan component on the structure under design is considered separately.

The amplitude of displacement, velocity, and acceleration at any moment are, of course, related, as each measures the change in the other over time. Given the record of how one parameter has changed over time (time history), the other two can be calculated. However, due to the direct relationship of force to acceleration (F-Ma) and also because acceleration is easiest to instrumentally measure, acceleration has become the standard measurement parameter. The characteristically spiked and jagged shape of the acceleration time history (accelergram, Figure 9-5) is universally recognized as being associated with earthquakes.

When any nonrigid structure, such as the pendulum or cart and spring of Figure 9-6(A) is subjected to a time history of base motion, the movement (D) of the mass (M) can be measured over time and this record of motions becomes the dynamic response (K). The dynamic response will be different than the input motion because of the inertial lag of the mass behind the base and the resultant energy stored by distorting the connecting structure. The dynamic response to any input motion, then, will depend on the size of the mass and the stiffness of the supporting structure.

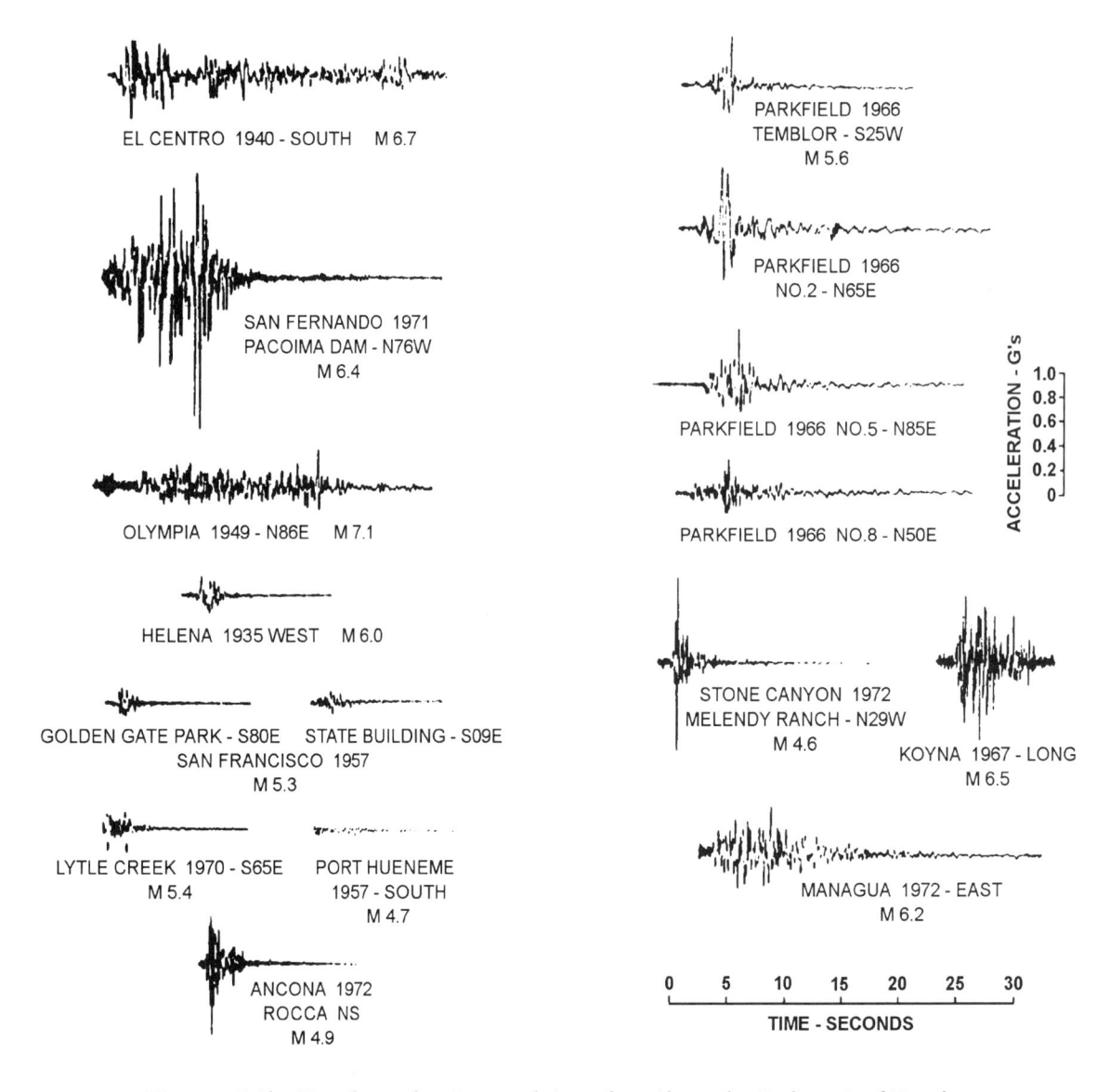

Figure 9-5  Earthquake Ground Accelerations in Epicentral Regions

## The Response Spectrum

Because of the difficulty of measuring all the variations of distortion in a normal structure at each moment of time, a shorthand measure of maximum response is often used. The maximum response of a series of simple pendulums (single-degree-of-freedom system) to a given time history of motion is calculated and the resulting set of maximums is known as a "response spectrum." (See Figure 9-7.) The response parameter could be displacement, velocity, or acceleration, although acceleration is most often used. The variation in dynamic characteristics of each pendulum in the infinite set is measured by the natural period of vibration. The natural period of any system is dependent on stiffness and mass and measures the length of one complete cycle of free (natural) vibration. Frequency, or the inverse of the period, is also often used in place of the period.

If the input motion (or forcing function) for a structure is of constant frequency and matches the natural frequency, resonance occurs, and the response is theoretically infinite. Damping that occurs to some degree in all real systems prevents infinite response, and the amplitude of the actual response will be proportional to the damping present. Damping is normally measured as

# Chapter 9 — Seismic Protection of Plumbing Equipment

a percentage of the amount of damping that would create zero response; that is, the pendulum when set in motion would simply return to its at-rest position. The damping in most structures is between 2 and 10%. For any input motion, the response would depend on the amount of damping present and therefore responses (and response spectra) are often presented as families of similar curves, each corresponding to a different damping value. (Refer to Figure 9-7.)

By the response spectrum technique, the maximum single response to a given base motion of a structure with a known period and damping can be predicted. It must be remembered that the response spectra eliminates the time element from consideration because the maximums plotted for each period are likely to have occurred at different times during the time history. Every ground motion will have its own distinct response spectrum, which will show on a gross basis which vibratory frequencies were predominant in motion. Since ground motions vary not only between earthquakes but between sites during the same earthquake, an infinite variety of response spectra must be considered possible. Fortunately, characteristics of wave transmission and physical properties of soil place upper bounds on spectral shapes. Using statistical analysis of many motions and curve fitting techniques, it is possible to create a design spectrum of energy stored by distorting the connecting structure. The spectrum that is theoretically most appropriate for a dynamic response to any input motion, then, will depend upon the region or even the given site.

With such a design spectrum for acceleration, measured in units of the acceleration of gravity (e.g., the maximum horizontal force in single degree of freedom), systems can be closely approximated using the ordinate as a percentage of the system.

Just as the response of a structure on the ground can be calculated by consideration of the ground motion time history, the response of a system on any floor of a building can similarly be calculated if the time history of the floor motion is known. Using computers, it is possible to calculate such floor motions in structures using base ground motion as input. Response spectra can then be calculated for each floor that would be appropriate for building contents or equipment. The vibratory response of the building is generally far more coherent than rock or soil, as the motion of floors is focused into the natural periods of the building. Floor response spectra are, therefore, often highly peaked around one or two frequencies, so responses nearer to theoretical resonance are more likely than they are on the ground. Responses 25 times greater than input acceleration can be calculated in such circumstances where response spectra for ground motion usually show response multiples of 25. (See US Department of Defense 1973.) These extreme responses are unlikely and are not considered in design, however, due to the many non-linearities in real structures and the low possibility of near-perfect resonance.

The response of multidegree-of-freedom systems [Figure 9-6(B)] cannot be simply calculated from a response spectrum, but spectra are often used to quickly approximate the upper limit of the total lateral force on the system. A "pseudo-

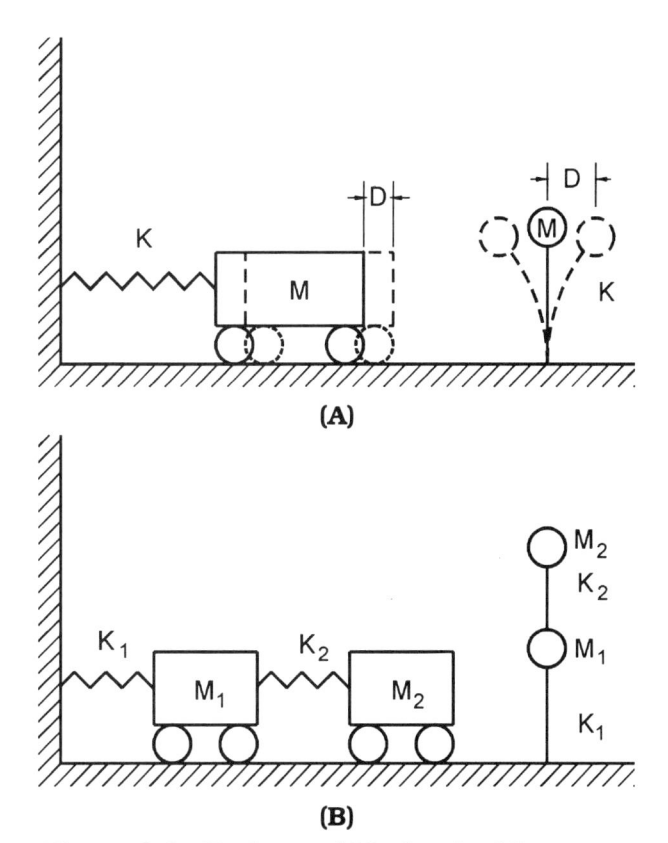

**Figure 9-6 Undamped Mechanical Systems: (A) Single-Degree-of-Freedom Systems; (B) Multiple-Degree-of-Freedom Systems.**

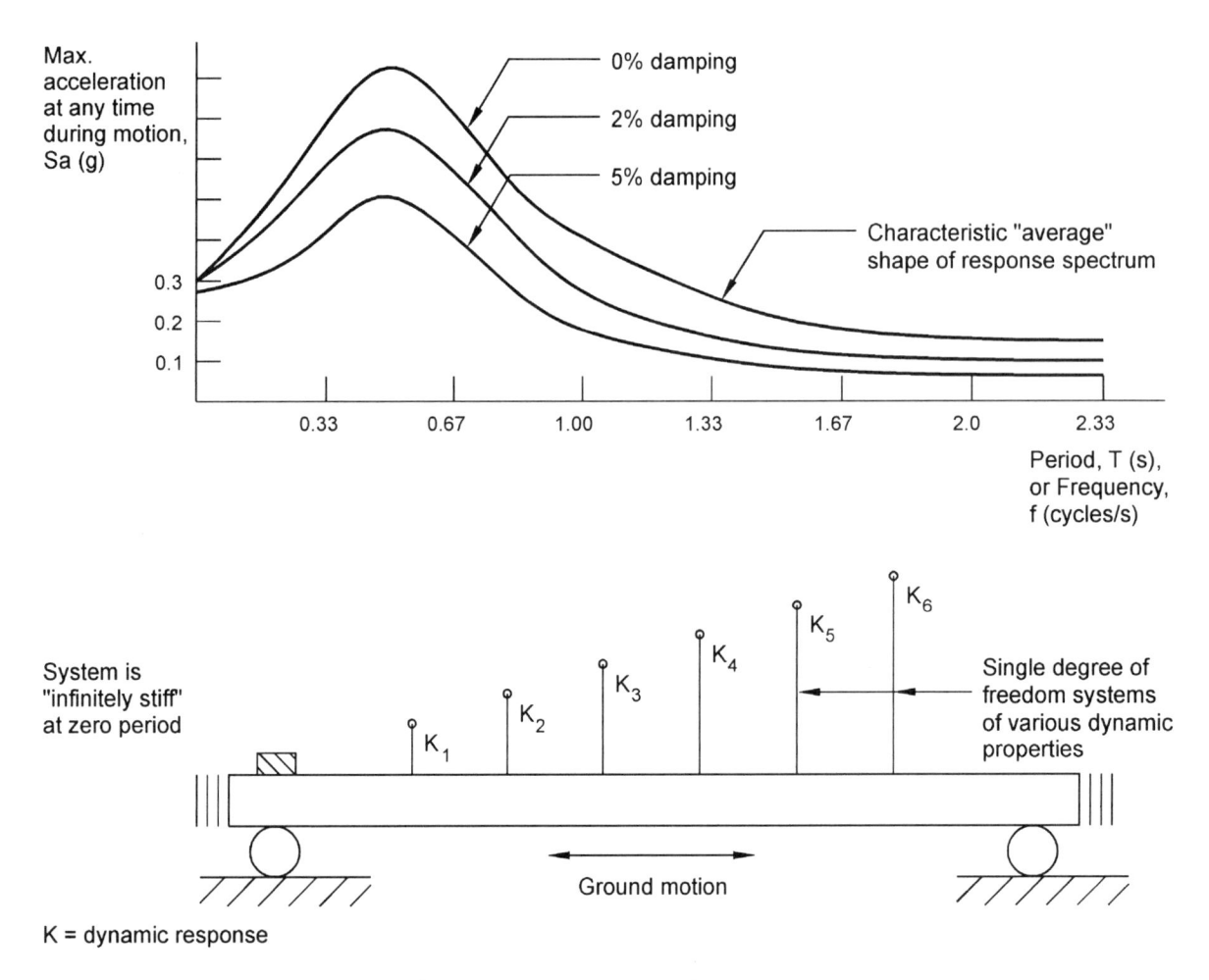

**Figure 9-7 Response Spectrum**

dynamic elastic analysis" can be done on any system using response spectra to obtain a close approximation of maximum forces or distortions. These analyses are typically done by an experienced engineer using a computer, as they can be labor intensive if performed manually.

## LEARNING FROM PAST EARTHQUAKES

### Damage to Plumbing Equipment

Damage to plumbing equipment or systems in earthquakes occurs in two ways:

1. Failure due to forces on the element resulting from dynamic response to ground or floor shaking. The most common example is the sliding or overturning of tanks.

2. Failure due to forced distortions on the element caused by differential movement of two or more supports. This can occur at underground utility entrances to buildings, at building expansion or seismic joints, or, on rare occasions, even between floors at a structure due to interstory drift.

An obvious method of determining failure modes and isolation elements susceptible to damage is to study the experience of past earthquakes.

Particularly useful are the following summaries.[1] (Concerning piping, it should be pointed out that both reports indicate that damage was light on an overall basis; the scattered damage found was as described below.)

---

[1] Ayres, Sun, et al. 1973 and Ayres and Sun 1973.

# Chapter 9 — Seismic Protection of Plumbing Equipment

## The 1964 Alaska Earthquake

### Damage summary

1. Most pipe failures occurred at fittings. Most brazed or soldered joints were undamaged, many screwed joints failed, and a few caulked joints were pulled apart or twisted.
2. Failures in screwed joints often occurred where long unbraced horizontal runs of pipe joined short vertical risers or were connected to equipment. Small branch lines that were clamped tightly to the building were torn from large horizontal mains if these were unbraced and allowed to sway.
3. Joints were loosened or pulled apart in long horizontal runs of unbraced cast-iron pipe, and hangers were bent, shifted, or broken.
4. Pipes crossing seismic joints were damaged if provisions were not made for the relative movements between structural units of buildings.
5. Thermal expansion loops and joints were damaged when the pipes were not properly guided.
6. Fire-sprinkler piping was practically undamaged because it was provided with lateral bracing.
7. Sand filter, water softener, domestic hot water, heating-hot-water expansion and cold-water-storage tanks shifted, toppled, or rolled over when they were not firmly anchored to buildings.
8. Hundreds of small, gas-fired and electric domestic water heaters fell over. Many of the legs on which heaters stood collapsed, and vent connectors were damaged.
9. Some plumbing fixtures were damaged by falling debris.
10. Vertical plumbing stacks in tall buildings were practically undamaged.

## The 1971 San Fernando Earthquake

### Damage summary

1. Unanchored heavy equipment and tanks moved and damaged the connected piping.
2. Heavy equipment installed with vibration isolation mounts moved excessively, often destroyed the isolators, and damaged the connected piping.
3. Cast-iron supports for heavy cast-iron boilers failed.
4. Pipes failed at threaded connections to screwed fittings. Some cast-iron fittings were fractured.
5. Pipes were damaged when crossing separations between buildings.
6. Screwed pipe legs under heavy tanks failed and angle iron legs were deformed.
7. Plumbing fixtures were loosened from mounts and enamel was chipped.
8. Domestic water heater legs were bent or collapsed.

The overall recommendations applicable to plumbing equipment from the Alaska report, made primarily as a response to observed damage, are worth relating:

1. Pipelines should be tied to only one structural system. Where structural systems change, and relative deflections are anticipated, movable joints should be installed in the piping to allow for the same amount of movement.
2. Suspended piping systems should have consistent freedom throughout; for example, branch lines should not be anchored to structural elements if the main line is allowed to sway.
3. If the piping system is allowed to sway, movable joints should be installed at equipment connections.
4. Pipes leading to thermal expansion loops or flexible pipe connections should be guided to confine the degree of pipe movement.
5. Whenever possible, pipes should not cross seismic joints. Where they must cross seismic joints, appropriate allowance for differential movements must be provided. The crossing should be made at the lowest floor possible, and all pipe deflections and stresses induced by the deflections should be carefully evaluated. Standards of the National Fire Protection Association (NFPA) for earthquake protection to fire-sprinkler systems should be referred to for successful, field-tested, installation details that are applicable to any piping system. The latest revision to FM data sheet 2-8 for sprinkler systems is also valuable as a reference guide.
6. Supports for tanks and heavy equipment should be designed to withstand earthquake forces and should be anchored to the floor or otherwise secured.

7. Suspended tanks should be strapped to their hanger systems and provided with lateral bracing.

8. Pipe sleeves through walls or floors should be large enough to allow for the anticipated movement of the pipes and ducts.

9. Domestic water heaters should be provided with legs that can withstand earthquake forces, and the legs should be anchored to the floor and/or strapped to a structurally sound wall.

10. Earthquake-sensitive shut-off valves on gas-service lines should be provided where maximum protection from gas leaks is required.

11. Vibrating and noisy equipment should, if possible, be located far from critical occupancies, so that the equipment can be anchored to the structure, and vibration isolation is not required.

Avoid mounting heavy mechanical equipment on the top or upper floors of tall buildings unless all vibration-isolation mounts and supports are carefully analyzed for earthquake-resistant design.

When equipment and the attached piping must be isolated from the structure by vibration isolators, constraints should be used.

## SEISMIC PROTECTION TECHNIQUES

### General

Assuming that the building in which the piping systems are supported is designed to perform safely in response to earthquake forces, the piping systems must be designed to resist the seismic forces through the strength of the building attachments.

The design professional must consider local, state, and federal seismic requirements, as applicable, in the area of consideration. Only those engineers with seismic experience should design the supports required for seismic zones. Close coordination with the structural engineer is required to ensure the structural system properly supports the mechanical systems and equipment.

### Equipment

Seismic protection of equipment in buildings, as controlled by the design professional, consists of preventing excessive movement that would either damage the equipment directly or break the connected services. The ability of the equipment housing or working parts to withstand earthquake vibration is generally not formally considered for one or more of the following reasons:

1. Such failure would not endanger life.

2. Continued functioning is not always required.

3. Most equipment will experience transportation shocks or working vibrations that are similar to earthquake motions and the housing and internal parts are therefore considered adequate.

4. The design professional has little control over the manufacturing process. Competitively priced equipment specially qualified to resist earthquake motion is not available.

5. Because of a lack of performance data for equipment that is anchored, the extent of the problem is unknown.

Movement to be prevented is essentially overturning and sliding, although these effects can take place with a variety of characteristics:

1. Overturning (moment).

    A. Overturn of equipment.

    B. Failure in tension or compression of perimeter legs, vibration isolators, hangers or their supports.

    C. Excessive foundation rotation.

2. Sliding (shear).

    A. Sliding of floor-mounted equipment.

    B. Swinging of hung equipment.

    C. Excessive sideways failure of legs, stands, tank mounts, vibration isolators, or other supports. Although these failures are often be described as local overturning of the support structure, they are categorized as a shear or sliding failure because they are caused by the straight lateral movement of the equipment rather than the tendency to overturn.

Prevention of overturning and sliding effects can best be discussed by considering the categories of mounting equipment, such as fixed or vibration-isolated, and floor-mounted or hung.

**Fixed, floor-mounted equipment** This group includes tanks, water heaters, boilers, and other equipment that can rest directly on the floor. Although anchoring the base of such equipment to the floor is obvious, simple, and inexpensive, it is commonly omitted. Universal base anchorage of equipment undoubtedly would be the single largest improvement and would yield the largest cost-benefit ratio in the entire field of seismic protection of plumbing equipment. This anchoring is almost always to concrete and is accomplished by cast-in-place anchor bolts or other inserts, or by drilled or shot-in concrete anchors. The connection to the equipment base is totally configuration dependent and may require angles or other hardware to supplement the manufactured base. For elements that have a high center of gravity, it may be most efficient to resist overturning by bracing at the top, either diagonally down to the floor, to the structure above, or to adjacent structural walls. Vertical steel beams, or "strongbacks," can also be added on either side of tall equipment to span from floor to floor; a vertical slip joint connection should be placed at the top of such beams to avoid unexpected interaction between the floor structures.

Tanks that are supported on cast-iron legs or threaded pipes have proven to be particularly susceptible to support failure. These types of legs should be avoided or should have supplemental bracing.

The horizontal earthquake loads from equipment mounted on or within concrete stands or steel frames should be braced from the equipment through the support structure and out the base. Concrete tank saddles often are not attached to the tank, are of inadequate strength (particularly in the longitudinal direction), are not anchored to the floor foundation, or have inadequate provisions for earthquake-generated forces in the floor or foundation. Steel equipment frames often have similar problems, some of which can be solved by diagonal bracing between legs.

**Fixed, suspended equipment** The most common element in this group is the suspended tank. Seldom are these heavy elements laterally braced. The best solution is to install the tank tightly against the structural member above, thus eliminating the need for bracing. However, even these tanks should be secured to the suspension system to prevent slipping. Where the element is suspended below the supporting member, cross-bracing should be installed in all directions to provide lateral stability. Where a tank is suspended near a structural wall, struts to the wall may prove to be simpler and more effective than diagonal bracing.

**Vibration-isolated, floor-mounted equipment** This group includes units containing internal moving parts, such as pumps, motors, compressors, and engines. The entire concept of vibratory isolation by flotation on a nontransmitting material (spring, neoprene, cork, etc.), although necessary for equipment operating movement, is at cross-purposes with seismic anchorage. The isolation material generally has poor lateral, force-carrying capacity in itself, plus the housing devices are prone to overturning. It is, therefore, necessary to either supplement conventional isolators with separate snubbing devices (Figure 9-8), or to install specially designed isolators that have built-in restraints and overturning resistance (Figure 9-9). Isolators with minimal lateral-force resistance used in exterior applications to resist wind are usually inadequate for large seismic forces and are also commonly made of brittle cast iron. The possibility of complete isolator unloading and ensuing tension forces due to overturning or vertical acceleration also must be considered. Manufacturers' ratings of lateral loads for isolators should be carefully examined, for often the capacity is limited by the anchorage of the isolators themselves, which is normally unspecified.

The containment surfaces in these devices must be hard connections to the equipment or its base to avoid vibratory short circuits. Because this requirement for complete operational clearance allows a small, ¼-⅜" (6.4-9.5 mm), movement before restraint begins, resilient pads are added to ease the shock load that could be caused by impact.

Because of the stored energy in isolation springs, it is more efficient to anchor the assembly, as restraint is built into the isolator rather than being a separate unit. In retrofit applications, or occasionally due to dimensional limitations, separate snubbers are preferable. Once snubbers are decided upon, those that restrain in three dimensions are preferred because that minimizes the number required. Although some unconfirmed rubber-in-shear isolators are intended to resist loads in several directions, there is little data to indicate adequacy to resist the concurrent large amplitude dynamic loading

166                                                                                        ASPE Data Book — Volume 1

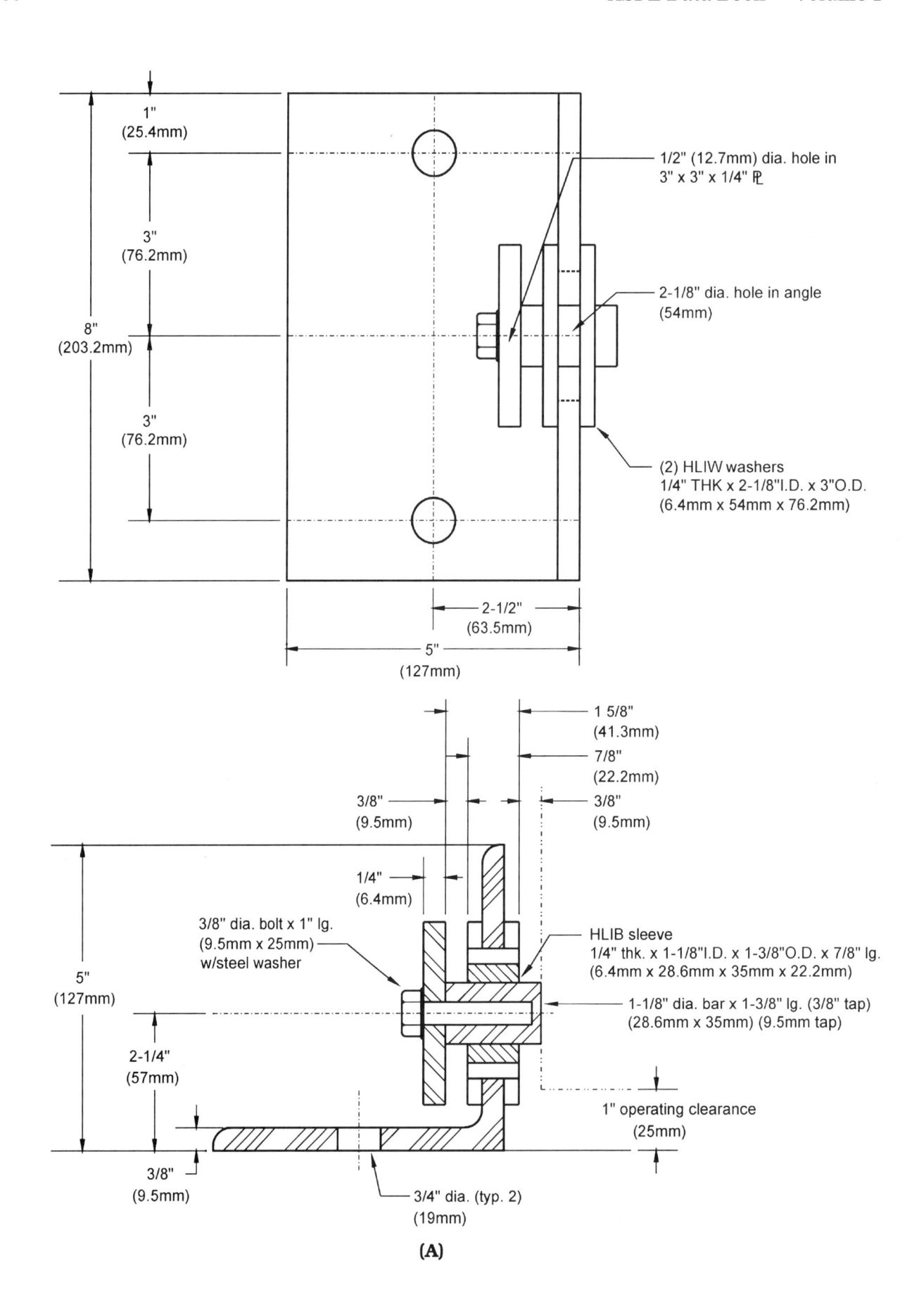

1"
(25.4mm)

3"
(76.2mm)

8"
(203.2mm)

3"
(76.2mm)

1/2" (12.7mm) dia. hole in
3" x 3" x 1/4" ℞

2-1/8" dia. hole in angle
(54mm)

(2) HLIW washers
1/4" THK x 2-1/8"I.D. x 3"O.D.
(6.4mm x 54mm x 76.2mm)

2-1/2"
(63.5mm)

5"
(127mm)

1 5/8"
(41.3mm)

7/8"
(22.2mm)

3/8"
(9.5mm)

3/8"
(9.5mm)

1/4"
(6.4mm)

3/8" dia. bolt x 1" lg.
(9.5mm x 25mm)
w/steel washer

HLIB sleeve
1/4" thk. x 1-1/8"I.D. x 1-3/8"O.D. x 7/8" lg.
(6.4mm x 28.6mm x 35mm x 22.2mm)

1-1/8" dia. bar x 1-3/8" lg. (3/8" tap)
(28.6mm x 35mm) (9.5mm tap)

5"
(127mm)

2-1/4"
(57mm)

3/8"
(9.5mm)

3/4" dia. (typ. 2)
(19mm)

1" operating clearance
(25mm)

**(A)**

# Chapter 9 — Seismic Protection of Plumbing Equipment

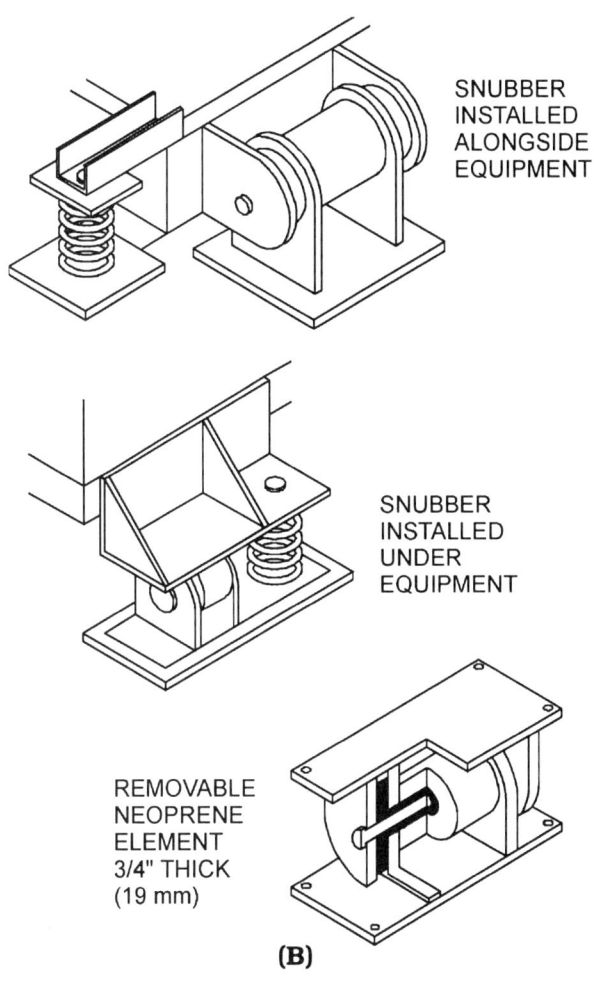

**Figure 9-8  Snubbing Devices:
(A) Three-Directional Angle Snubbers;
(B) Three-Dimensional Cylinder Snubber**

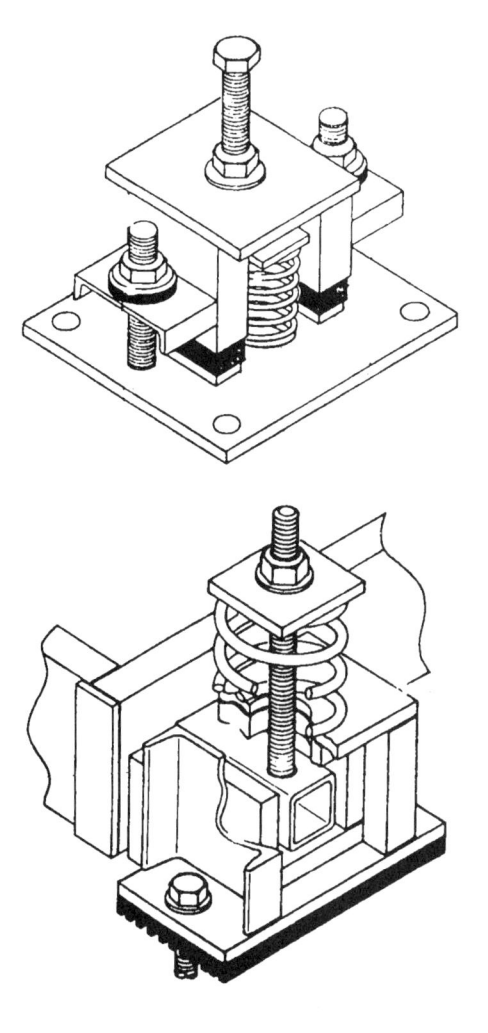

**Figure 9-9  Isolators with Built-In Seismic Restraint**

that could occur in an earthquake. Unless such isolators are considered for real earthquake loading (as opposed to code requirements) with a suitable safety factor, additional snubbing is recommended. Rubin-in-shear isolators with metal housing are more likely to have the overload capacity that may be needed to resist seismic loading, but unless they are specifically tested and rated for this loading, ultimate capacities should be compared with expected real seismic loads.

**Vibration-isolated, suspended equipment**  This is by far the most difficult type of equipment to restrain, particularly if only a small movement can be tolerated. The best method is to place an independent, laterally stable frame around the equipment with proper operating gaps padded with resilient material, similar to a snubber. However, this frame and its support system can be elaborate and awkward. An alternate method is to provide a self-contained, laterally stable, suspended platform upon which conventional seismic isolators or snubbers can be mounted.

For smaller equipment, or when a limited degree of lateral movement can be tolerated, the unit can be installed as close as possible to the supporting structure above without additional lateral restraint. Isolators within hangers should always be installed tight against the supporting structural member. When hanger rods are used to lower the unit, cross bracing or diagonal bracing should be installed.

Slack cables used as diagonal tension braces at each corner of suspended isolated equipment have also been employed. Adjustment of such systems to both avoid short circuits in the isolation system and appropriately limit the movement is critical. Shock loadings that could occur can be partially avoided by neoprene cushioning within cable connections.

## Piping Systems

Normally, piping suspended by hangers less than 12 in. (305 mm) in length, as measured from the top of the pipe to the bottom of the support where the hanger is attached, do not require bracing. The following piping shall be braced:

1. Fuel oil, gas, medical gas, and compressed air piping 1-in. (25.4-mm) nominal diameter and larger.

2. Piping in boiler rooms, mechanical rooms, and refrigeration mechanical rooms 1¼-in. (31.8-mm) nominal diameter and larger.

3. All piping 2½-in. (63.5-mm) nominal diameter and larger.

Conventionally installed piping systems have survived earthquakes with minimal damage. Fitting failures generally occur at or near equipment connectors where equipment is allowed to move, or where a main is forced to move and small branches connected to the main are clamped to the structural elements. In theory then, a few well-placed pipe restraints in the problem areas could provide adequate seismic protection. In practice, however, the exact configuration of piping is seldom known to the designer and even if it was, the key brace locations are not easy to determine. Often partial restraint in the wrong location is worse than no restraint at all. Correct practice is therefore to provide complete restraint when seismic protection of piping systems is advisable. This restraint can be applied throughout the system or in local, well-defined areas such as mechanical or service rooms.

Although there are many variables to consider when restraining pipe against seismic movement, the techniques to do so are simple and similar to those used for hanging equipment. Fixing pipe directly to structural slabs, beams, columns, or walls is, of course, the simplest method. Many codes and guidelines consider hangers of less than 12 in. (305 mm) as being equivalent to direct attachment. For pipes suspended more than 12 in. (305 mm), diagonal braces to the structure above or horizontal struts to an adjacent structure are normally installed at vertical hanger locations. Vertical

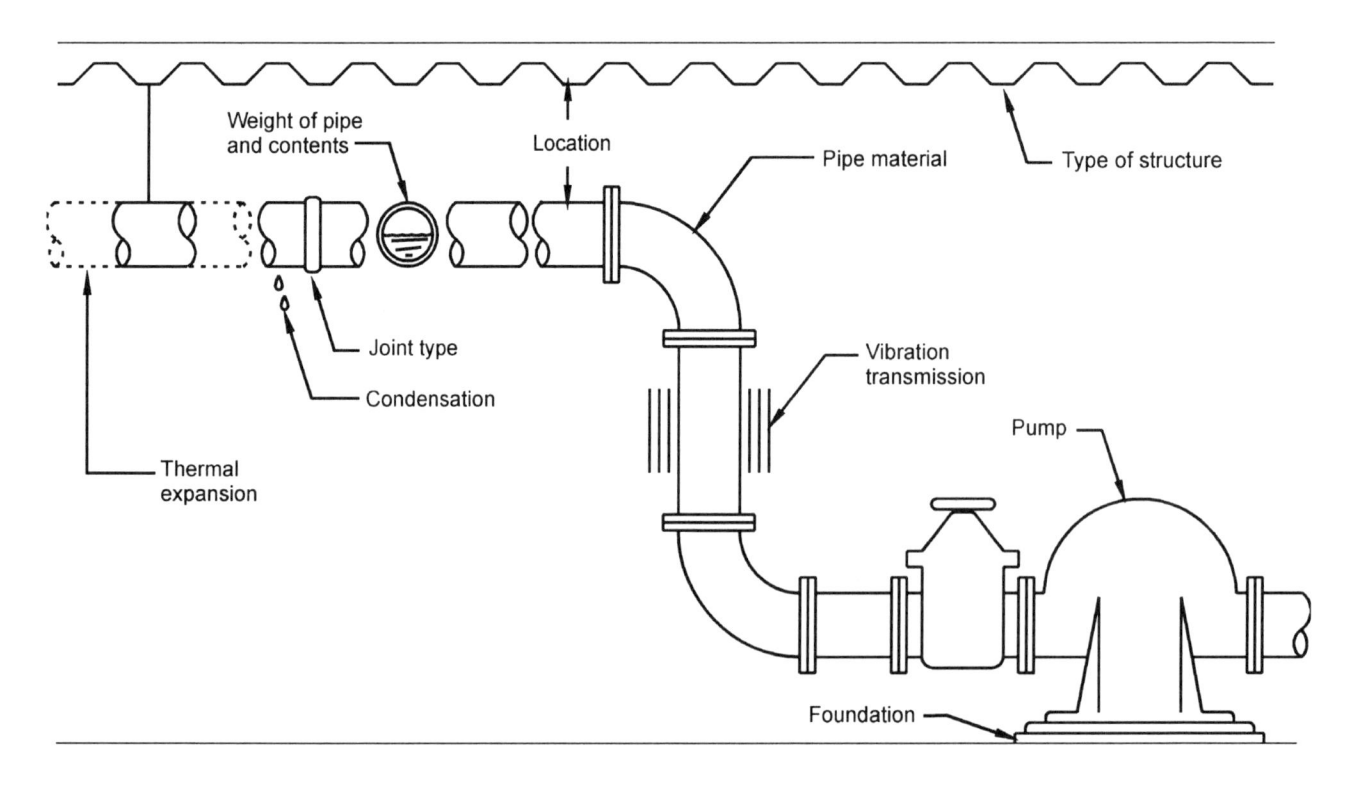

**Figure 9-10 Parameters to Be Considered for Pipe Bracing**

# Chapter 9 — Seismic Protection of Plumbing Equipment

suspension hardware is usually incorporated into braces, both for efficiency and because it is readily available.

Connection to the pipe at transverse braces is accomplished by bearing the pipe or insulation on the pipe clamp or hanger. Attachment to the pipe at longitudinal brace points is not as simple. For small loads, tight-fitting clamps (such as riser clamps) dependent on friction are often used. For larger loadings, details commonly used for anchor points in high-temperature systems with welded or brazed direct connections to the piping may be necessary. Welding should be done by certified welders in accordance with American Welding Society (AWS) D 1.1 and shall use either the shielded or submerged arc method.

Transverse bracing shall be at 40 ft (12.2 m) maximum spacing, except that fuel oil and gas piping shall be at 20 ft (6.1 m) maximum spacing. Longitudinal bracing shall be at 80 ft (24.4 m) maximum spacing, except that fuel oil and gas piping shall be at 40 ft (12.2) maximum spacing.

The many parameters that must be considered before the exact details and layout of a pipe bracing system can be completed are shown schematically in Figure 9-10. These parameters are discussed in more detail below:

1. *Weight of pipe and contents* Since the motion being restrained is a dynamic response, the forces that must be resisted in each brace are proportional to the tributary weight.

2. *Location of pipe* The strength of structural members, particularly compression members, is sensitive to length, so a pipe that must run far from a structural support may require more or longer braces. In boiler service rooms, a horizontal grid of structural beams has sometimes been placed at an intermediate height to facilitate bracing of pipes.

3. *Type of structure* The connection of hangers and braces to the structure is an important factor in determining a bracing system, as demonstrated by the following considerations: Many light roof-deck systems cannot accept point loads except at beam locations; pipe locations and brace layout are thereby severely limited unless costly cross beams are placed at every brace. Other roof and floor systems have significant limitations on the magnitude of point loads, which limit brace spacing.

   It is often unacceptable to have anchors drilled or shot into the underneath of pre-stressed concrete floors. Limitations on depth and location also exist in the bottom flange of steel or reinforced concrete beams and in the bottom chord of joists.

   Many steel floor-deck styles have down flutes 1½ in. (38.1 mm) or less in width; the strength of drilled or shot-in anchors installed in these locations is questionable.

   The structures of buildings that employ interstitial space may have the capacity to brace pipe to either the top or the bottom of the space, which greatly increases bracing layout flexibility.

4. *Piping material* The strength and ductility of the material will affect brace spacing. The stiffness will affect dynamic response and therefore loading.

5. *Joint type* The joint has proven to be the element most likely to be damaged in piping systems; threaded and bell-and-spigot joints have been particularly susceptible. The joint type also determines, in conjunction with the pipe material, the length of the span between braces. Brazed and soldered joints perform acceptably. Most no-hub joints, however, have virtually no stiffness; effective bracing of such systems is nearly impossible. Mechanical joints exhibit the most complex behavior, with spring-like flexibility (when pressurized) within a certain rotation and then rigidity. In addition, the behavior of such systems under earthquake conditions, which cause axial loadings necessary to transmit forces to longitudinal braces, is unknown. As a minimum, cast iron and glass pipe, and any other pipe joined with a shield-and-clamp assembly, where the top of the pipe exceeds 12 in. (305 mm) from the supporting structure, shall be braced on each side of a change of direction of 90° or more. Riser joints shall be braced or stabilized between floors. For hubless, pipe-riser joints unsupported between floors, additional bracing is required. All pipe vertical risers shall be laterally supported with a riser clamp at each floor.

6. *Vibration* Traditionally, unbraced pipe systems seldom cause vibration transmission problems because of their inherent flexibility. Many engineers are concerned that completely braced "tight" piping systems could cause unpredictable sound and vibration problems.

7. *Temperature movement* Pipe anchors and guides used in high-temperature piping systems must be considered and integrated into a seismic bracing system. A misplaced longitudinal brace can become an unwanted anchor and cause severe damage. Thermal forces at anchor points, unless released after the system is operational, are additive to tributary seismic forces. Potential interference between seismic and thermal support systems is particularly high near pipe bends where a transverse brace can become an anchor for the perpendicular pipe run.

8. *Condensation* The need to thermally insulate high-temperature and chilled water lines from hanging hardware makes longitudinal brace attachment difficult. In some configurations of short runs with bends, transverse braces can be utilized near elbows to brace the system in both directions. Friction connections, using wax-impregnated oak or calcium-silicate sleeves as insulators, have been used.

Several bracing systems have been developed that contain some realistic and safe details governing a wide range of loading conditions and

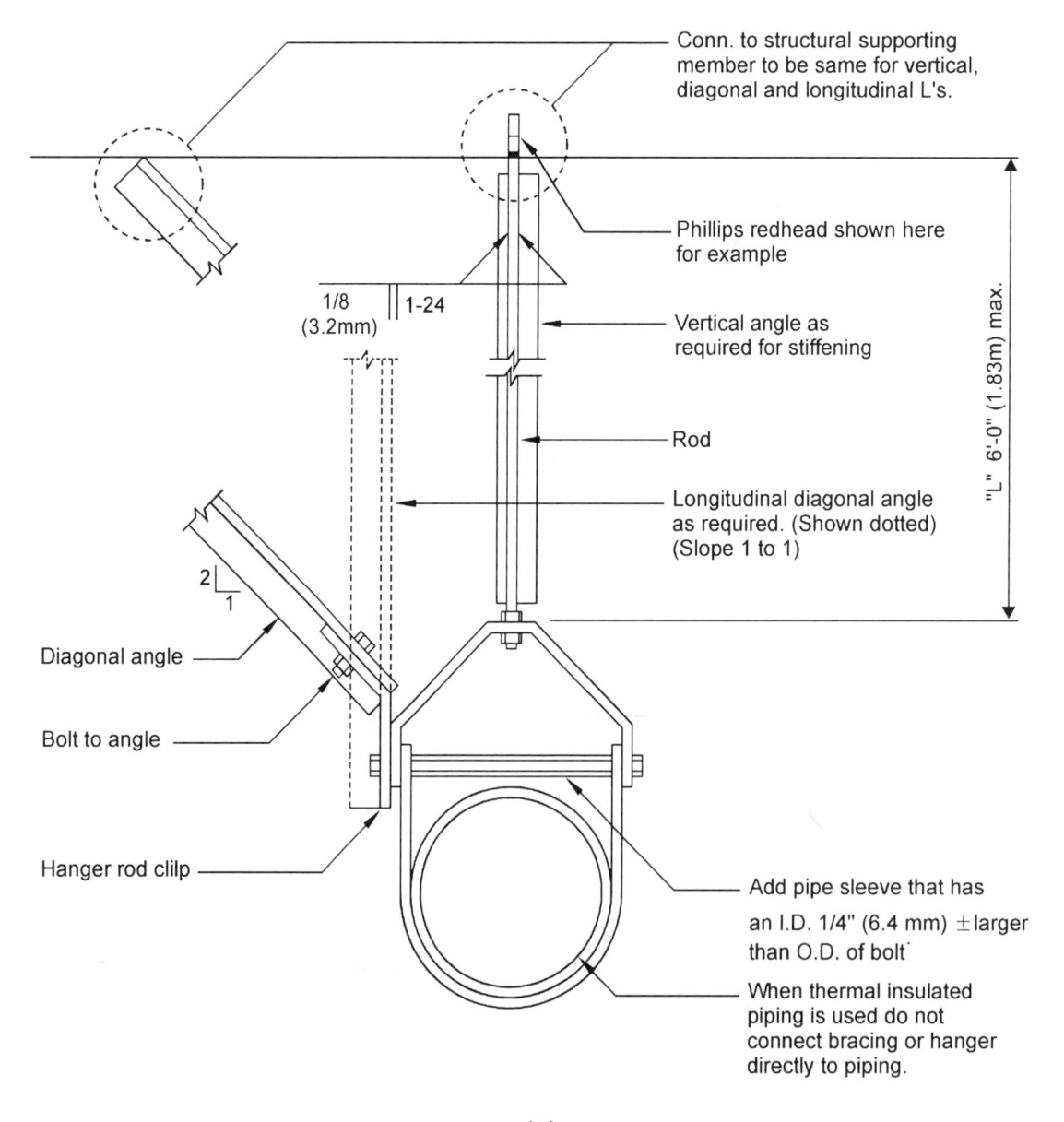

**(A)**

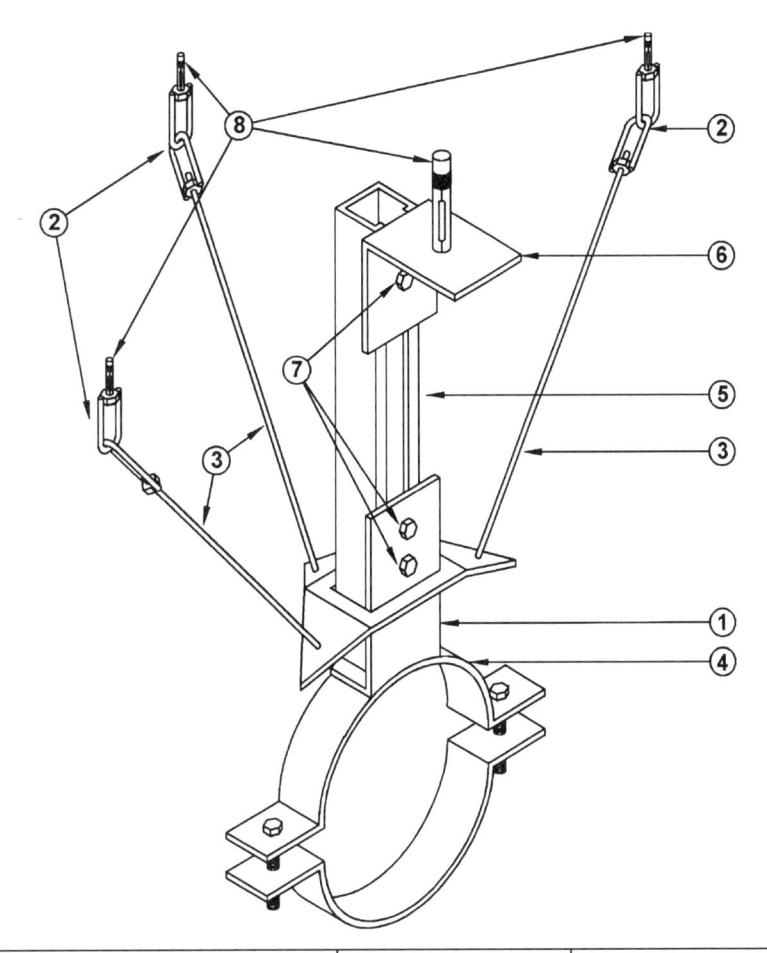

# Chapter 9 — Seismic Protection of Plumbing Equipment

| | | | | |
|---|---|---|---|---|
| ① | Triangle Plate<br>1/4" (6.4 mm) Plate<br>1/4"x 1-1/4" Flat Bar<br>(6.4 x 31.8 mm) | ⑤ | 1-1/4"x1-1/4"x12 Ga. Channel<br>(31.8 x 31.8 mm)<br>Length Varies |
| ② | 1/2" Flexible Connector<br>(12.7 mm) | ⑥ | 1/4" (6.4 mm) Angle Clip |
| ③ | 1/2" (12.7 mm) All Thread Rod<br>Length Varies<br>with Nylon Lock Nut | ⑦ | 1/4" (6.4 mm) Dia Mach. Bolt<br>With Clamp Nut<br>30 Ft-lb Torque |
| ④ | Pipe Hanger<br>Size and type can vary | ⑧ | Phillips Sleeve Anchor |

**(B)**

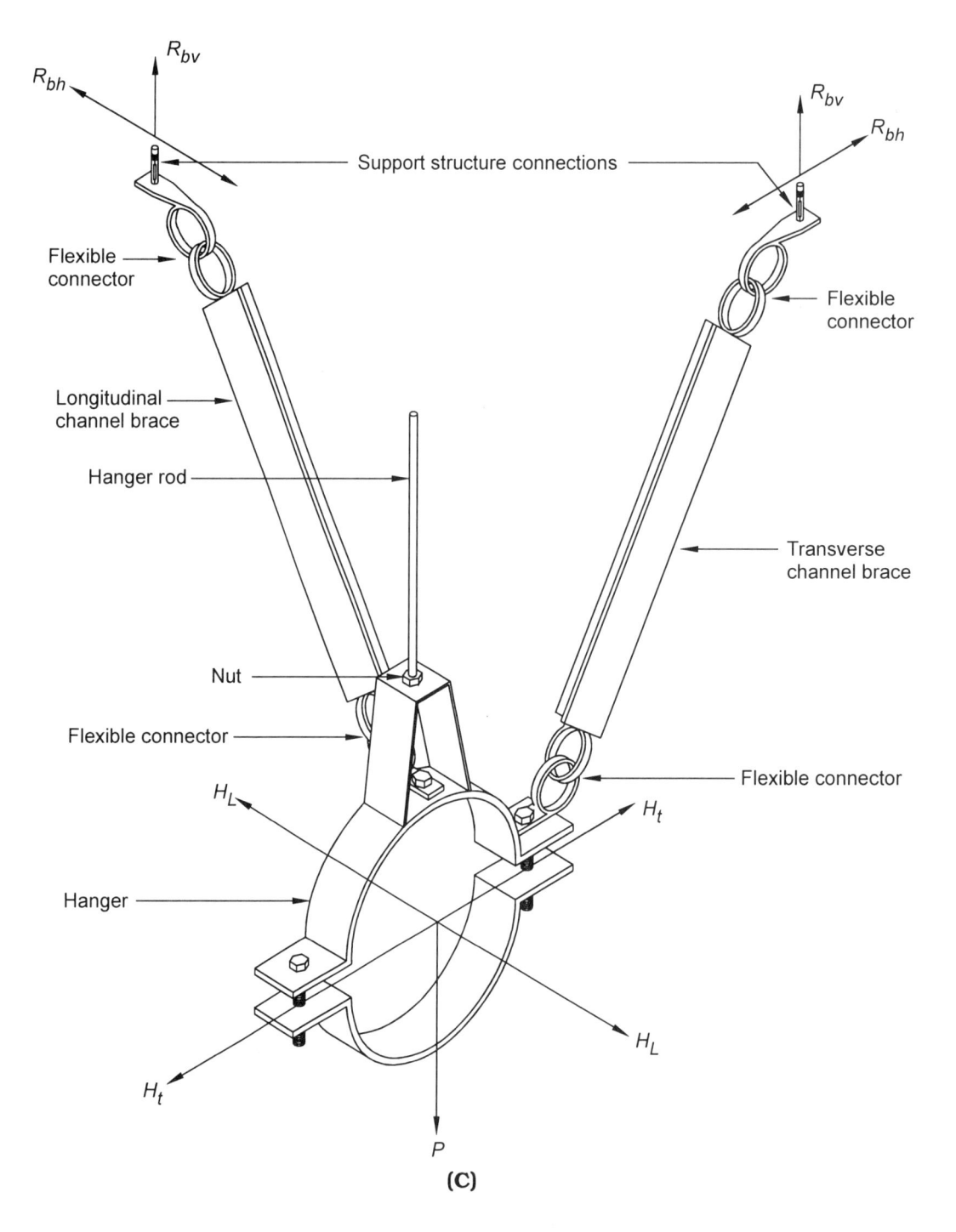

**Figure 9-11** Pipe Bracing Systems: (A) Typical Pipe Bracing; (B) Tension 360; (C) Superstrut.
*Source*: (A) SMACNA.

configurations. For example, SMACNA (Sheet Metal and Air Conditioning Contractors' National Association) and PPIC (Plumbing and Piping Industry Council) have prepared some guidelines on bracing systems for use by engineers, architects, contractors, and approving authorities. Some of these details for construction of seismic restraints are seen in Figures 9-11 and 9-12.

The guidelines set forth by SMACNA and PPIC utilize three pipe bracing methods:

# Chapter 9 — Seismic Protection of Plumbing Equipment

1. The structural angle.
2. The structural channel.
3. The aircraft cable method. (See Figure 9-12.)

Several manufacturers have developed their own seismic bracing methods. (See Figures 9-13 and 9-14.)

Whatever method is used, one should determine the adequacy of the supporting structure by properly applying acceptable engineering procedures.

Pipe risers seldom pose a problem because they are normally clamped at each floor and movement due to temperature changes are routinely considered. Very large or stiff configurations, which could be affected by interstory drift, or situations where long, free-hanging horizontal runs could be inadvertently "braced" by a riser are possible exceptions. The effect of mid-span couplings with less strength or rigidity than the pipe itself must also be considered.

The techniques for handling the possible differential movement at locations of utility entrances to buildings or at building expansion joints are well developed because of the similarity to nonseismic problems of settlement, temperature movement, and wind drift. Expansion loops or combinations of mechanically flexible joints are normally employed. For threaded piping, flexibility may be provided by the installation of swing joints. For manufactured ball joints, the length of piping offset should be calculated using seismic drift of 0.015 feet per foot (0.0046 meter per meter) of height above the base where seismic separation occurs. The primary consideration in seismic applications is to recognize the possibility of repeated, large differential movements.

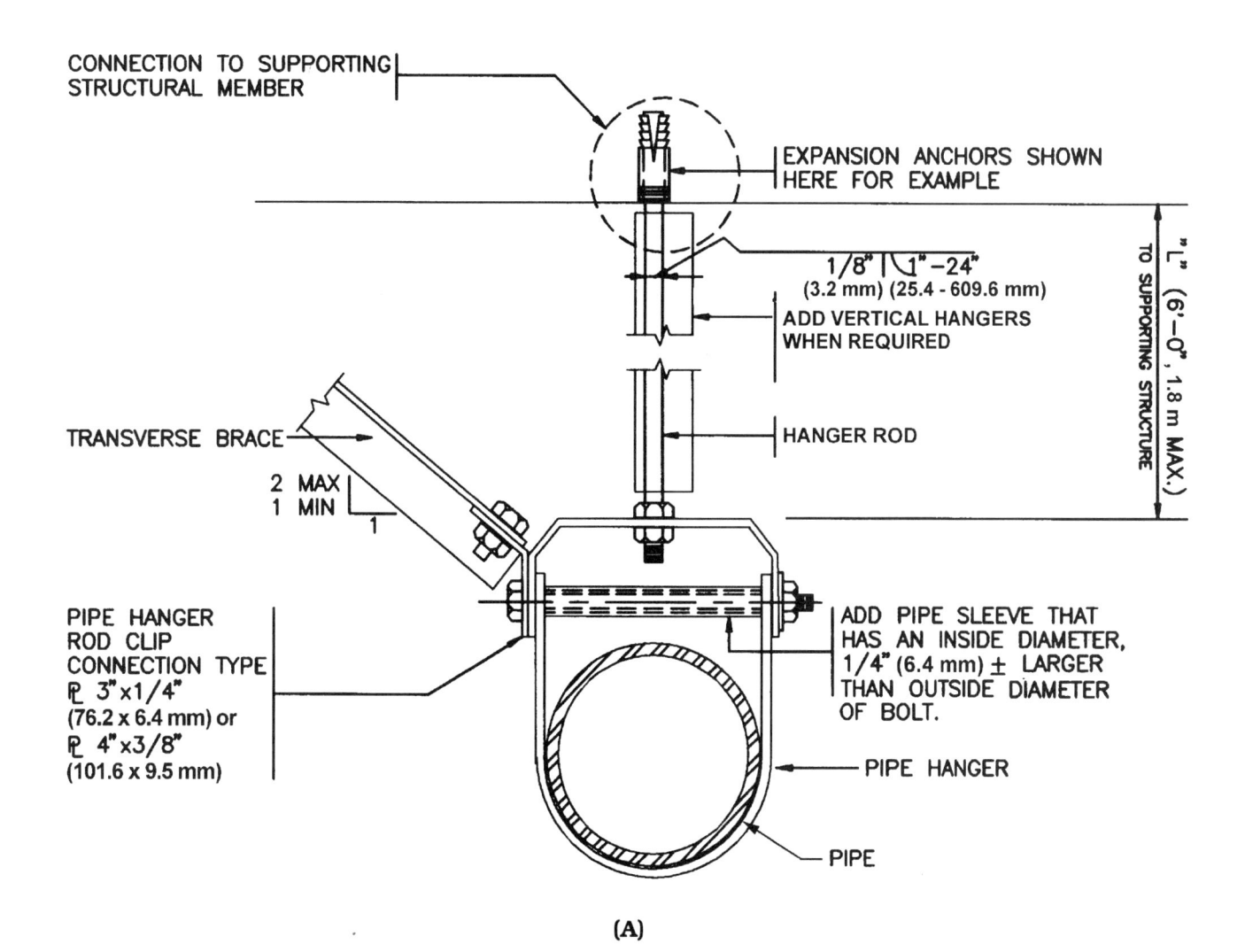

(A)

**(B)**

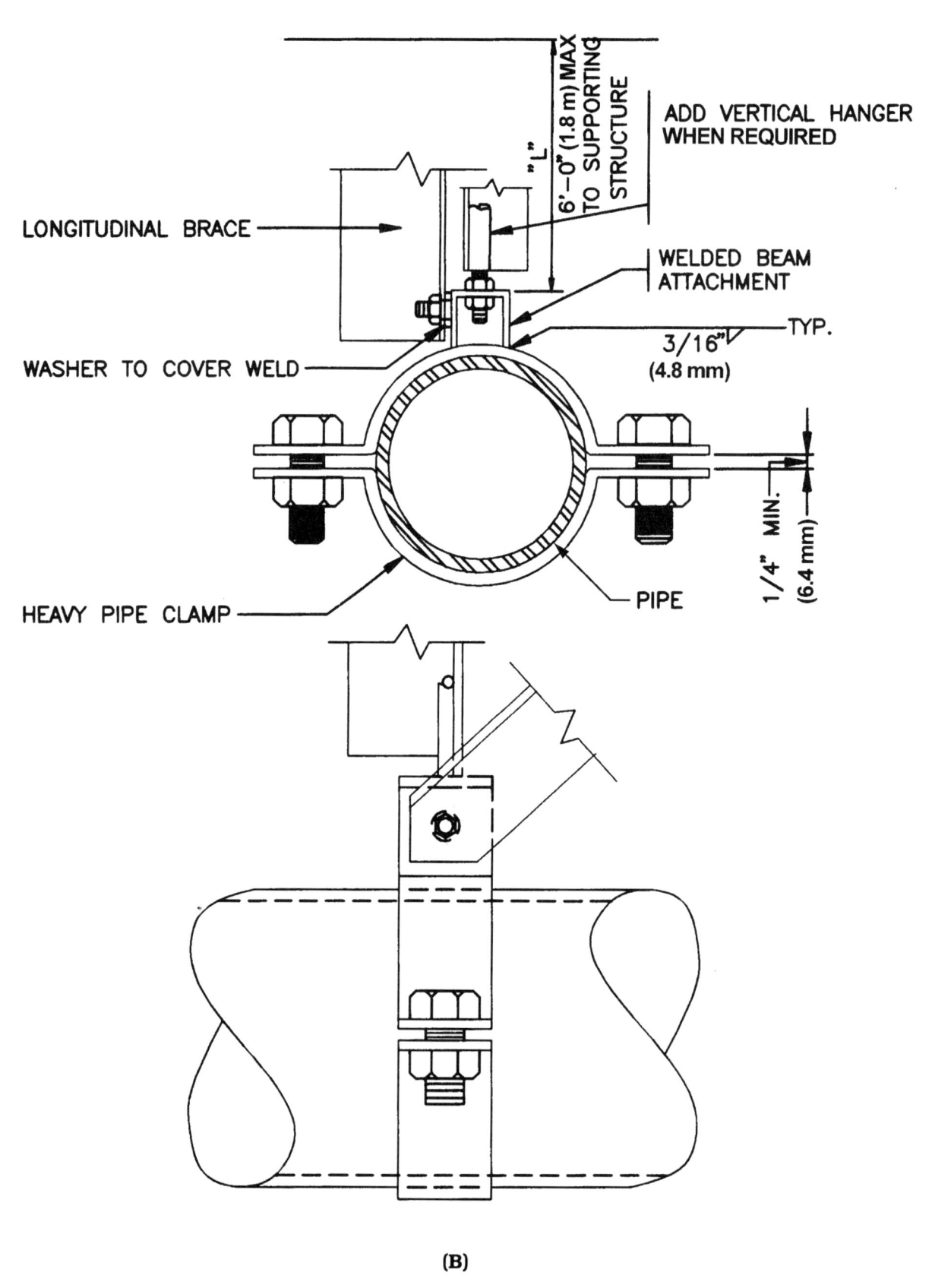

*Note:* Movement due to temperature has been neglected in this example.

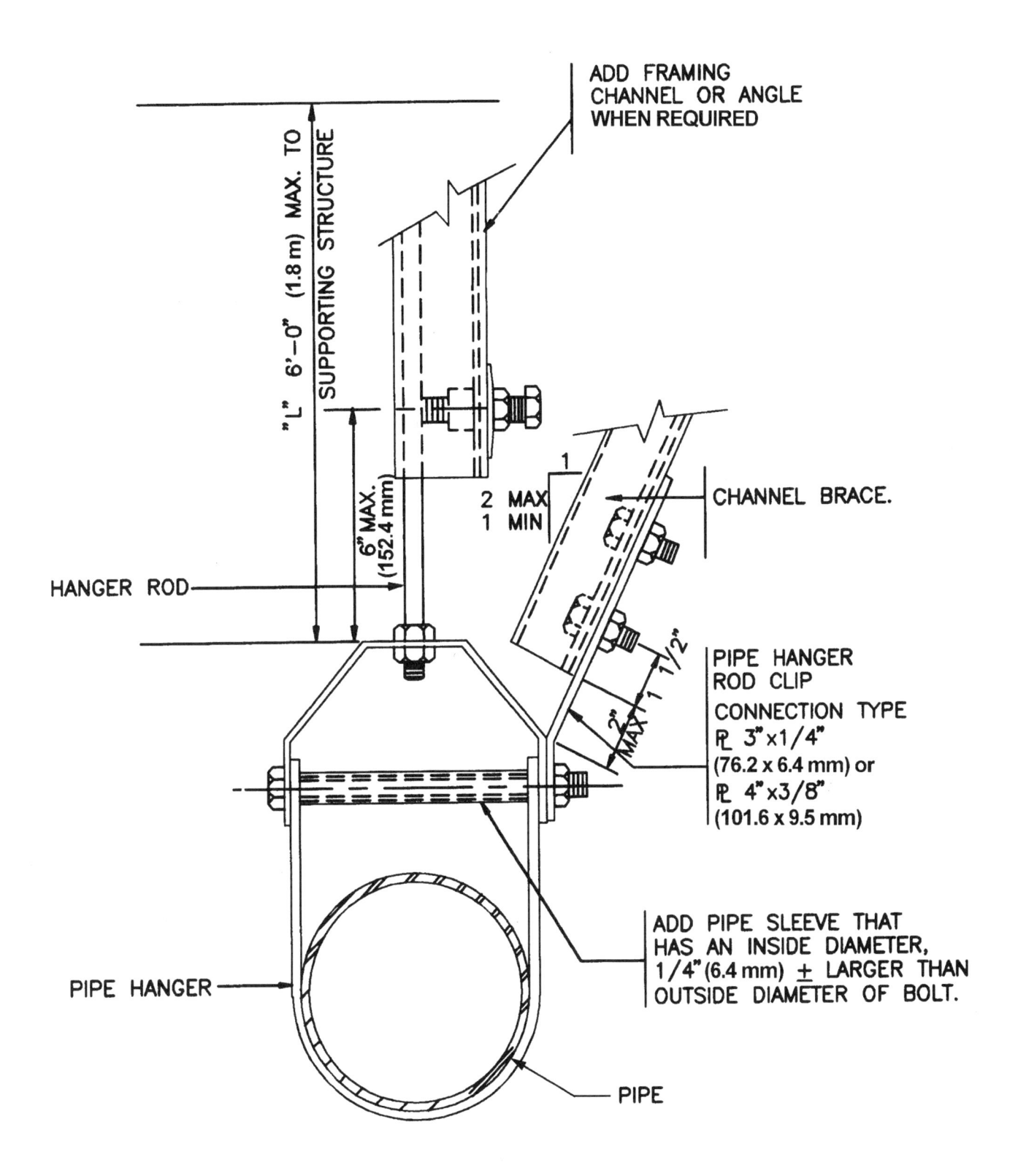

# Chapter 9 — Seismic Protection of Plumbing Equipment

(C)

**(D)**

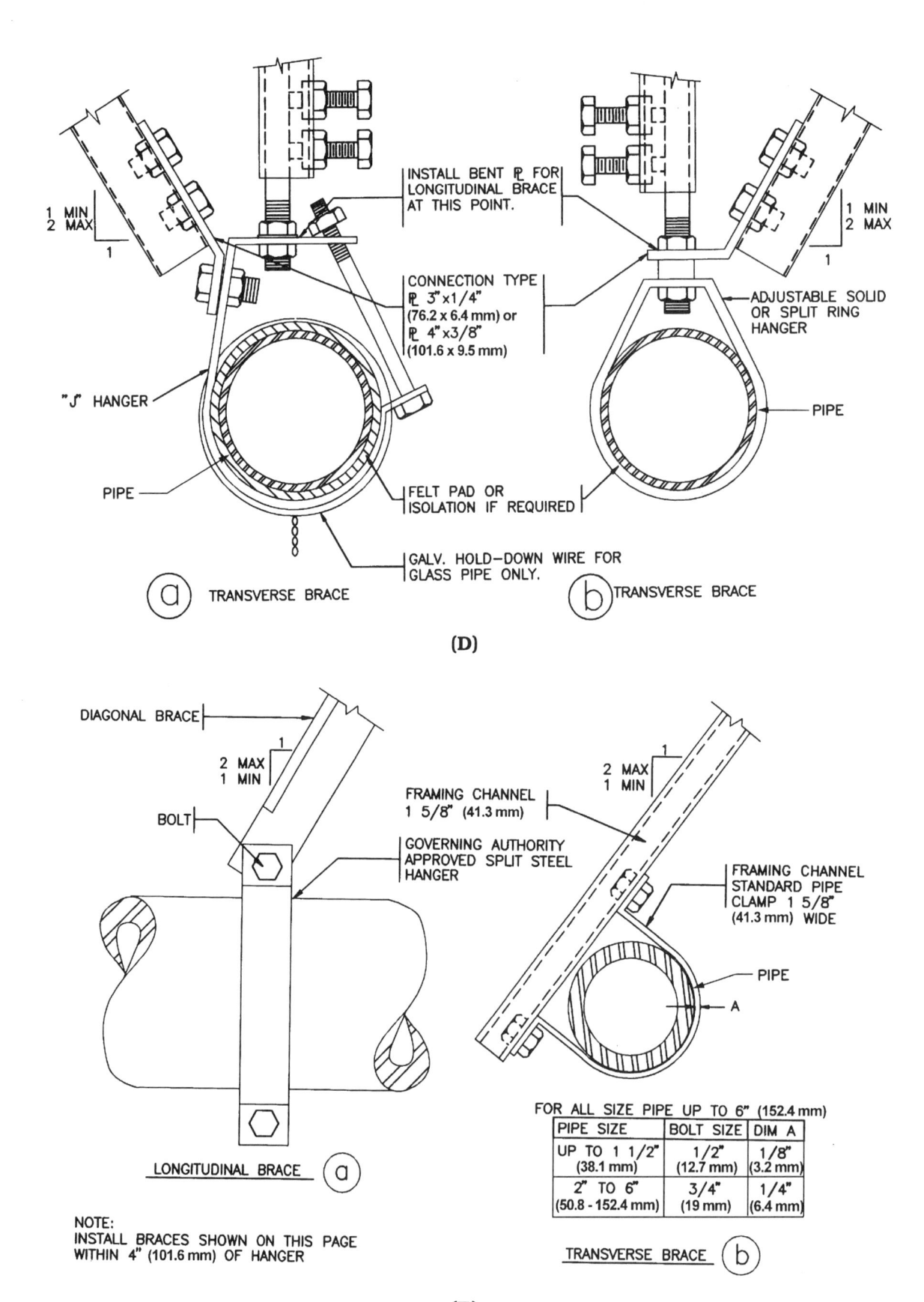

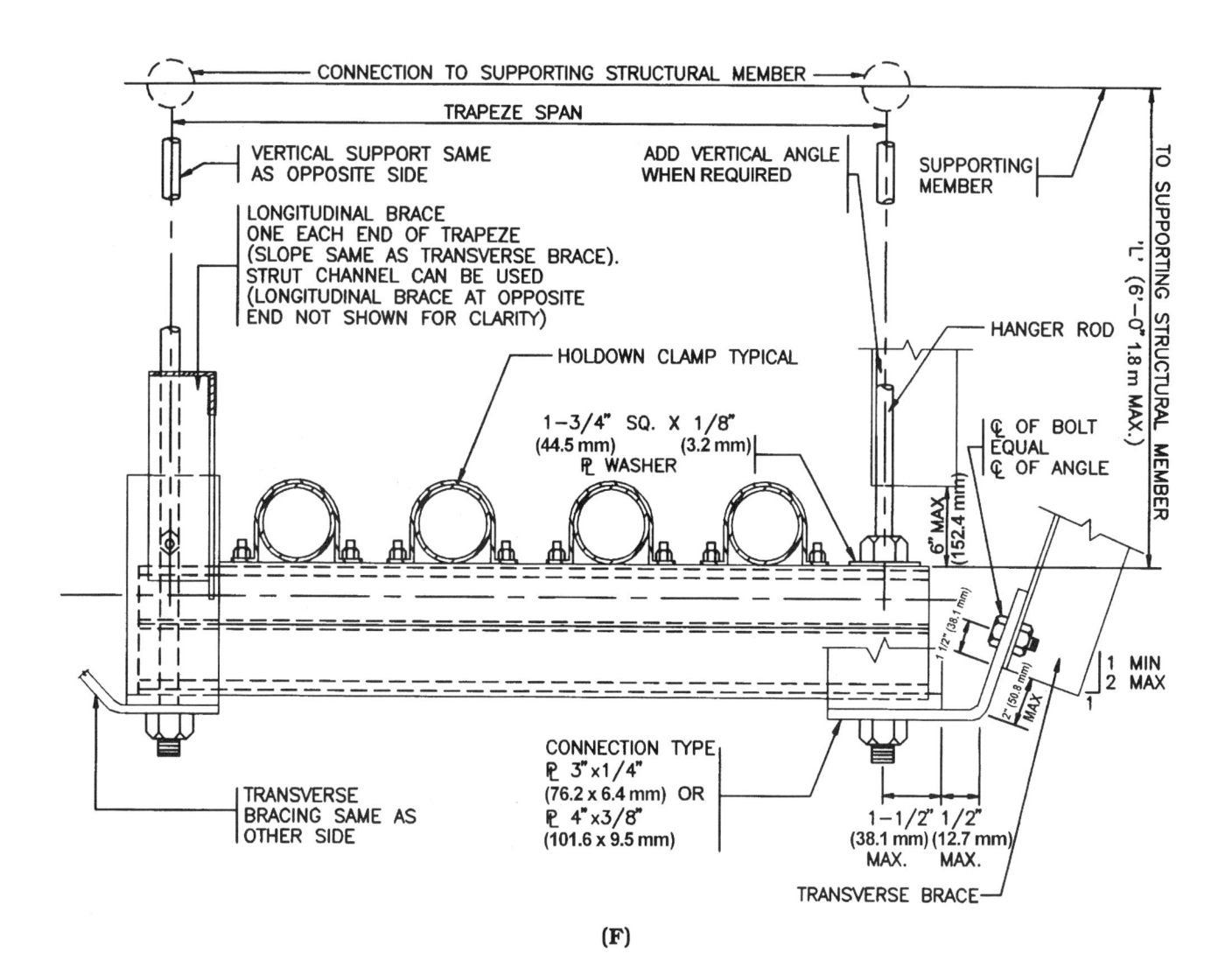

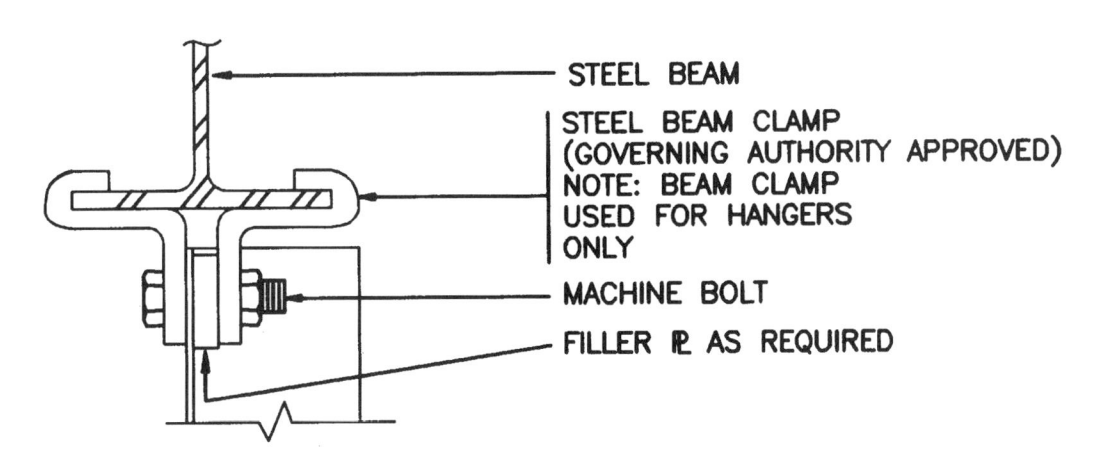

# Chapter 9 — Seismic Protection of Plumbing Equipment

**(F)**

**(G)**

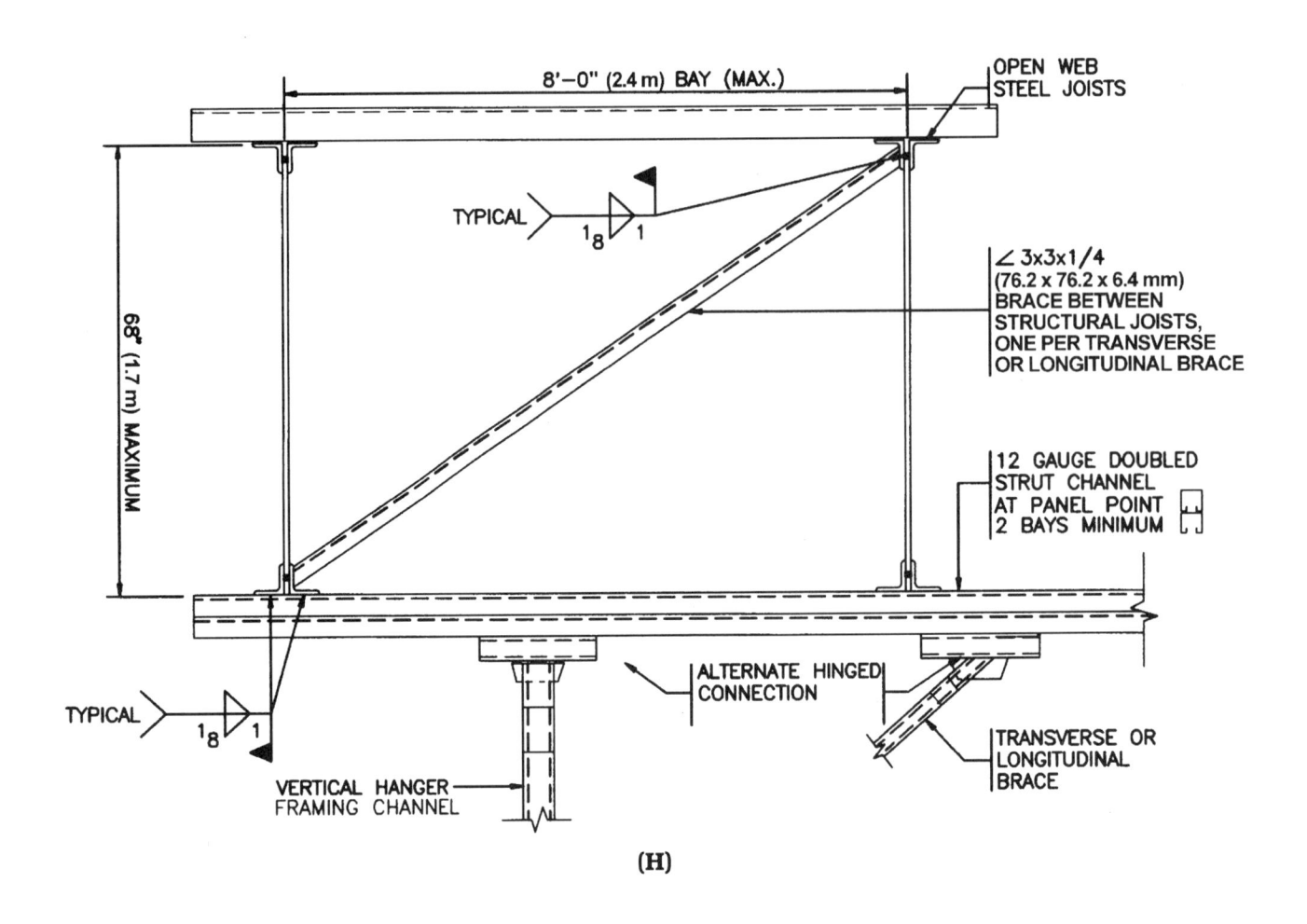

**(H)**

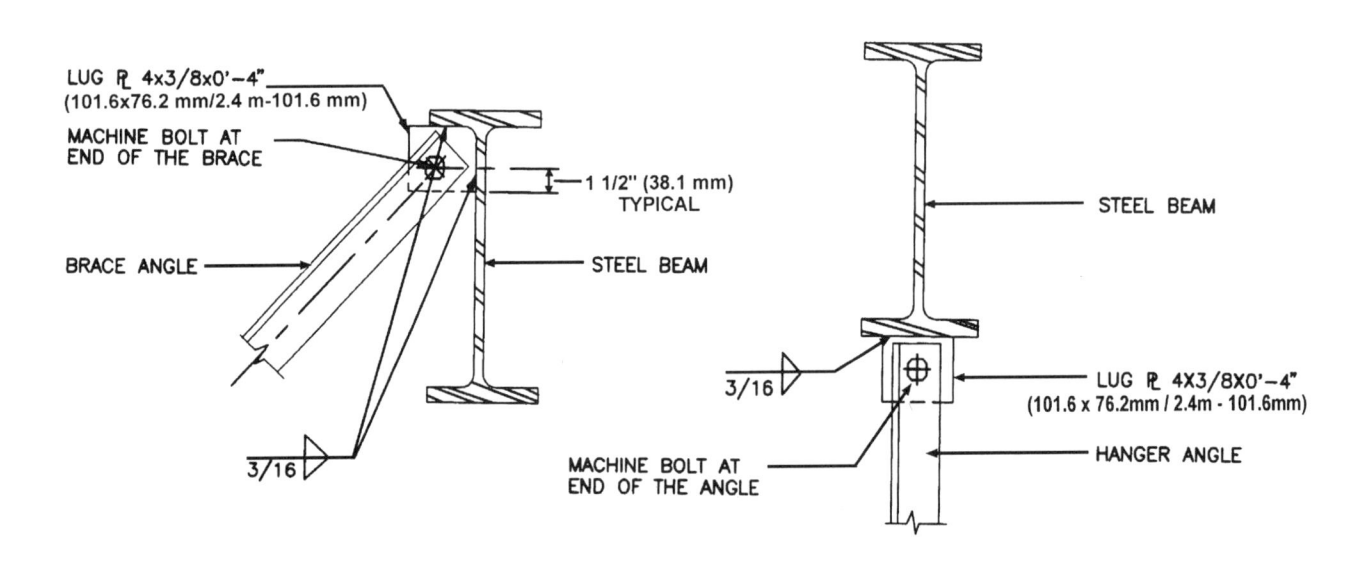

- 8'-0" (2.4 m) BAY (MAX.)
- 68" (1.7 m) MAXIMUM
- OPEN WEB STEEL JOISTS
- TYPICAL 1/8 1
- ∠ 3x3x1/4 (76.2 x 76.2 x 6.4 mm) BRACE BETWEEN STRUCTURAL JOISTS, ONE PER TRANSVERSE OR LONGITUDINAL BRACE
- 12 GAUGE DOUBLED STRUT CHANNEL AT PANEL POINT 2 BAYS MINIMUM
- ALTERNATE HINGED CONNECTION
- TRANSVERSE OR LONGITUDINAL BRACE
- VERTICAL HANGER FRAMING CHANNEL

**(I)**

- LUG ℞ 4x3/8x0'-4" (101.6x76.2 mm/2.4 m-101.6 mm)
- MACHINE BOLT AT END OF THE BRACE
- 1 1/2" (38.1 mm) TYPICAL
- BRACE ANGLE
- STEEL BEAM
- 3/16
- STEEL BEAM
- 3/16
- LUG ℞ 4X3/8X0'-4" (101.6 x 76.2mm / 2.4m - 101.6mm)
- MACHINE BOLT AT END OF THE ANGLE
- HANGER ANGLE

# Chapter 9 — Seismic Protection of Plumbing Equipment

**CONNECTION TO STRUCTURE ABOVE**

1/2" (12.7 mm) TYPICAL CLR. MAX.

1 1/4" (31.8 mm)

PL 4x1/4 (101.6 x 6.4 mm)

CLEVIS

HANGER ROD

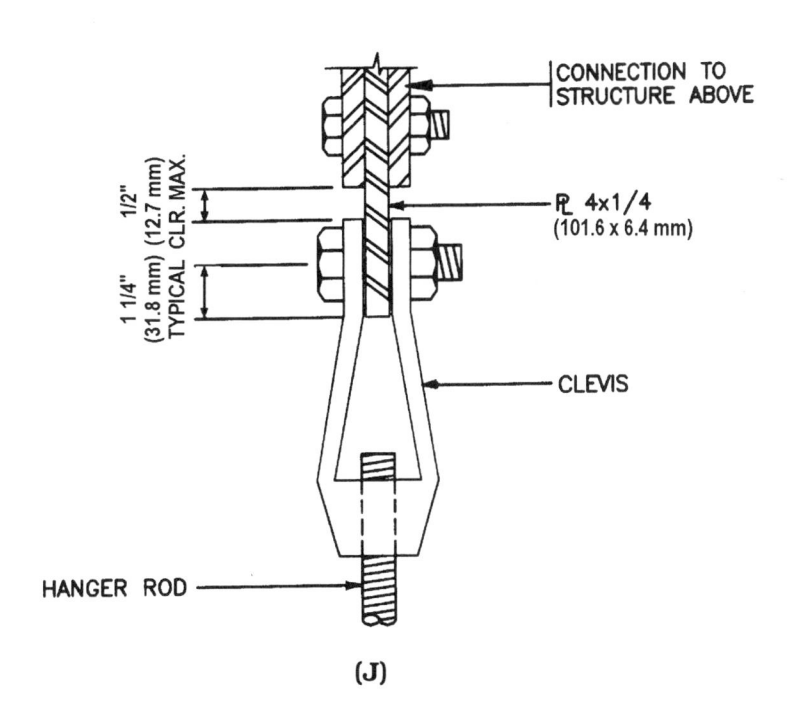

**(J)**

WHERE MULTIPLE SHIELD AND CLAMP JOINTS OCCUR IN A CLOSELY SPACED ASSEMBLY (I.E. FITTING–FITTING–FITTING, ETC.) A 16 GAUGE HALF SLEEVE MAY BE INSTALLED UNDER THE ASSEMBLY WITH A PIPE HANGER AT EACH END OF THE SLEEVE.

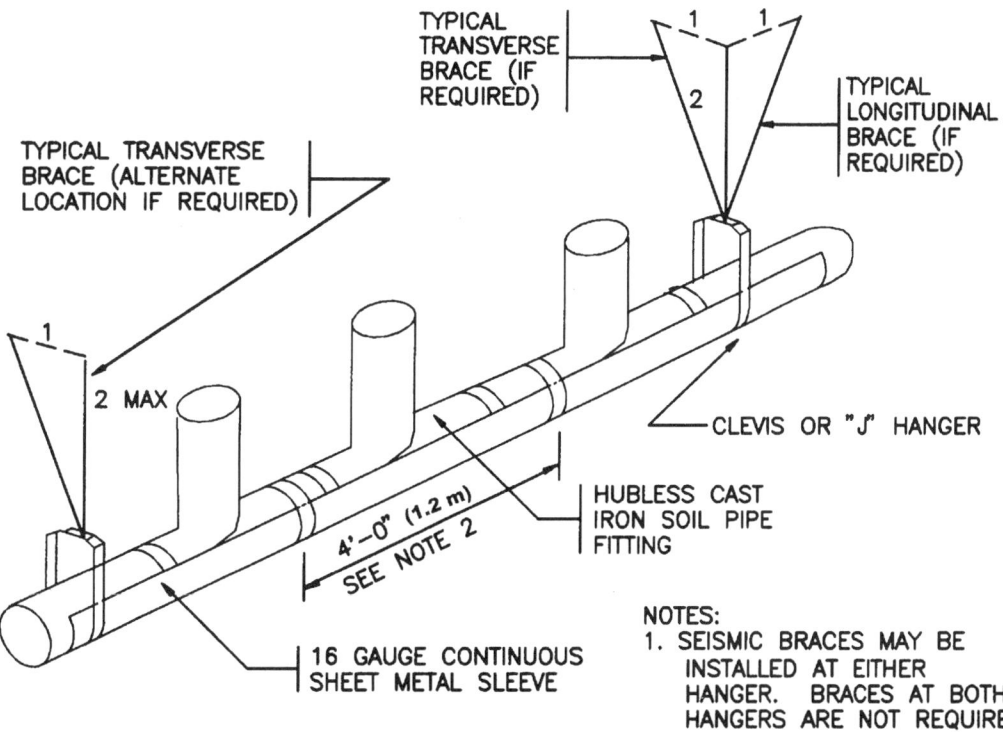

TYPICAL TRANSVERSE BRACE (IF REQUIRED)

TYPICAL LONGITUDINAL BRACE (IF REQUIRED)

TYPICAL TRANSVERSE BRACE (ALTERNATE LOCATION IF REQUIRED)

1 / 2 MAX

CLEVIS OR "J" HANGER

HUBLESS CAST IRON SOIL PIPE FITTING

4'-0" (1.2 m) SEE NOTE 2

16 GAUGE CONTINUOUS SHEET METAL SLEEVE

MAXIMUM HANGER SPACING 10'-0" (3 m)

NOTES:
1. SEISMIC BRACES MAY BE INSTALLED AT EITHER HANGER. BRACES AT BOTH HANGERS ARE NOT REQUIRED.
2. STRAP SLEEVE TO PIPE AT 4'-0" (1.2 m) CENTERS.

**(K)**

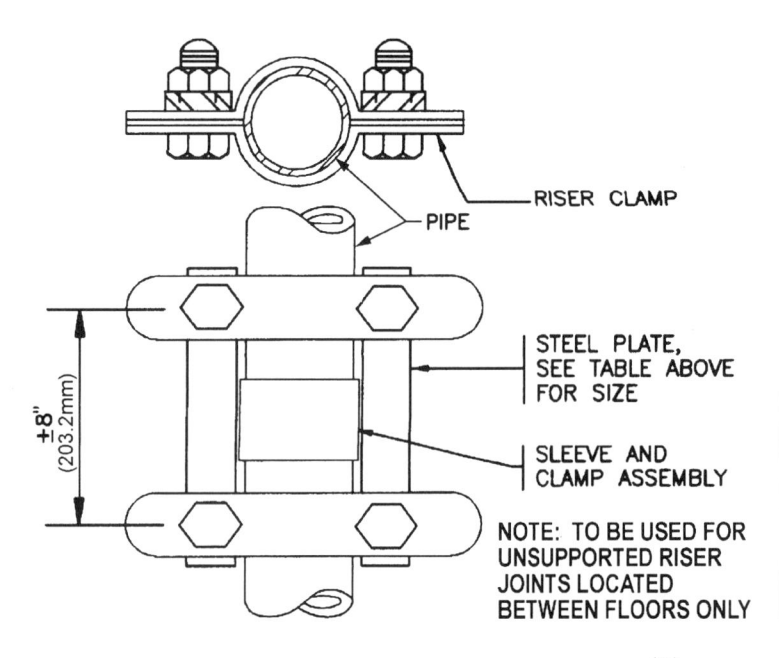

| PIPE SIZE | PLATE SIZE | BOLT SIZE |
|---|---|---|
| Up To 2" (50.8mm)" | ¼" x 1¼" (6.4 x 31.8mm) | ¼" (6.4mm) |
| 2½" to 3" (63.5 - 76.2mm) | ¼" x 1¼" (6.4 x 31.8mm) | ⅜" (9.5mm) |
| 4" and 5" (101.6 & 127mm) | ¼" x 1¼" (6.4 x 31.8mm) | ½" (12.7mm) |
| 6" (152.4mm) | ⅜" x 1½" (9.5 x 38.1mm) | ½" (12.7mm) |
| 8" (203.2mm) | ⅜" x 1½" (9.5 x 38.1mm) | ⅝" (15.9mm) |

(L)

(M)

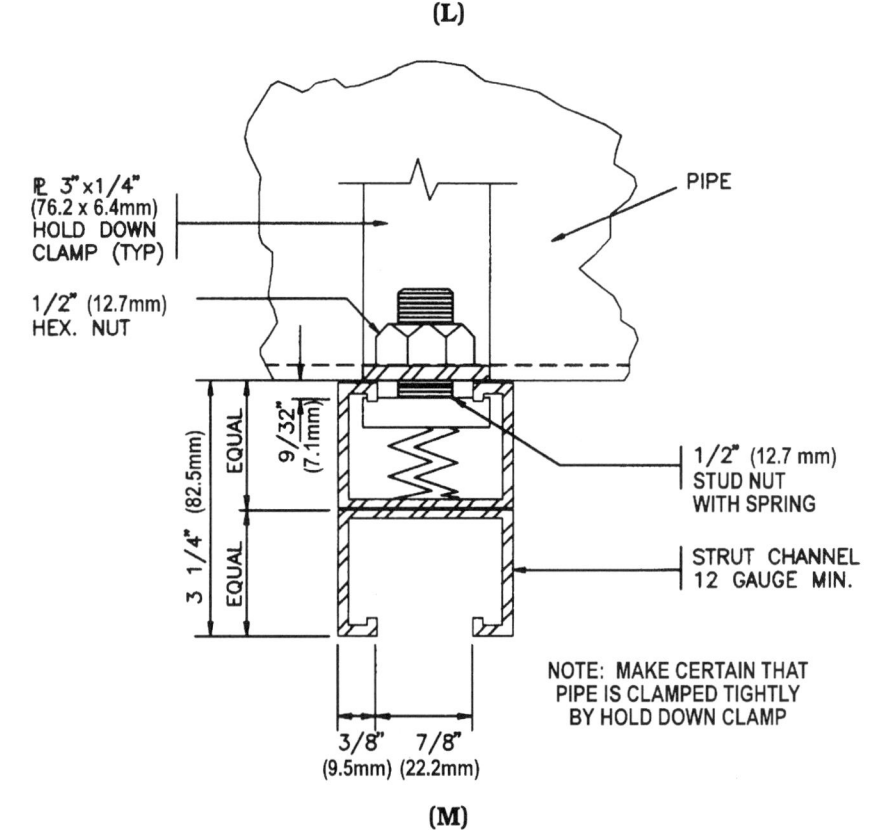

**Figure 9-12 Construction Details of Seismic Protection for Pipes:**
**(A)** Transverse Bracing for Pipes; **(B)** Longitudinal Bracing for Pipes; **(C)** Strut Bracing for Pipes; **(D)** Alternate Attachment to Hanger for Pipe Bracing; **(E)** Alternate Bracing for Pipes; **(F)** Strut Bracing for Pipe Trapeze; **(G)** Connections to Steel Beams; **(H)** Connections to Open-Web Steel Joists; **(I)** Connections to Steel; **(J)** Hanger Rod Connections; **(K)** Hubless Cast-Iron Pipe; **(L)** Riser Bracing for Hubless Pipes; **(M)** Connections for Pipes on Trapeze.
*Source:* SMACNA 1991. *Note:* For additional information, refer to SMACNA 1991.

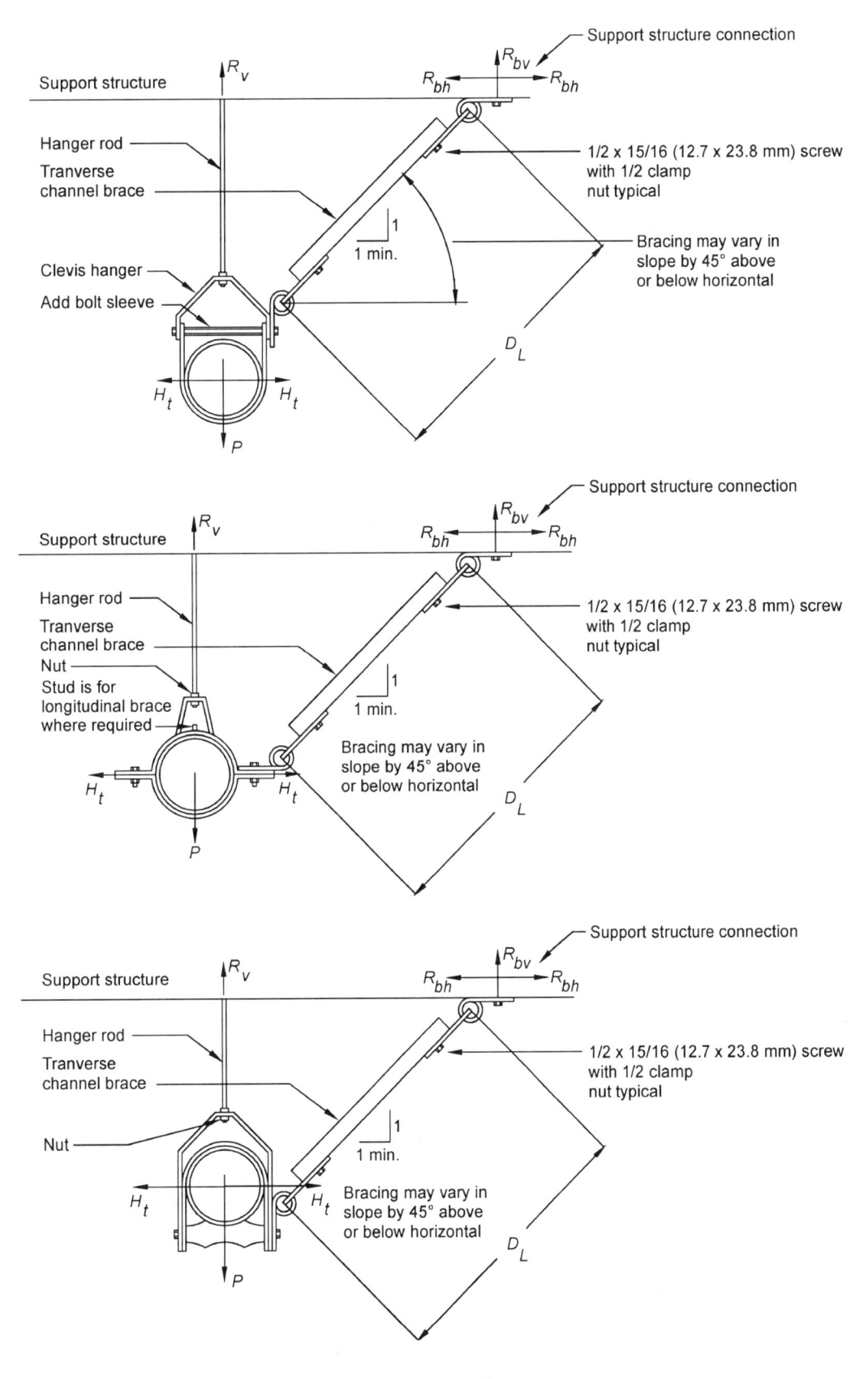

# Chapter 9 — Seismic Protection of Plumbing Equipment 181

*Transverse Only*

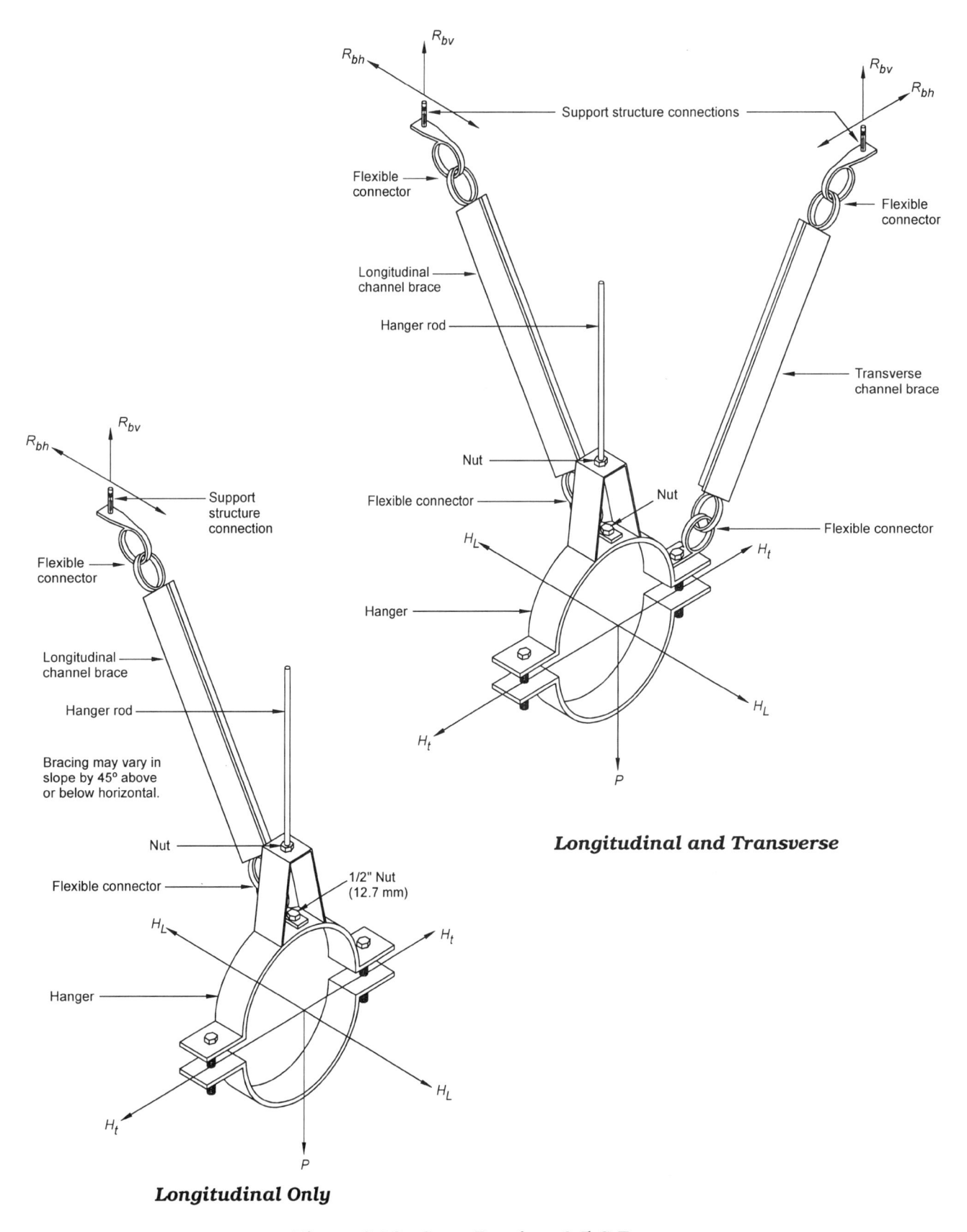

**Figure 9-13 Sway Bracing, 0.5 G Force**

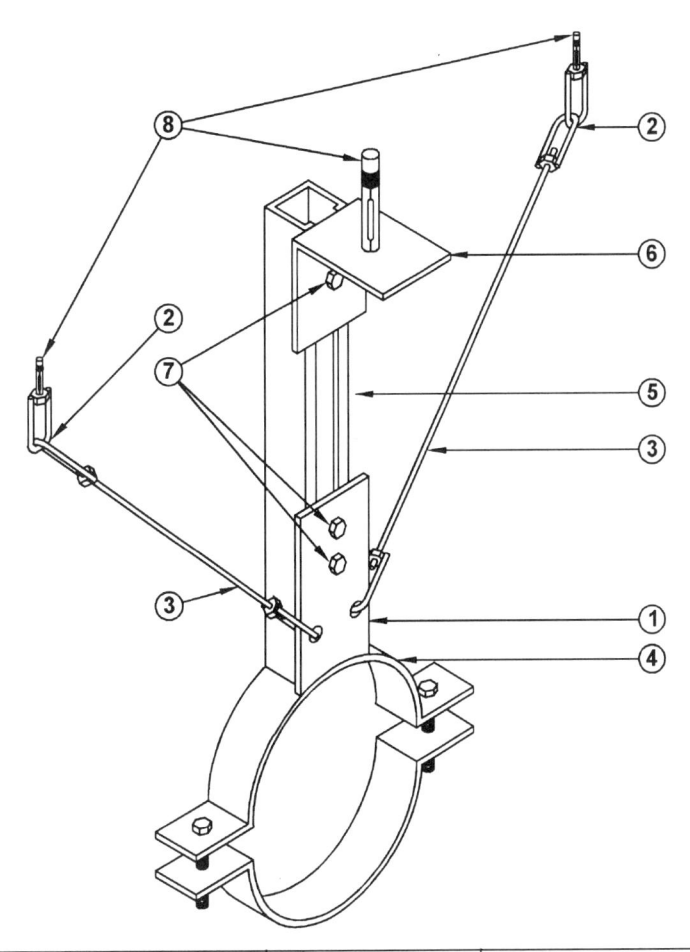

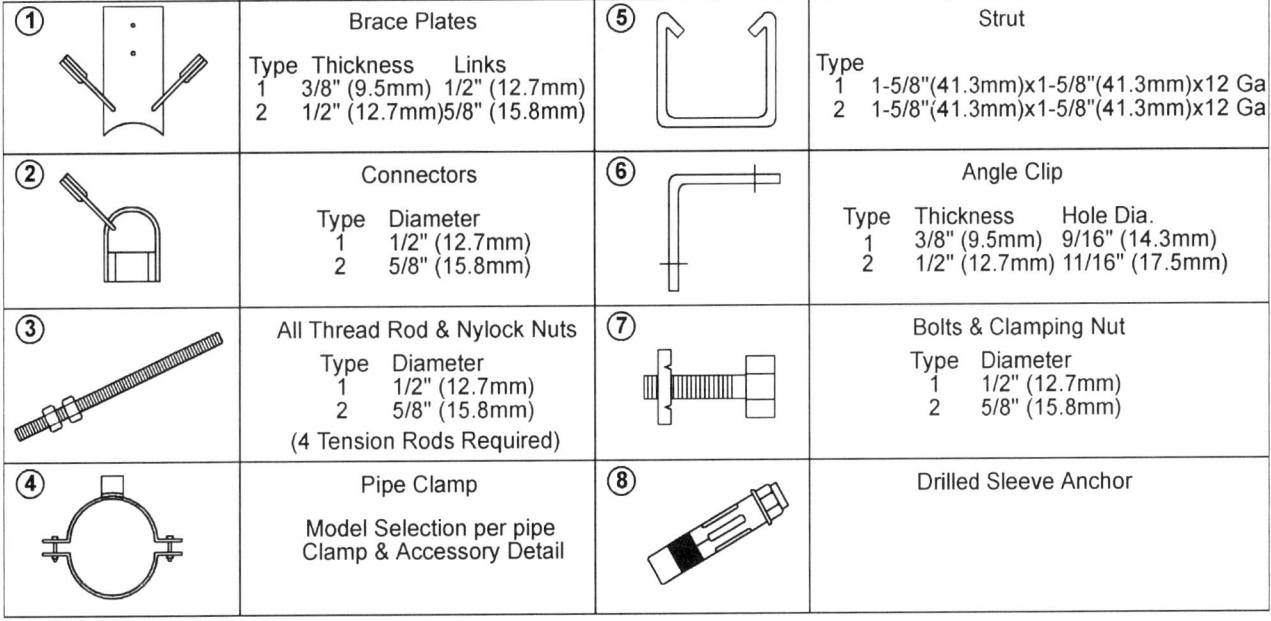

# Chapter 9 — Seismic Protection of Plumbing Equipment

| | | | | |
|---|---|---|---|---|
| ① | Brace Plates<br><br>Type  Thickness     Links<br>1      3/8" (9.5mm)   1/2" (12.7mm)<br>2      1/2" (12.7mm) 5/8" (15.8mm) | ⑤ | Strut<br><br>Type<br>1   1-5/8"(41.3mm)x1-5/8"(41.3mm)x12 Ga<br>2   1-5/8"(41.3mm)x1-5/8"(41.3mm)x12 Ga |
| ② | Connectors<br><br>Type  Diameter<br>1      1/2" (12.7mm)<br>2      5/8" (15.8mm) | ⑥ | Angle Clip<br><br>Type  Thickness       Hole Dia.<br>1      3/8" (9.5mm)    9/16" (14.3mm)<br>2      1/2" (12.7mm)   11/16" (17.5mm) |
| ③ | All Thread Rod & Nylock Nuts<br><br>Type  Diameter<br>1      1/2" (12.7mm)<br>2      5/8" (15.8mm)<br>(4 Tension Rods Required) | ⑦ | Bolts & Clamping Nut<br><br>Type  Diameter<br>1      1/2" (12.7mm)<br>2      5/8" (15.8mm) |
| ④ | Pipe Clamp<br><br>Model Selection per pipe<br>Clamp & Accessory Detail | ⑧ | Drilled Sleeve Anchor |

Figure 14a    Sway bracing - Lateral only

**(A)**

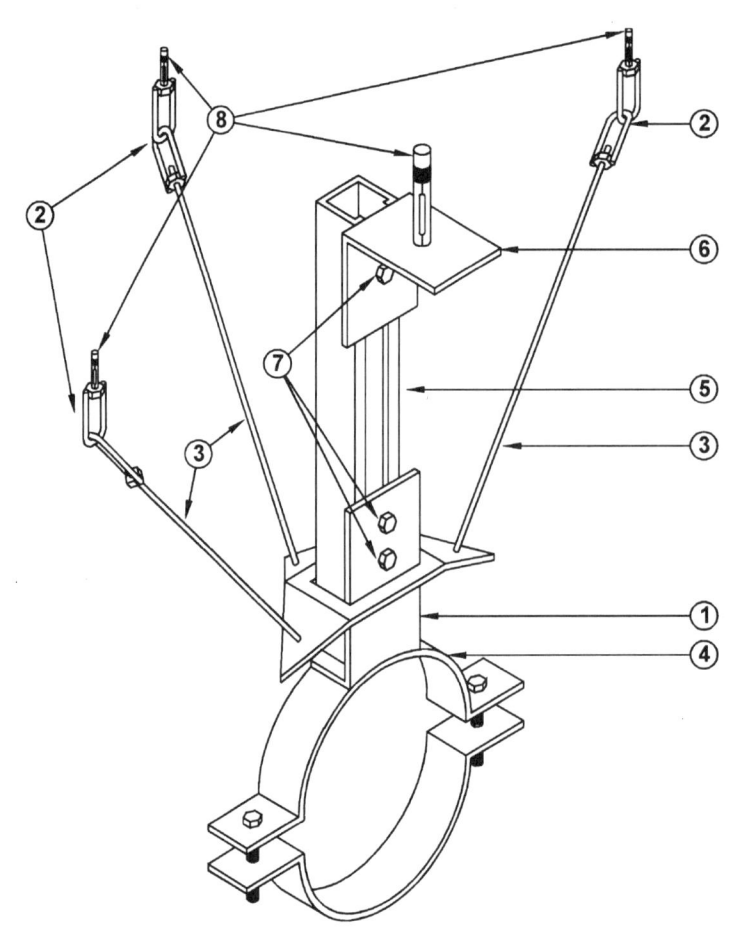

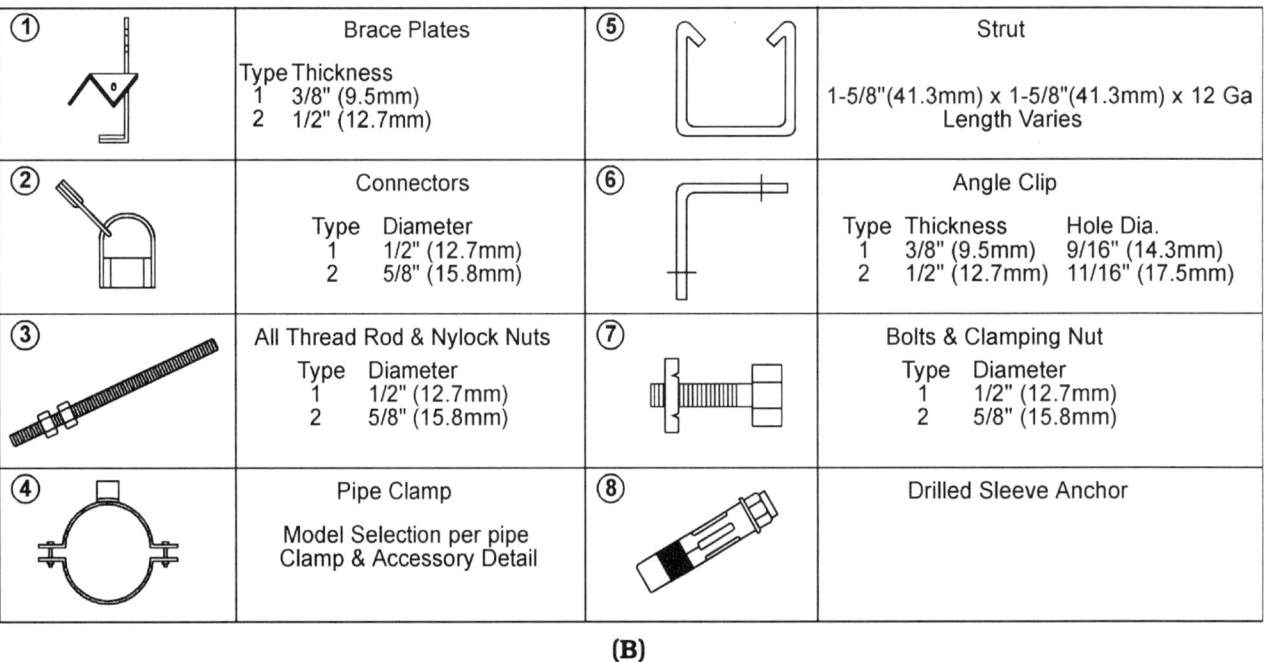

| | | | |
|---|---|---|---|
| ① | Brace Plates<br><br>Type Thickness<br>1    3/8" (9.5mm)<br>2    1/2" (12.7mm) | ⑤ | Strut<br><br>1-5/8"(41.3mm) x 1-5/8"(41.3mm) x 12 Ga<br>Length Varies |
| ② | Connectors<br><br>Type    Diameter<br>1    1/2" (12.7mm)<br>2    5/8" (15.8mm) | ⑥ | Angle Clip<br><br>Type  Thickness       Hole Dia.<br>1    3/8" (9.5mm)     9/16" (14.3mm)<br>2    1/2" (12.7mm)   11/16" (17.5mm) |
| ③ | All Thread Rod & Nylock Nuts<br><br>Type    Diameter<br>1    1/2" (12.7mm)<br>2    5/8" (15.8mm) | ⑦ | Bolts & Clamping Nut<br><br>Type    Diameter<br>1    1/2" (12.7mm)<br>2    5/8" (15.8mm) |
| ④ | Pipe Clamp<br><br>Model Selection per pipe<br>Clamp & Accessory Detail | ⑧ | Drilled Sleeve Anchor |

(B)

**Figure 9-14   A Seismic Bracing Method:**
**(A) Lateral Sway Bracing; (B) Lateral and Longitudinal Sway Bracing.**

# Chapter 9 — Seismic Protection of Plumbing Equipment

## CODES

### Design Philosophy

The process of the seismic design for buildings has had a reasonably long time to mature. Beginning in the 1920s, after engineers observed heavy building damage from earthquakes, they began to consider lateral forces on buildings in this country and Japan. Today's procedures are based on analytical results as well as considerable design experience and observed performance in earthquakes of varying characteristics. Lateral forces for buildings specified in most codes are much lower than could be calculated from structural dynamics for a variety of reasons, including:

1. Observed acceptable performance at low design levels.
2. Expected ductile action of building systems (ability to continue to withstand force and distort after yielding). Redundancy of resisting elements in most systems.
3. High damping as distortions increase, which creates a self-limiting characteristic on response.
4. Less-than-perfect compliance of the foundation to the ground motion.
5. Economic restraints on building codes.

The fact that the actual response of a building during an earthquake could be 3 or 4 times that represented by code forces must be understood and considered in good seismic design. Traditionally, this is done by rule of thumb and good judgment to ensure that structural yielding is not sudden or does not produce a collapsed mechanism. More recently, the response of many buildings to real earthquake input is being considered more specifically using computer analysis.

Design of seismic protection for nonstructural elements, including plumbing components and equipment, has neither the tradition nor a large number of in-place tests by actual earthquakes to enable much refinement of design force capability or design technique. Unfortunately, few of the effects listed above that mitigate the low force level for structures apply to plumbing or piping. Equipment and piping systems are generally simple and have low damping, and their lateral force resisting systems are usually nonredundant. It is imperative, therefore, when designing seismic protection for these elements, to recognize whether force levels being utilized are arbitrarily low for "design" or realistic predictions of actual response. Even when predictions of actual response are used, earthquake forces are considered sufficiently unpredictable that friction is not allowed as a means of "anchorage," and often less than full dead load is used to both simulate vertical accelerations and to provide a further safety factor against overturning or swinging action.

### Code Requirements

All current building codes require most structures and portions of structures to be designed for a horizontal force based on a certain percentage of its weight. Each code may vary in the method of determining this percentage, based on factors including the seismic zone, the importance of the structure, and the type of construction.

It is difficult to consider specific code requirements out of context. The code documents themselves should be consulted for specific usage. Most codes currently in use, or being developed, can generally be discussed by considering these four:

1. *Uniform building code 1988* (UBC). (See International Conference of Building Officials 1982.)
2. *California administrative code of regulations*, Chap. 2, Div. 122, Part 6 of Title 24 (Title 24, CAL). (See California, State of, 1988.)
3. Seismic design for buildings. *Tri-Services Manual.* (See US Department of Defense 1973.)
4. *Tentative provisions for the development of seismic regulations for buildings* (ATC-3). (See Applied Technology Council 1978.).

All of these codes require consideration of a lateral force that must be placed at the center of gravity of the element. The lateral force, or "equivalent static force," is calculated using some or all of the following parameters:

1. *Zone* Similar to Figure 9-2, the zone category affects the lateral force calculated by considering the size and frequency of potential earthquakes in the region.
2. *Soils* The effect of specific site soils on ground motion.
3. *Force factor* This considers the basic response of the element to ground motion and

is affected by subparameters, which could include location within the building and possible resonance with the structure.
4. *Importance* A measure of the desirability of protection for a specific element.
5. *Element weight* All codes require calculation of a lateral force that is a percentage of the element weight.

It is of critical importance that the various building codes and their requirements be obtained and adhered to.

**Sprinkler systems: NFPA 13** Because of the potential for fire immediately after earthquakes, sprinkler piping has long received special attention. The reference standard for installation of sprinkler piping, NFPA 13 (National Fire Protection Association 1996), is often cited as containing prototype seismic bracing for piping systems. In fact, in those cases observed, sprinkler piping has performed well. The bracing guidelines followed for some time in seismically active areas are actually contained in Appendix A of NFPA 13. However, good earthquake performance by sprinkler piping is also due to other factors, such as limited pipe size, use of steel pipe, coherent layouts, and conservative suspension (for vertical loads).

Use of NFPA 13 guidelines for pipe bracing is not discouraged, but it should not be considered a panacea for all piping systems. Other organizations, such as Factory Mutual (FM), have developed guidelines for properties insured by them and in many cases are more restrictive.

For reference, the following three tables provide good information for the engineer. Table 9-1 provides weights of steel pipes filled with water for determining horizontal loads. Table 9-2 provides load information for the spacing of sway bracing, and Table 9-3 provides maximum horizontal loads for sway bracing.

## ANALYSIS TECHNIQUES

### Determination of Seismic Forces

As discussed in the previous section, the most common method of defining seismic forces is by use of code static equivalents of dynamic earthquake forces. Regardless of the parameters used, this procedure reduces to the following formula:

***Equation 9-1***

$$F_p = K_g W_p$$

*where:*

$F_p$ = Lateral (seismic) force applied at element center of gravity

$K_g$ = Coefficient considering the parameters discussed above, under "Codes." The final percentage of the element weight is

**Table 9-1 Piping Weights for Determining Horizontal Load**

| Schedule 40 Pipe, in. (mm) | Weight of Water-Filled Pipe, lb/ft (kg/m) | ½ Weight of Water-Filled Pipe, lb/ft (kg/m) |
|---|---|---|
| 1 (25.4) | 2.05 (0.28) | 1.03 (0.14) |
| 1¼ (31.8) | 2.93 (0.40) | 1.47 (0.20) |
| 1½ (38.1) | 3.61 (0.50) | 1.81 (0.25) |
| 2 (50.8) | 5.13 (0.70) | 2.57 (0.35) |
| 2½ (63.5) | 7.89 (1.08) | 3.95 (0.54) |
| 3 (76.2) | 10.82 (1.48) | 5.41 (0.74) |
| 3½ (88.9) | 13.48 (1.85) | 6.74 (0.92) |
| 4 (101.6) | 16.40 (2.25) | 8.20 (1.12) |
| 5 (127) | 23.47 (3.22) | 11.74 (1.61) |
| 6 (152.4) | 31.69 (4.35) | 15.85 (2.17) |
| 8[a] (203.2) | 47.70 (6.54) | 23.85 (3.27) |

| Schedule 10 Pipe, in. (mm) | | |
|---|---|---|
| 1 (25.4) | 1.81 (0.25) | 0.91 (0.12) |
| 1¼ (31.8) | 2.52 (0.35) | 1.26 (0.17) |
| 1½ (38.1) | 3.04 (0.42) | 1.52 (0.21) |
| 2 (50.8) | 4.22 (0.58) | 2.11 (0.29) |
| 2½ (63.5) | 5.89 (0.81) | 2.95 (0.40) |
| 3 (76.2) | 7.94 (1.09) | 3.97 (0.54) |
| 3½ (88.9) | 9.78 (1.34) | 4.89 (0.67) |
| 4 (101.6) | 11.78 (1.62) | 5.89 (0.81) |
| 5 (127) | 17.30 (2.37) | 8.65 (1.19) |
| 6 (152.4) | 23.03 (3.16) | 11.52 (1.58) |
| 8 (203.2) | 40.08 (5.50) | 20.04 (2.75) |

[a]Schedule 30.

# Chapter 9 — Seismic Protection of Plumbing Equipment

### Table 9-2  Assigned Load Table for Lateral and Longitudinal Sway Bracing

| Spacing of Lateral Braces, ft (m) | Spacing of Longitudinal Braces, ft (m) | 2 | 2½ | 3 | 4 | 5 | 6 | 8 |
|---|---|---|---|---|---|---|---|---|
| | | \multicolumn{7}{c}{Assigned Load for Pipe Size to Be Braced, lb (kg)} |
| 10 (3.0) | 20 (6.0) | 380 (171) | 395 (177.8) | 410 (184.5) | 435 (195.8) | 470 (211.5) | 655 (294.8) | 915 (411.8) |
| 20 (6.0) | 40 (12.2) | 760 (342) | 785 (353.3) | 815 (366.8) | 870 (391.5) | 940 (423) | 1,305 (587.3) | 1,830 (823.5) |
| 25 (7.6) | 50 (15.2) | 950 (427.5) | 980 (441) | 1,020 (459) | 1,090 (490.5) | 1,175 (528.8) | 1,630 (733.5) | 2,290 (1030.5) |
| 30 (9.1) | 60 (18.3) | 1,140 (513) | 1,180 (531) | 1,225 (551.3) | 1,305 (587.3) | 1,410 (634.5) | 1,960 (882) | 2,745 (1235.3) |
| 40 (12.2) | 80 (24.4) | 1,515 (681.8) | 1,570 (706.5) | 1,630 (733.5) | 1,740 (783) | 1,880 (846) | 2,610 (1174.5) | 3,660 (1647) |
| 50 (15.2) | | 1,895 (852.8) | 1,965 (884.3) | 2,035 (915.8) | 2,175 (978.8) | 2,350 (1057.5) | 3,260 (1467) | 4575 (2058.8) |

*Note*: Table based on half the weight of a water-filled pipe.

### Table 9-3  Maximum Horizontal Loads for Sway Bracing

| Shape and Size, in. (mm) | Least Radius of Gyration | Maximum Length for 1/r = 200 | 30-44° Angle from Vertical | 45-59° Angle from Vertical | 60-90° Angle from Vertical |
|---|---|---|---|---|---|
| | | | \multicolumn{3}{c}{Maximum Horizontal Load, lb (kg)} |
| **Pipe (Schedule 40)** | $=\dfrac{\sqrt{r_o^2 + r_i^2}}{2}$ | | | | |
| 1    (25.4) | 0.42  | 7 ft 0 in. (2.1 m)  | 1,767 (801.5)  | 2,500 (1134.0) | 3,061 (1 388.4) |
| 1¼  (31.8) | 0.54  | 9 ft 0 in. (2.7 m)  | 2,393 (1085.4) | 3,385 (1535.4) | 4,145 (1 880.1) |
| 1½  (38.1) | 0.623 | 10 ft 4 in. (3.1 m) | 2,858 (1296.4) | 4,043 (1833.9) | 4,955 (2 241.5) |
| 2    (50.8) | 0.787 | 13 ft 1 in. (4 m)   | 3,828 (1736.3) | 5,414 (2455.7) | 6,630 (3 007.3) |
| **Pipe (Schedule 10)** | $=\dfrac{\sqrt{r_o^2 + r_i^2}}{2}$ | | | | |
| 1    (25.4) | 0.43  | 7 ft 2 in. (2.2 m)  | 1,477 (670.0)  | 2,090 (948.0)  | 2,559 (1 160.7) |
| 1¼  (31.8) | 0.55  | 9 ft 2 in. (2.8 m)  | 1,900 (861.8)  | 2,687 (1218.8) | 3,291 (1 492.8) |
| 1½  (38.1) | 0.634 | 10 ft 7 in. (3.2 m) | 2,194 (995.2)  | 3,103 (1407.5) | 3,800 (1 723.6) |
| 2    (50.8) | 0.802 | 13 ft 4 in. (4.1 m) | 2,771 (1256.9) | 3,926 (1780.8) | 4,803 (2 178.6) |
| **Angles** | | | | | |
| 1½ x 1½ x ¼ (38.1 x 38.1 x 6.4) | 0.292 | 4 ft 10 in. (1.5 m) | 2,461 (1116.3) | 3,481 (1578.9) | 4,263 (1 933.7) |
| 2 x 2 x ¼   (50.8 x 50.8 x 6.4) | 0.391 | 6 ft 6 in.  (2 m)   | 3,356 (1522.2) | 4,746 (2152.7) | 5,813 (2 636.7) |
| 2½ x 2 x ¼  (63.5 x 50.8 x 6.4) | 0.424 | 7 ft 0 in.  (2.1 m) | 3,792 (1720.0) | 5,363 (2432.6) | 6,569 (2 979.6) |
| 2½ x 2½ x ¼ (63.5 x 63.5 x 6.4) | 0.491 | 8 ft 2 in.  (2.5 m) | 4,257 (1930.9) | 6,021 (2731.1) | 7,374 (3 344.8) |
| 3 x 2½ x ¼  (76.2 x 63.5 x 6.4) | 0.528 | 8 ft 10 in. (2.7 m) | 4,687 (2126.0) | 6,628 (3006.4) | 8,118 (3 682.2) |
| 3 x 3 x ¼   (76.2 x 76.2 x 6.4) | 0.592 | 9 ft 10 in. (3 m)   | 5,152 (2336.9) | 7,286 (3304.9) | 8,923 (4 047.4) |
| **Rods** | $=\dfrac{r}{2}$ | | | | |
| ⅜  (9.5)  | 0.094 | 1 ft 6 in. (0.5 m) | 395   (179.2) | 559   (253.6)  | 685   (310.7)   |
| ½  (12.7) | 0.125 | 2 ft 6 in. (0.8 m) | 702   (318.4) | 993   (450.4)  | 1,217 (552.0)   |
| ⅝  (15.9) | 0.156 | 2 ft 7 in. (0.8 m) | 1,087 (493.1) | 1,537 (697.2)  | 1,883 (854.1)   |
| ¾  (19.1) | 0.188 | 3 ft 1 in. (0.9 m) | 1,580 (716.7) | 2,235 (1013.8) | 2,737 (1 241.5) |
| ⅞  (22.2) | 0.219 | 3 ft 7 in. (1.1 m) | 2,151 (975.7) | 3,043 (1380.3) | 3,726 (1 690.1) |

*(Continued)*

*(Table 41-3 continued)*

| Shape and Size, in. (mm) | | Least Radius of Gyration | Maximum Length for 1/r = 200 | | 30-44° Angle from Vertical | | 45-59° Angle from Vertical | | 60-90° Angle from Vertical | |
|---|---|---|---|---|---|---|---|---|---|---|
| **Flats** | | = 0.29 h (where h is smaller of two side dimensions) | | | | | | | | |
| 1½ x ¼ | (38.1 x 6.4) | 0.0725 | 1 ft 2 in. | (0.4 m) | 1,118 | (507.1) | 1,581 | (717.1) | 1,936 | (878.2) |
| 2 x ¼ | (50.8 x 6.4) | 0.0725 | 1 ft 2 in. | (0.4 m) | 1,789 | (811.5) | 2,530 | (1147.6) | 3,098 | (1405.2) |
| 2 x ⅜ | (50.8 x 9.5) | 0.109 | 1 ft 9 in. | (0.5 m) | 2,683 | (1217.0) | 3,795 | (1721.4) | 4,648 | (2108.3) |
| **Pipe (Schedule 40)** | | $= \dfrac{\sqrt{r_o^2 + r_i^2}}{2}$ | | | | | | | | |
| 1 | (25.4) | 0.42 | 3 ft 6 in. | (1.1 m) | 7,068 | (3206.0) | 9,996 | (4534.1) | 12,242 | (5 552.8) |
| 1¼ | (31.8) | 0.54 | 4 ft 6 in. | (1.4 m) | 9,567 | (4339.5) | 13,530 | (6137.1) | 16,570 | (7 516.0) |
| 1½ | (38.1) | 0.623 | 5 ft 2 in. | (1.6 m) | 11,441 | (5189.5) | 16,181 | (7339.5) | 19,817 | (8 988.8) |
| 2 | (50.8) | 0.787 | 6 ft 6 in. | (2 m) | 15,377 | (6974.9) | 21,746 | (9863.8) | 26,634 | (12 080.9) |
| **Pipe (Schedule 10)** | | $= \dfrac{\sqrt{r_o^2 + r_i^2}}{2}$ | | | | | | | | |
| 1 | (25.4) | 0.43 | 3 ft 7 in. | (1.1 m) | 5,910 | (2680.7) | 8,359 | (3791.6) | 10,237 | (4 643.4) |
| 1¼ | (31.8) | 0.55 | 4 ft 7 in. | (1.4 m) | 7,600 | (3447.3) | 10,749 | (4875.6) | 13,164 | (5 971.1) |
| 1½ | (38.1) | 0.634 | 5 ft 3 in. | (1.6 m) | 8,777 | (3981.2) | 12,412 | (5630.0) | 15,202 | (6 895.5) |
| 2 | (50.8) | 0.802 | 6 ft 8 in. | (2 m) | 11,105 | (5037.1) | 15,705 | (7123.6) | 19,235 | (8 724.8) |
| **Rods** | | $= \dfrac{r}{2}$ | | | | | | | | |
| ⅜ | (9.5) | 0.094 | 0 ft 9 in. | (0.2 m) | 1,580 | (716.7) | 2,234 | (1013.3) | 2,737 | (1 241.5) |
| ½ | (12.7) | 0.125 | 1 ft 0 in. | (0.3 m) | 2,809 | (1274.1) | 3,972 | (1801.7) | 4,865 | (2 206.7) |
| ⅝ | (15.9) | 0.156 | 1 ft 3 in. | (0.4 m) | 4,390 | (1991.3) | 6,209 | (2816.3) | 7,605 | (3 449.6) |
| ¾ | (19.1) | 0.188 | 1 ft 6 in. | (0.5 m) | 6,322 | (2867.6) | 8,941 | (4055.5) | 10,951 | (4 967.3) |
| ⅞ | (22.2) | 0.219 | 1 ft 9 in. | (0.5 m) | 8,675 | (3934.9) | 12,169 | (5519.7) | 14,904 | (6 760.3) |
| **Pipe (Schedule 40)** | | $= \dfrac{\sqrt{r_o^2 + r_i^2}}{2}$ | 1/r = 300 | | | | | | | |
| 1 | (25.4) | 0.42 | 10 ft 6 in. | (3.2 m) | 786 | (356.5) | 1111 | (503.9) | 1,360 | (616.9) |
| 1¼ | (31.8) | 0.54 | 13 ft 6 in. | (4.1 m) | 1,063 | (482.2) | 1,503 | (681.7) | 1,841 | (835.1) |
| 1½ | (38.1) | 0.623 | 15 ft 7 in. | (4.7 m) | 1,272 | (577.0) | 1,798 | (815.5) | 2,202 | (998.8) |
| 2 | (50.8) | 0.787 | 19 ft 8 in. | (6 m) | 1,666 | (755.7) | 2,355 | (1068.2) | 2,885 | (1 308.6) |
| **Pipe (Schedule 10)** | | $= \dfrac{\sqrt{r_o^2 + r_i^2}}{2}$ | | | | | | | | |
| 1 | (25.4) | 0.43 | 10 ft 9 in. | (3.3 m) | 656 | (297.8) | 928 | (420.9) | 1,137 | (515.7) |
| 1¼ | (31.8) | 0.55 | 13 ft 9 in. | (4.2 m) | 844 | (383.2) | 1,194 | (541.6) | 1,463 | (663.6) |
| 1½ | (38.1) | 0.634 | 15 ft 10 in. | (4.8 m) | 975 | (442.3) | 1,379 | (625.5) | 1,194 | (541.6) |
| 2 | (50.8) | 0.802 | 20 ft 0 in. | (6.1 m) | 1,234 | (559.7) | 1,745 | (791.5) | 2,137 | (969.3) |
| **Rods** | | $= \dfrac{r}{2}$ | | | | | | | | |
| ⅜ | (9.5) | 0.094 | 2 ft 4 in. | (0.7 m) | 176 | (79.8) | 248 | (112.5) | 304 | (137.9) |
| ½ | (12.7) | 0.125 | 3 ft 1 in. | (0.9 m) | 312 | (141.5) | 441 | (200.0) | 540 | (244.9) |
| ⅝ | (15.9) | 0.156 | 3 ft 11 in. | (1.2 m) | 488 | (221.4) | 690 | (313.0) | 845 | (383.3) |
| ¾ | (19.1) | 0.188 | 4 ft 8 in. | (1.4 m) | 702 | (318.4) | 993 | (450.4) | 1,217 | (552.0) |
| ⅞ | (22.2) | 0.219 | 5 ft 6 in. | (1.7 m) | 956 | (433.6) | 1,352 | (613.3) | 1,656 | (751.1) |

often described in units of g, the acceleration of gravity, e. g., "0.5 g." This is equivalent to specifying a percentage of the weight; thus 0.5 = 50% of W.

$W_p$ = Weight tributary to anchorage (pipe and contents weight)

Since $F_p$ is a representation of vibratory response, it can be applied in a plus or minus sense.

In piping systems, since vertical supports will probably be placed more frequently than lateral braces, $W_p$ will be greater than the dead load supported at that point. This mismatching of $F_p$ and available dead load often causes uplift on the pipe, which should be taken into consideration.

The loading ($F_p$) can also be calculated using a response spectrum determined for the appropriate floor or by modeling the equipment or piping as part of the structure and, by computer, inputting an appropriate time history of motion at the base. In practice, these techniques are seldom used except in buildings of extreme importance, or when the mass of the equipment becomes a significant percentage of the total building mass (10% is sometimes used as the limit).

Vertical seismic load, $F_{pv}$, of equipment or piping is normally considered by specifying a percentage of the horizontal force factor to be applied to the weight concurrently. In several codes the factor is taken as 30% $K_g$; therefore, $F_{pg}$ = 0.3 $K_g W$, where W is the tributary vertical load.

The three generalized loadings that must be considered in the design of seismic restraints, $F_p$, $F_{pv}$, and W, are shown schematically in Figure 9-15.

### Determination of Anchorage Forces

In most cases, anchorage or reaction forces, $R_h$ and $R_v$ [Figure 9-16(A)], created by the loading described above, are calculated by simple statistics. Although trivial for a professional familiar with statistics, calculations to find all maximums become numerous when the center of gravity is off one or both plan centerline axes or if the base support is nonsymmetrical.

In typical pipe braces [Figure 9-16(B)], it is important to note that R, the gravity force in the hanger rod, is significantly affected by the addition of the brace and is not equal to W; as indicated previously. A tension rod hanger commonly goes into compression in such a situation.

## COMPUTER ANALYSIS OF PIPING SYSTEMS

Computers programs have been used to analyze piping systems for stress for some time. These programs were initially developed to consider thermal stresses and anchor point load, but software is now commonly available that can consider seismic and settlement loading, spring or damping supports, snubbers (similar to equipment snubbers), differing materials, and nonrigid couplings. The seismic loading normally can be figured by using a full-time history, as a response spectrum, or equivalent static forces. The time history has the inherent problem of requiring a search of each time increment for worst-case stresses and brace loadings. The computer time and man-hours required are seldom justified. In fact, for seismic loading alone, computer analysis is almost never performed because brace loadings can easily be determined by tributary length methods, and rule-of-thumb pipe spans (brace spacing) are contained in several publications (see National Fire Protection Association 1996; Hillman, Biddison, and Loevenguth; and US Dept. of Defense 1973). Computer analysis may be appropriate, however, when it is necessary to combine seismic loading with several of the following considerations:

1. Temperature changes and anchorage.
2. Nonlinear support conditions (springs, snubbers, etc.).
3. Complex geometry.
4. Several loading conditions.
5. Piping materials other than steel or copper.
6. Joints or couplings that are significantly more flexible or weaker than the pipe itself.

Because of the variety of computer programs available and because many have proprietary restrictions, specific programs are not listed here. Piping analysis programs are available at most computer service bureaus, many universities, and national computer program clearinghouses.

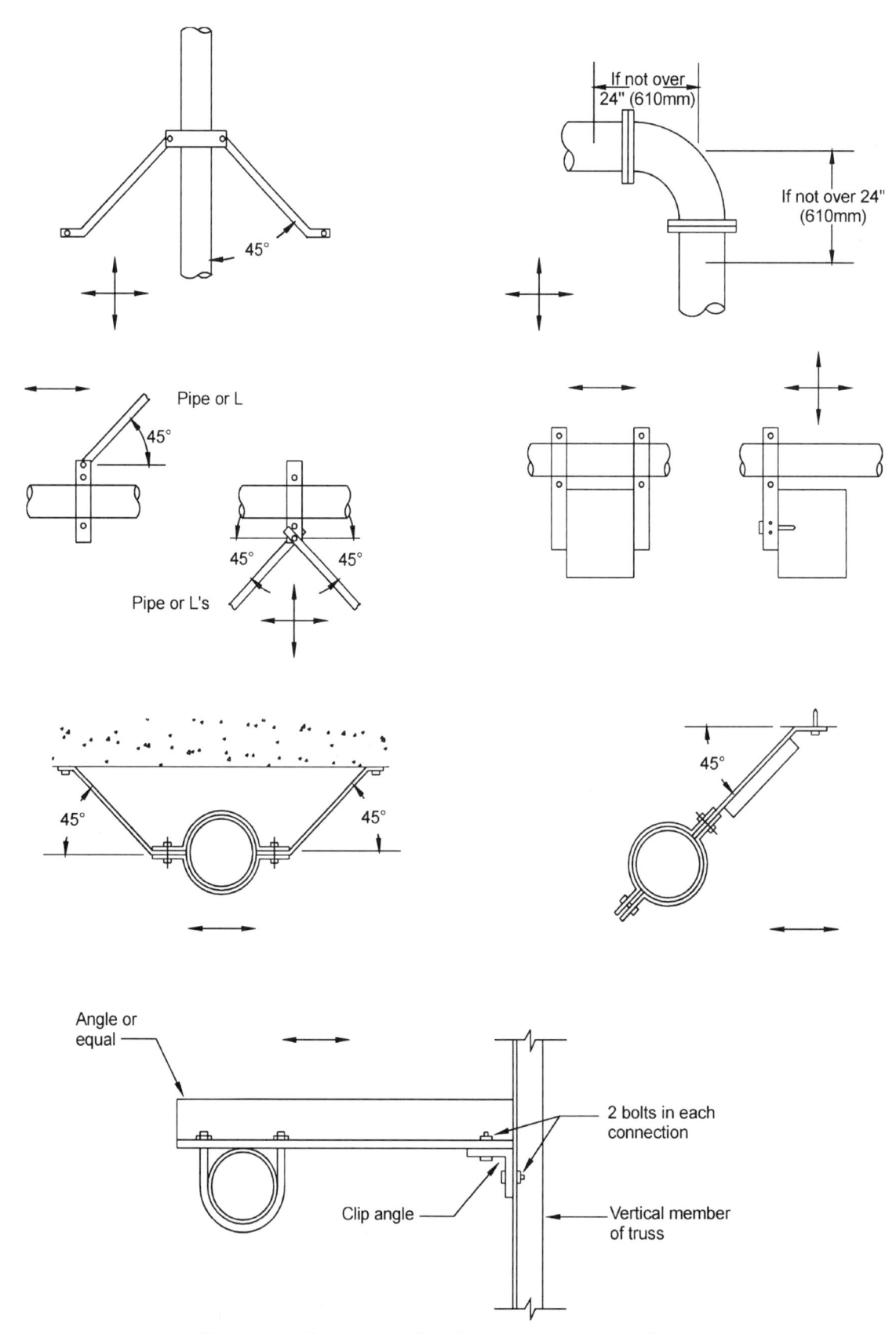

**Figure 9-15 Acceptable Types of Sway Bracing**

# Chapter 9 — Seismic Protection of Plumbing Equipment

## DESIGN CONSIDERATIONS

### Loads in Structures

It is always important to identify unusual equipment and piping loads during the first stages of project design to assure that the structural system being developed is adequate. Consideration of seismic effects makes this coordination even more important because seismic forces produce unusual reactions. During an earthquake, not only must horizontal forces be taken into the structure, but vertical load effects are intensified due to vertical accelerations and overturning movements. These reactions must be acceptable to the structure locally (at the point of connection) and globally (by the system as a whole).

If the structural system is properly designed for the appropriate weights of equipment and piping, seismic reactions will seldom cause problems to the overall system. However, local problems are not uncommon. Most floors are required by code to withstand a 2000-lb (908 kg) concentrated load, so this is a reasonable load to consider acceptable without special provisions. However, seismic reactions to structures can easily exceed this figure; for example:

1. A longitudinal brace carrying a tributary load of 80-ft (24.8-m) of 8-in. (203-mm) steel pipe filled with water will generate reactions of this magnitude.
2. Transverse or longitudinal braces on trapezes often have larger reactions.
3. A 4000-lb (1816-kg) tank on legs could also yield such a concentrated load. In addition, possible limitations on attachment methods due to structure type could reduce the effective maximum allowable concentration.

Roof structures have no code-specified concentrated load requirement and often are the source of problems, particularly concerning piping systems, because of the random nature of hanger-and-brace locations. Many roof-decking systems cannot accept concentrations greater than 50 lb (22.7 kg) without spreaders or strengthening beams. Such limitations should be considered both in the selection of a structural system and in the equipment and piping layout.

If equipment anchorage or pipe bracing is specified to be contractor supplied, attachment load limitations or other structural criteria should be given. Compliance with such criteria should be checked to assure that the structure is not being damaged or overloaded.

## POTENTIAL PROBLEMS

It would be impractical to cover the details of structural design for seismic anchorage and bracing in this chapter. The engineer can get design information and techniques from standard textbooks and design manuals or, preferably, obtain help from a professional experienced in seismic and/or structural design. Simple, typical details are seldom appropriate and all-encompassing, seismic-protection "systems" quickly become complex. Certain common situations that have the potential to create problems can be identified, however; these are shown schematically in Figure 9-17 and discussed below.

Condition 1 in Figure 9-17 occurs frequently in making attachments to concrete. Often an angle is used, as indicated. The seismic force, P, enters the connector eccentric to the reaction, R, by the distance e; this is equivalent to a concentric force plus the moment $P_e$. In order for the connector to perform as designed, this moment must be resisted by the connection of the angle either to the machine or to the concrete. To use the machine to provide this moment, the machine base must be adequate and the connection from angle to base must be greatly increased over that required merely for P. Taking this moment into the concrete significantly increases the tension in the anchorage, R, which is known as "prying action." The appropriate solution must be decided on a case-by-case basis, but eccentricities in connection should not be ignored.

Legs 18 in. (457 mm) or longer supporting tanks or machines clearly create a sideways problem and are commonly cross-braced. However, shorter legs or even rails often have no strength or stiffness in their weak direction, as shown in Condition 2, and should also be restrained against base failure.

Conditions 3 and 4 point out that spring isolators typically create a significant height, h, through which lateral forces must be transmitted. This height, in turn, creates conditions similar to the problems shown in 1 or 2 and must be treated in the same manner.

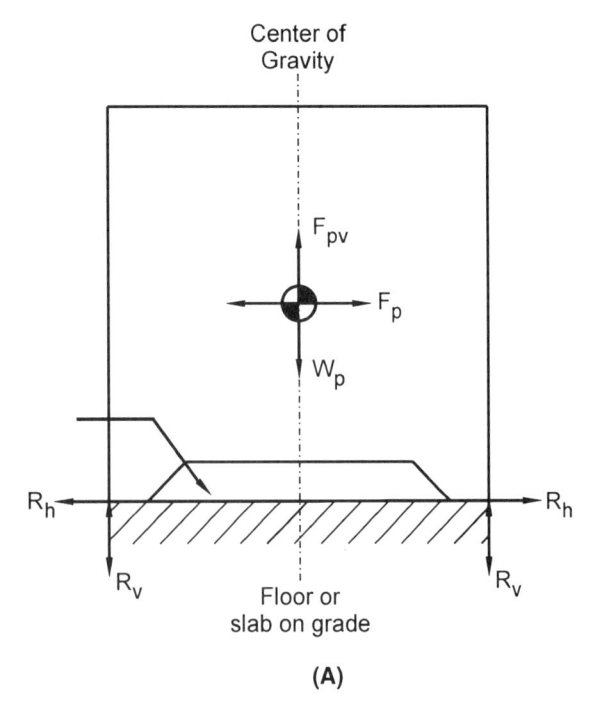

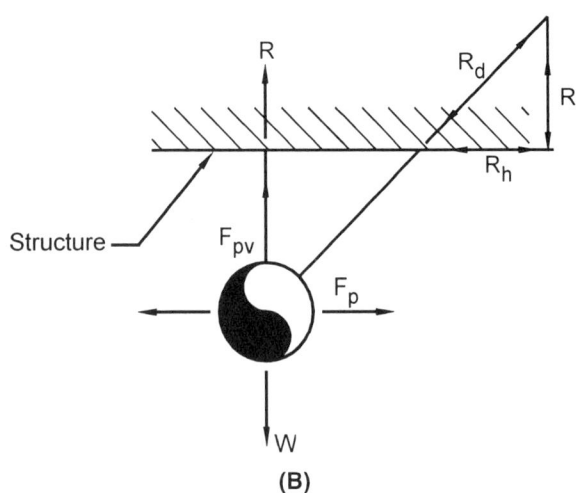

**Figure 9-16 Forces for Seismic Design: (A) Equipment; (B) Piping.**

cant stresses can be introduced into the beam if the load is large or the beam small. Considering the variability and potential overload characteristics of seismic forces, this condition should probably be avoided. Condition 7 also shows a connector in common use, which is probably acceptable in a nonseismic environment but which should be secured in place as shown under dynamic conditions.

Most pipe bracing systems utilize bracing members in pure tension or compression for stiffness and efficiency. This truss-type action is only possible when bracing configurations make up completed triangles, as shown on the right under condition 8. The brace configuration on the far left is technically unstable and the eccentric condition shown produces moment in the vertical support.

As previously indicated, "typical" details must be carefully designed and presented to prevent their misuse. Condition 9 shows the most common deficiency: a lack of limiting conditions.

Condition 10 shows a situation often seen in the field where interferences may prevent placement of longitudinal braces at the ends of a trapeze and either one is simply left out or two are replaced by one in the middle. Both of these "substitutions" can cause an undesirable twist of the trapeze and subsequent pipe damage. All field revisions to bracing schemes should be checked for adequacy.

Other potential problems that occur less frequently include incompatibility of piping systems with differential movement of the structure (drift) and inadvertent "self bracing" of piping through short, stiff service connections or branches that penetrate the structure. If the possibility of either is apparent, pipe stresses should be checked or the self-bracing restraint eliminated.

A few problems associated with making connection to a structure were discussed above, in relation to 9-17. When connecting to structural steel, in addition to manufactured clip devices, bolting and welding are also used. Holes for bolting should never be placed in structural steel without the approval of the structural engineer responsible for the design. Field welding should consider the effects of elevated temperatures on loaded structural members.

The preferred method of connecting to concrete is through embedments, but this is seldom

Condition 5 is meant to indicate that seldom can the bottom flange of a steel beam resist a horizontal force; diagonal braces, which are often connected to bottom flanges, create such a horizontal force. This condition can be rectified by attaching the diagonal brace near the top flange or adding a stabilizing element to the bottom flange.

Condition 6 depicts a typical beam connection device (beam clamp), which slips over one flange. Although this is often acceptable, signifi-

# Chapter 9 — Seismic Protection of Plumbing Equipment

practical. Since the location of required anchorages or braces is often not known when concrete is poured, the use of drilled-in or shot-in anchors is prevalent for this purpose. Although these anchors are extremely useful and practically necessary connecting devices, their adequacy has many sensitivities and they should be applied with thorough understanding and caution. The following items should be considered in the design or installation of drilled or shot-in anchors:

1. Manufacturers often list ultimate (failure) values in their literature. Normally, factors of safety of 4 or 5 are applied to these values for design.

2. Combined shear and tension should be considered in the design. A conservative approach commonly used is the following equation:

### Equation 9-2

$$\frac{\text{shear}}{\text{shear allowable}} + \frac{\text{tension}}{\text{tension allowable}} \leq 1$$

3. Edge distances are important because of the expansive nature of these anchors. Six (6) diameters are normally required.

4. Review the embedments required for design values. It is difficult to install an expansion bolt over ½-in. (12.7-mm) diameter in a typical floor system of 2½-in. (63.5-mm) concrete over steel decking.

5. Bolt sizes over ¼-in. (6.4-mm) diameter have embedments sufficient to penetrate the reinforcing envelope. Bolts should therefore not be placed in columns, the bottom flange of beams, or the bottom chord of joists. Bolts in slabs or walls are less critical, but the possibility of special and critical reinforcing bars being cut should always be considered. The critical nature of each strand of tendon in prestressed concrete as well as the stored energy generally dictates a complete prohibition of these anchors.

6. Installation technique has been shown to be extremely important in developing design strength. Field testing of a certain percentage of anchors should be considered.

## Additional Considerations

Seismic anchorage and bracing, like all construction, should be thoroughly reviewed in the field. Considering the lack of construction tradition, the likelihood of field changes or interferences, and other potential problems (discussed above), seismic work probably should be more clearly controlled, inspected, and/or tested than normal construction.

Another result of the relative newness of seismic protection of equipment and piping is the lack of performance data for the design and detailing techniques now being used. Considerable failure data were collected in Anchorage and San Fernando, but essentially no field data are available to assure that our present assumptions, although scientifically logical and accurate, will actually provide the desired protection. Will firm anchorage of equipment cause damage to the internal workings? Will the base cabinet, or framework (which is now seldom checked), of equipment be severely damaged by the anchorage? In contrast, the present requirements for structures are largely the result of observations of damage to structures in actual earthquakes over 75 years.

The net result of current standards in seismic protection can only be positive. The fine-tuning of scope, force levels, and detailing techniques must wait for additional, full-scale testing in real earthquakes.

## GLOSSARY

***Anchor*** A device, such as an expansion bolt, for connecting pipe-bracing members into the structure of a building.

***Attachment*** See "positive attachment."

***Bracing*** Metal channels, cables, or hanger angles that prevent pipes from breaking away from the structure during an earthquake. See also "longitudinal bracing" and "transverse bracing." Together, these resist lateral loads from any direction.

***Dynamic properties of piping*** The tendency of pipes to change in weight and size because of the movement and temperature of fluids in them. This does not refer to movement due to seismic forces.

***Essential facilities*** Buildings that must remain safe and usable for emergency purposes after an earthquake in order to preserve the health and safety of the general public. Examples include hospitals, emergency shelters, and fire stations.

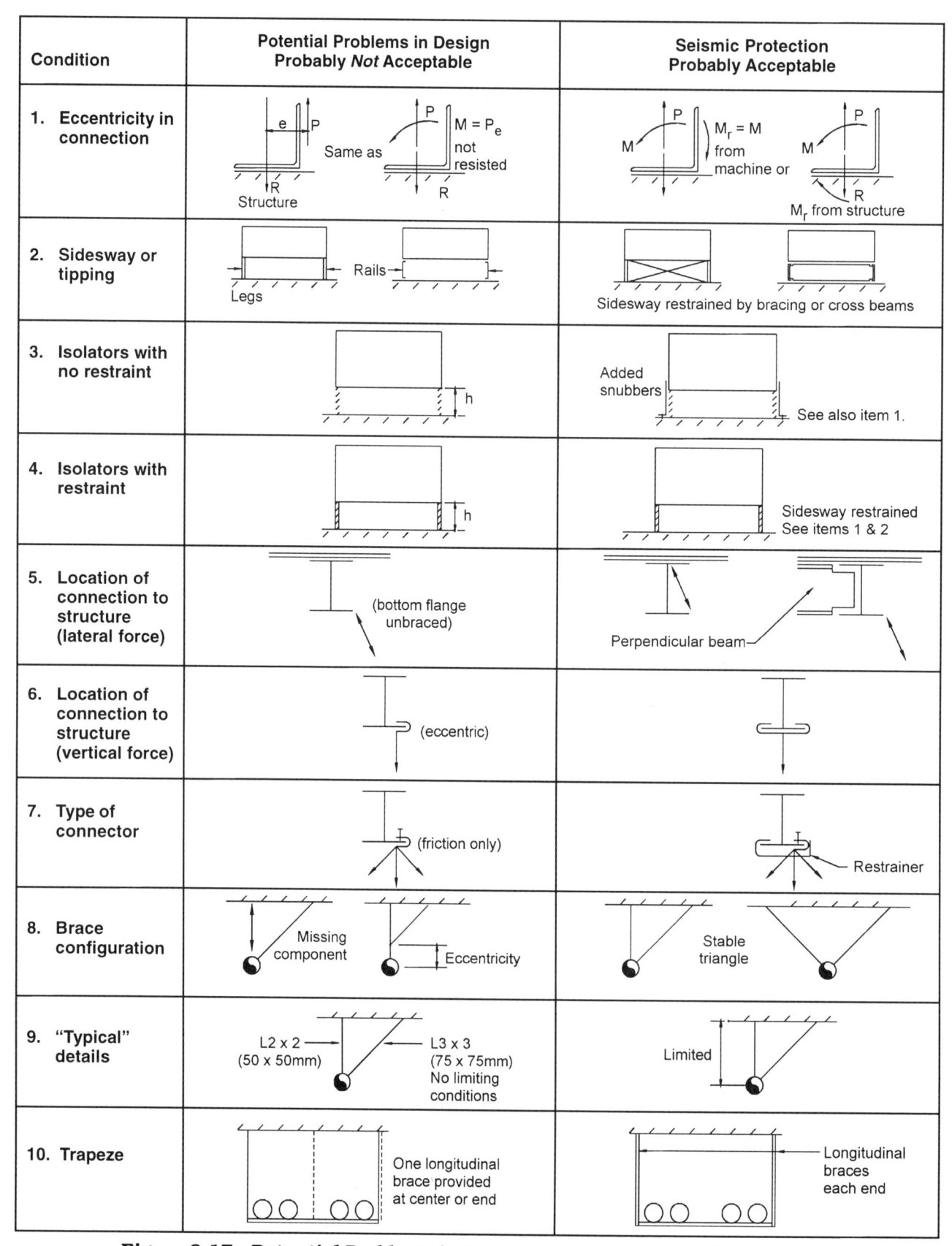

| Condition | Potential Problems in Design Probably *Not* Acceptable | Seismic Protection Probably Acceptable |
|---|---|---|
| 1. Eccentricity in connection | $M = Pe$ not resisted | $M_r = M$ from machine or $M_r$ from structure |
| 2. Sidesway or tipping | Legs, Rails | Sidesway restrained by bracing or cross beams |
| 3. Isolators with no restraint | | Added snubbers. See also item 1. |
| 4. Isolators with restraint | | Sidesway restrained. See items 1 & 2 |
| 5. Location of connection to structure (lateral force) | (bottom flange unbraced) | Perpendicular beam |
| 6. Location of connection to structure (vertical force) | (eccentric) | |
| 7. Type of connector | (friction only) | Restrainer |
| 8. Brace configuration | Missing component; Eccentricity | Stable triangle |
| 9. "Typical" details | L2 x 2 (50 x 50mm); L3 x 3 (75 x 75mm) No limiting conditions | Limited |
| 10. Trapeze | One longitudinal brace provided at center or end | Longitudinal braces each end |

**Figure 9-17  Potential Problems in Equipment Anchorage or Pipe Bracing**

# Chapter 9 — Seismic Protection of Plumbing Equipment

*Equipment*  For the purposes of this chapter, "equipment" refers to the mechanical devices associated with pipes that have significant weight. Examples include: pumps, tanks, and electric motors.

*Gas pipe*  For the purposes of this chapter, "gas pipe" is any pipe that carries fuel gas, fuel oil, medical gas, vacuum, or compressed air.

*Lateral force*  A force acting on a pipe in the horizontal plane. This force can be in any direction.

*Longitudinal bracing*  Bracing that prevents a pipe from moving in the direction of its run.

*Longitudinal force*  A lateral force that happens to be in the same direction as the pipe.

*OSHPD*  Office of Statewide Health Planning and Development (California).

*Positive attachment*  A mechanical device designed to resist seismic forces that connects a nonstructural element, such as a pipe, to a structural element, such as a beam. Bolts and screws are examples of positive attachments. Glue and friction due to gravity do not create positive attachments.

*Seismic*  Related to an earthquake. Seismic loads on a structure are caused by wave movements in the earth during an earthquake.

*Transverse bracing*  Bracing that prevents a pipe from moving from side to side.

## REFERENCES

1. American National Standards Institute. Draft. ANSI-ASSI: Building code requirements for minimum design loads in buildings and other structures. New York.

2. Applied Technology Council. 1978. *Tentative provisions for the development of seismic regulation for buildings (ATC-3)*. Washington, DC: US Department of Commerce, National Bureau of Standards.

3. Ayres, J. M., and T. Y. Sun. 1973. Non-structural damage. *The San Fernando, California, Earthquake of February 9, 1971*. Washington, DC: National Oceanic and Atmospheric Administration.

4. Ayres, J. M., T. Y. Sun, and F. R. Brown. 1973. Non-structural damage to buildings. *The Great Alaska Earthquake of 1964: Engineering*. Washington, DC: National Academy of Sciences.

5. California, State of. 1988. *California Code of Regulations*. Division 122 of Title 24, Building Standards.

6. Hillman, Biddison, and Loevenguth. *Guidelines for seismic restraints of mechanical systems*. Los Angeles: Sheet Metal Industry Fund.

7. Hodnott, Robert M. *Automatic sprinkler systems handbook*. Boston, MA: NFPA.

8. International Conference of Building Officials. 1988. *Uniform Building Code 1988*. Whittier, California: International Conference of Building Officials.

9. National Fire Protection Association (NFPA). 1996. *Standard for the installation of sprinkler systems*. NFPA no. 13. Boston, MA: NFPA.

10. Sheet Metal and Air Conditioning Contractors' National Association, Inc. (SMACNA). 1991. *Seismic restraint manual guidelines for mechanical systems*. Chantilly, VA: SMACNA.

11. The Sheet Metal Industry Fund of Los Angeles, CA, and the Plumbing and Piping Industry Council, Inc. 1982. *Guidelines for seismic restraint of mechanical systems*. Los Angeles, CA.

12. US Department of Defense. April 1973. Seismic design for buildings. In *Department of Defense Tri-Services Manual*. (TM-5-809-10. NAVFAC P-355, AFM SB-S Ch. 13) Washington, DC: Department of the Army, the Navy, and the Air Force.

13. US General Services Administration Public Buildings Service. Design guidelines. *Earthquake Resistance of Buildings*. Vol. 1. Washington, DC: Government Printing Office.

14. US Veterans Administration. Earthquake resistant design requirements handbook (H-08-8). Washington, DC: Veterans Administration Office of Construction.

# The New Aqua-Saver From Eljer®
## *Performance Under Pressure*

*Eljer introduces the most comprehensive line of pressure-assisted toilets in the industry! We provide the products you need for those typical and not so typical installations.*

**091-7015**
Aqua-Saver

**091-7925**
Walford

**091-7025**
Aqua-Saver

*The improved PF/2 pressure unit is so compact, it fits into a smaller tank for a more streamlined appearance.*

**091-7875**
Hylando

**091-7045**
Aqua-Saver

*Old System*

*You can count on Eljer for the features and selection of models that are second to none. The New Aqua-Saver:*

- Requires 60% less force to actuate than our previous system

- Is factory tested to ensure flawless performance

- Redesigned system provides a quieter flush

- Boasts a large 11-1/2" x 15" water surface

PF/2™ ENERGIZED FLUSH

*Award Winning Technology*

14801 Quorum Drive • Dallas, Texas 75240-7584 • 1-800-435-5372 • Visit our website at www.eljer.com

# 10 Acoustics in Plumbing Systems

## INTRODUCTION

The plumbing system can be the source of one of the most intrusive, unwanted noises in highrise apartment buildings, hospitals, hotels, and dormitories. It is essential, therefore, that plumbing engineers understand the terminology and theory of the field of acoustics in order to reduce the acoustical impact of plumbing.

## ACCEPTABLE ACOUSTICAL LEVELS IN BUILDINGS

Acceptable acoustical levels in buildings are usually assessed in a number of ways, depending upon the classification of a building occupancy (or normal usage), the time of the day (or night), the extent of the intrusion of external noises from other sources (including traffic), and the socioeconomic nature of a building (or of the areas in which it is located).

Typical sound levels are normally established in terms of their relationship with a preexisting background sound level, which is often specified in standards. Thus, for example, the background sound levels for broadcast studios would be specified in the range of 10 to 25 dB(A); those for sleeping quarters would be specified in the range of 20 to 35 dB(A); and those for offices would be specified in the range of 30 to 50 dB(A).

## ACOUSTICAL PERFORMANCE OF BUILDING MATERIALS

### Insulation Against Airborne Sound

The noise reduction provided by a barrier, partition, or wall is dependent on the transmission loss of that particular barrier, partition, or wall, together with the acoustical characteristics (and, specifically, the amount of sound absorption) existing on each side of the element.

For damped, single-leaf barriers, this transmission loss will depend primarily on the product of the surface weight of the barrier and the frequency of the signal being attenuated. This phenomena is described as the "mass law." Doubling the surface weight of the barrier only results in a 3-dB improvement in transmission loss.

For double-leaf barriers, the transmission loss is determined by the spacing between the leaves at the edges of the barrier system and the respective surface weights of the two leaves. Maximizing the spacing between the leaves has the result that the performance of the barrier tends to be the highest possible value at all frequencies. At minimum spacing between the leaves, the maximum improvement is at the highest frequencies while the typical improvement may be as little as 3 dB at the low frequencies.

In any barrier system, maximum performance requires the closing off and effective sealing of all holes and gaps, particularly around penetrations of the type required for pipe and pipe fittings. Such penetrations usually require an effective flexible sealing in order to accommodate the thermal movement while simultaneously minimizing the extent of vibration transmission from the pipe into the surrounding barrier system. The preferred type of sealing system should incorporate fire-rated flexible fiberglass, mineral wool, or ceramic wool wrapping retained by sealant and, where required, metal sleeving for protection or to span between the cavity access on the opposite sides of thick walls or large cavities.

Barrier systems used to surround or enclose piping should incorporate acoustic-absorbing linings or retained fiberglass or mineral wool together with effectively sealed external barriers of high-mass drywall construction fixed to steel stud framework, or masonry, as required. Barriers in close contact with pipe or fittings should provide noise reduction or have a sound absorption capability less than that indicated by laboratory tests carried out on large-scale samples evaluated under normal conditions.

## ACOUSTICAL RATINGS OF PLUMBING FIXTURES AND APPLIANCES

The acoustical rating tests for fixtures are still in their infancy and have not yet been internationally standardized. While some countries, most notably Germany, do have useful standards, the United States has yet to formalize any plumbing acoustic tests. The problem of adequately defining the direct airborne and structure-borne components of vibration still constitutes a major problem in performing acoustic rating tests. Only the German standard DIN 52218, *Laboratory Testing on the Noise Emitted by Valves, Fittings and Appliances Used in Water Supply Installations (Part 1)*, has so far addressed this problem. Also, the International Organization for Standardization (ISO) has published standard 3822/1, *Laboratory Tests on Noise Emission by Appliances and Equipment Used in Water Supply Installations*, and the American Society for Testing and Materials (ASTM) has established a project E-33.08B, Plumbing Noise, to investigate this problem.[1]

[1] Copies of the DIN and ISO standards are available from the American National Standards Institute (ANSI), 1430 Broadway, New York, NY 10018.

The airborne sound radiated by showers, dishwashers, waste-disposal units, washing machines, water closets and bathtubs is specified by the fixture/appliance manufacturer. The sound ratings for fittings are normally expressed in terms of sound power or A-weighted and octave-band levels measured in a reverberant toilet room or kitchen-type environment at a distance of 3 ft (0.9 m). Because of the differences in the reverberations between one environment and another, the characteristics of the test environment should ideally have a reverberation time lying in the range of 1 to 2 seconds and be independent of the frequency.

**Valves** The sound levels from valves are dependent upon the size of the fitting, the mass flow rate, and the pressure differential across the fitting. Sound levels from taps and valves at a distance of 3 ft (0.9 m) may range between 30 and 50 dB(A) for well designed and properly installed fittings, 50 and 70 dB(A) for adequately designed and adequately installed fittings, and 70 and 90 dB(A) for poorly designed and poorly installed fittings. Improvements in the performance of faucets are most notably achieved through the incorporation of aerators, which may result in reductions in noise levels of as much as 15 or more dB.

**Water closets** The noise from water closets can be subdivided into:

- The noise of the water flushing the closet bowl.
- The noise of the water refilling the tank.
- The noise of a flush-valve operation, including water discharge into the fixture and the ejection of materials from the closet bowl.

The noise of the water flushing the closet bowl is a function of the specific flow rate from the tank, the proximity of the tank to the closet bowl, and the method of mounting the tank itself. While sound levels as high as 90 dB(A) at 3 ft (0.9 m) are possible in older style fittings with the tank located as much as 6 ft (1.8 m) above the bowl, modern, close-coupled tanks, when properly installed (with bowl cover down), may be as low as 55 dB(A).

The noise of the tank refilling is a function of its design, which includes the type of construction of its envelope, the method of mounting to the wall (or closet bowl), the type of tank valve used, the water, and the time required for the refill cycle. There are many cases where the noise of the tank refilling is far more annoying than the noise produced by the toilet flushing. The noise of toilet tanks refilling may be as low as 40 dB(A) at 3 ft (0.9 m) in well-designed units incorporating quiet valves and silenced, tail-pipe assemblies. In poorly designed installations operating at the maximum flow rate, this noise may be as high as 95 dB. Flush-valve operation can be as high as 95 dB, while blowout-type fixtures have been recorded at as high as 120 dB.

**Urinals** The noise associated with urinals is a function of the wall-mounting method used to install the fixture and the flushing action of the urinal—water discharge into the fixture and the ejection of the materials. The flush-valve operation may be as high as 95 dB, with blowout units as high as 110 dB.

**Bathtubs** The noise from bathtubs is usually caused by the impact of a high-velocity water stream into a glazed metallic, fiberglass, or acrylic bathtub. While this noise varies during the filling cycle, it may also be significant during the drainage cycle.

In both cases, this noise is a function of the bathtub material and its structural design as well as its method of installation, particularly the extent of its structural decoupling from the walls and floor (in order to reduce the acoustical impact on adjacent rooms or other apartments). In many European countries, the building regulations specify stringent decoupling procedures to minimize the structure-borne noise propagation. Outside these countries, such procedures are relatively unknown and are not generally utilized. The noise of a bathtub filling typically lies in the range of 60 to 100 dB(A) at 3 ft (0.9 m), depending on the flow rate. The point of impact of the water stream with the side of the bathtub is generally reduced by using an aerator on the spout. Good design practice calls for the water spout to be installed so that the water stream is not directed to strike the bottom of the bathtub.

**Showers** Shower noise is mostly a function of the floor surface in the shower enclosure and the type of shower head. The constant-temperature controller is only significant when the water pressure drop across it is unusually high or the method of supporting the pipe from the walls results in the generation of resonant noise. Shower noise typically lies in the range of 60 to 90 dB(A) at 3 ft (0.9 m).

**Dishwashers** Dishwasher noise is a function of the basic design of the unit, the choice of the mounting procedures employed, and the extent to which the installer provides additional thermal and acoustical insulation. The noise from dishwashers can be minimized by mounting the units on rubber isolation devices or by providing layers of fiberglass or mineral wool insulation on their tops, rear and sides. Significant noise is created by the activation of solenoids and solenoid-activated valves that create water hammer and sound propagation through the piping.

The use of flexible connections and the incorporation of surge eliminators to minimize the water hammer are highly recommended. Typical noise levels from dishwashers are in the range of 65 to 85 dB(A) at 3 ft (0.9 m) with peaks as high as 105 dB(A) created by solenoid operation.

**Waste-disposal units** Sink or waste-disposal-unit noise is a function of the design of the unit as well as the design of the sink or basin to which it is attached. Lightly constructed, stainless-steel sinks or basins will tend to amplify the sound energy. The supporting cupboards or fixtures may also have a similar effect. The sinks and basins used to support such fittings should be designed to incorporate an effective dampening of the bowl through the application of damping materials or framework.

The plumbing connection to the waste-disposal units should be flexible at the inlet and the discharge. The noise levels produced by these units can vary widely and most manufacturers do not publish the noise-rating data. Noise levels can vary between 75 and 105 dB(A) at 3 ft (0.9 m), depending on the method of mounting and the extent to which the protective covers are used.

**Washing machines** Washing-machine noise is a function of the design of the unit and, to a lesser extent, the method of mounting or type of plumbing connections used. The airborne noise levels can be somewhat reduced through the application of dampening material on the inside surfaces of the enclosing panels.

The incorporation of isolation mounts does not normally reduce the direct airborne sound but may drastically reduce the structure-borne component audible in adjacent apartments or rooms. The noise levels produced by washing machines range between 65 and 90 dB(A) at 3 ft

(0.9 m), and few manufacturers publish the sound-rating data.

An indirect discharge of the wastewater into a trough or hub drain may be very loud in the room but may have a reduced noise transmission through the piping system.

## GENERAL ACOUSTICAL DESIGN

### Water Pipes

**Origin and spread of noise** The causes of noise are the surge due to the sudden opening or closing of valves and flow where the cross sections of such valves are greatly restricted. In addition, because of the high velocity, cavitation and turbulence are also created by the sudden changes in direction. The higher the pressure head of the fittings, the louder the noise. Water hammer arrestors (shock absorbers) may be beneficial in eliminating sound noise generated by these problems.

Noises originate when a stream of water strikes the base of the bathtub, sink, or lavatory. In the emptying operation, gurgling noises often occur because of the whirlpool action. These noises are conducted partly along the pipe and partly by the column of water. The pipes induce the walls and ceilings to vibrate and radiate sound.

**Reducing the noise at its origin** Fittings of satisfactory design with a low noise level should be employed whenever possible.

Flush tanks, in particular, can be substantially quieter than pressure flush valves, especially when insulated. Low-flush, gravity-flush valve fixtures will operate more quietly than high-velocity (blowout) type fixtures.

The largest possible cross sections for the pipes should be used and the water supply pipes in all critical areas should be designed for a maximum velocity of 4 ft/s (1.2 m/s).

The emptying noise can be reduced by using waste fittings to ensure that the air is simultaneously and uninterruptedly sucked out of the stream of water and carried away with it.

The pressure in the pipes inside of the building can be reduced to the extent that the operating conditions allow closed-circuit pressure.

**Reducing the spread of noise** The designer must make a distinction among the pipes laid on a wall, those in a wall, and those in shafts.

In the case of pipes installed on a wall or in pipe shafts, structure-borne, sound-damping packing (e.g., cork, felt, profiled strips of rubber, or other elastic materials) should be inserted between the fastenings and the pipe. The packing should not be compressed by excessive tightening of the pipe clips. Instead of packing, structure-borne, sound-damping fasteners may be used.

Pipes in the wall should be wrapped with sound-damping materials (e.g., felt, bituminized felt, or fibrous damping materials) without leaving any gaps. The same effect may be achieved by having pipes elasticity mounted in a firm outer casing.

Several pipes running in the same direction in shafts can be fastened to a single common rack. This rack should not have any structural, noise-conducting connections with the walls. Common racks should be fastened to the wall with rubber/metal connections interposed.

When pipes pass through ceilings or walls, they should be taken through sound-damping sleeves of fibrous damping materials, bituminized felt, or granulated cork. This approach must not adversely affect the airborne sound damping (for instance, through joints), in particular in the case of party ceilings and walls of separate tenants. In the case of ceilings and walls that have to be fire resistant, this approach must be complied with when deciding on the fire-rated sleeves.

In the case of apparatus and equipment, such as washing machines, spin dryers, bathtubs, and wash sinks that generate noise or in which noise occurs during filling and emptying, sound-damping materials should be used at the places where they touch or are attached to the structure. In the case of bathtubs, a solid joint between the bathtub outlet and the waste pipe should be avoided. Rubber pads under the bathtub supports are recommended.

Bear in mind, in the case of water pipes and apparatus in or on walls that border occupied spaces, that the permissible loudness levels are not exceeded. In such cases, conventional water-closet flush valves should be avoided and quiet-acting siphon jet actions should be used.

# Chapter 10 — Acoustics in Plumbing Systems

## Occupied Domestic Spaces

Keeping within the maximum allowable loudness levels in occupied rooms requires that steps be taken during the planning and construction stages of the building. The term "occupied domestic spaces" generally covers hotels, motels, dormitories, and other locations where, in addition to domestic appliances, the elevators, incinerators, ventilation equipment, switch gear, boilers, and refuse disposal installations can cause unacceptable noise levels in habitable areas, particularly sleeping quarters. As early as the planning stage, the various points requiring consideration must be taken into account by the plumbing designer.

Because of the multiplicity of influences involved, no simple or standard rules can be given for keeping within the permissible loudness levels. For some groups of installations, the following criteria apply:

1. Apparatus and machines in which the noise is predominantly transmitted as structure-borne sound (e.g., motors, pumps, pressure-increasing installations, ventilation machinery, drives for elevators, gearing, and heavy switch gear) must be sound insulated from the building.

2. In order to reduce the structure-borne sound transmission from the heating installations into occupied rooms, a heavy floating screed (attuned to low frequencies) should be laid in the rooms where the solid fuel is stored and where there is heating equipment. The boilers must stand on special foundations and be separated from other components and from the screed. The pipes can be supported by collector blocks on the floating screed. Rigid fastening to ceilings, floors, or walls should be avoided.

3. Ceilings over rooms where there is heating equipment should be provided with a floating screed in order to increase the airborne soundproofing.

4. In the case of refuse disposal installations, the inside shaft should be constructed in such a way that the building is insulated against structure-borne sounds. Whenever possible, low-noise materials should be used. Metal sheeting should be provided with a sound-insulating covering. The roof of the shaft should be made of sound-absorbent material.

5. Refuse bins should stand on a floating screed and be enclosed by walls and ceilings complying with the requirements for party walls and ceilings of apartments. If deflector plates are provided, these devices must be fixed in a flexible manner and with structure-borne sound insulation.

## Pumps

**Sources of noise** The following items are some of the major sources of noise from pumps in plumbing systems:

1. Unbalanced motors.

2. Pulsation of the air mass flow from electric fans (This is a major source of noise in 2-pole, fan-cooled, electric motors. The noise from the fan is usually so dominant that all other sources of noise in the electric motor can be neglected.)

3. Pulsation of the magnetic field in the electric motor.

4. Motor/gear/pump journal and thrust bearings.

5. Contacting of the components in parallel-shaft and epicyclic gears.

6. Imbalance of the pump impellers.

7. Pulsation in the pumps. (Hydrodynamic noise generation is inherent in all types of pumps; the fundamental frequency of noise, when the pump runs at the design point, is governed by the number of blades and their interaction with the volute cut-water or diffuser guide vane ring. The intensity of the noise generated and the relative strength of the various harmonics produced are determined by the velocity profile shape leaving the impeller passages, vortex wakes shed by vanes, and the impulsive effect as they pass under the volute cut-water. The impulse wave form, although very complex, can be resolved into a fundamental equal to the speed times the number of blades and a series of harmonics. Manufacturing errors, which produce angle or pitch variations between the blades, are instrumental in generating a less prominent series of harmonics with fundamental frequency equal to the speed with the amplitude and/or the frequency modulation of the blade-passing frequency. At off-design operation, unsteady flow conditions can arise

due to flow separation and rotating stall effects.)

8. Cavitation. (Air is entrained in the solution, which can damage the pump; impellers constructed with open-grain material, such as cast iron, may disintegrate because of the implosive effect of cavitation.)

**Possible modifications** Obviously, if the overall noise level of the pumping plant is considered to be too high to comply with the accepted specifications, identifying and reducing the noise output from the components and equipment in the plumbing system contributing the most noise will yield the most dramatic results.

Some possible modifications the plumbing engineer should consider in order to reduce the noise output from the system are as follows:

1. *Gearbox.* A silencing enclosure or cladding should be provided.

2. *Motor fan.* A silencer should be provided at the air inlet and outlet or, if possible, the design should be modified.

3. *Motor rotor.* The number of slots should be changed or, if possible, its design should be modified.

4. *Pump and pump bearings.* Sleeve types should be employed.

5. *Pump operation.* The pump should be operated near design flow conditions in order to achieve the correct system matching. (Modifying the characteristics of the system or altering the diameter of the impeller, resulting in operation at fewer rpms, will make the pump operate more quietly.)

6. *Pump impeller blades.* The clearance between the tip of the impeller and cut-water should be increased (a minimum of 85% impeller diameter to volute diameter is recommended).

7. *Impeller and guide-vane tips.* Should be dressed in order to reduce the thickness and intensity of the trailing wakes.

8. *Out of balance.* Should be balanced to fine limits. Impeller, motor, blades, and rotor should be balanced at all rpms to eliminate—or minimize—vibrations.

9. *Cavitation.* The suction characteristics of the installation/system should be improved. (Ideally, the pump should always be under positive head at the pump suction.)

**Plant noise** The resulting noise output of the plant, as installed on the site, is dependent upon all of the above factors coupled with the induced resonance of the adjacent parts (such as pipes, bedplates, fabricated stools, tanks, and panels). These factors form part of the final environment of the pump and are discussed under the section "Noise and Vibration Control," which follows. The effects are best investigated by determining the natural frequency of these parts, by separate excitation, or by operating the pump through its service-speed range. The natural frequency of the part can be modified by a simple trial-and-error stiffening or damping.

Other effects are also likely to appear for the first time on the site. One of them is the interaction of the pump with the intake sump. Testing a model of the sump intake before installation can prevent air-entraining vortices, eddies, and distorted flow distributions, which cause a mismatch at the impeller leading edges. Vortex formations generated entirely below the water level can be particularly troublesome in actual practice, since these formations are caused by water spinning at high velocities, which causes submerged cavitating vapor cores to be generated and drawn into the sump intake. These vortices cannot be observed on the site; however, they can be prevented at the outset by testing the sump model with observation windows fitted below the surface level.

Associated venturi-meters, valves, and pipes, in the final installation, contribute to raising the general noise level of the station. Water at high velocities passing through partially closed valves, particularly in high-pressure systems, can produce severe cavitation noise, which is generated by the rapid collapse of the vapor bubbles against the walls of the valve and downstream pipe. Sound-pressure levels of 110 dB have been recorded.

These high sound-pressure levels can be greatly reduced by using multistage pressure breakdown systems and by paying special attention to the valve's port design. Thick-walled pipe and external acoustic installation are only effective for localized noise reduction; they do not reduce the noise in the fluid stream but shroud it where treatment is used. Much of the noise is still carried downstream and, at times, upstream as well, depending on the system. Poor pipe design, involving many sharp bends and sudden expansions and contractions, can induce considerable turbulence and noise.

# Chapter 10 — Acoustics in Plumbing Systems

**Estimating the noise level of a pump** Small, motor-driven pump sets, which are commonly employed in plumbing systems, can be conveniently tested in an anechoic chamber with accurate results. However, the most important sources of sounds in large pumping stations are usually associated with custom-built units.

Pump noise levels are measured by the near-field technique 3 ft (0.9 m) from the unit in order to minimize the sound transmissions from pipe coupled to the pump. The sound-pressure level, at 3 ft (0.9 m) from the pump, can be estimated by using the following equation (presented in the International System of Units, or SI units, which is the best means for available test data):

### Equation 10-1

Pump sound-pressure level = 163.9 +

$$8.5 \log \frac{\text{(volume flow rate} \cdot \text{head)}}{\text{(rpm} \cdot \text{specific speed} \cdot \text{impeller diam.} \cdot \text{impeller width)}}$$

*where*

Pump sound-pressure level = dB(A)

Volume flow rate = L/s

Head = stages/m

Specific speed = m/s

Width and diameter of impeller = mm

Where the noise characteristics of a particular pump are already known, the change of the noise level with the pump speed can readily be determined by using the following equation:

### Equation 10-2

$$dB = 50 \log \left(\frac{N_1}{N_2}\right)$$

*where*

$N_1$ and $N_2$ = Pump speeds

## Flow Velocity and Water Hammer

In simple terms, the magnitude of the pressure increase due to water hammer is a function of the velocity of the pressure wave and the rate of exchange of the flow velocity. The velocity of the pressure wave (which is the same as the velocity of sound in the water contained in the pipe) depends on the physical properties of the water and of the pipe material. For all commercially available copper pipes, the velocity of the wave propagation has a value in the range of 3000 to 4000 ft/s (915 to 1220 m/s).

If the flow velocity changes abruptly (e.g., by the sudden closing of a tap or valve), the pressure increase can be determined by using the following equation:

### Equation 10-3

$$Pr = \frac{WaV}{144g}$$

*where*

Pr = Pressure rise, lb/ft² · s

W = Specific weight of liquid, lb/ft³

a = Velocity of pressure wave, ft/s²

V = Change in flow velocity, ft/s

g = Acceleration due to gravity, ft/s²

The pressure generated by water hammer may cause straight pipe lengths to vibrate. If the pipes are in close contact with the walls and not fixed at sufficiently rigid short intervals, they may strike against the walls with a succession of blows.

If such fittings (solenoid valves, foot-action valves, spring-loaded taps, and check valves) could be eliminated, then the incidence of water hammer could be greatly reduced. However, as these fittings are integral parts of a plumbing installation, the designer must allow for this condition. The following guidelines are recommended to the plumbing engineer:

- Maintain a water velocity in the range of 4 ft/s (1.2 m/s) at the appliance.
- Secure the piping so that it does not come into contact with the building structure.
- Use rubber isolators.

The use of air vessels (chambers) will reduce the effects of water hammer. A vessel with a flexible membrane to separate the air chamber from the water (water hammer arrestor or shock absorber) is recommended to prevent loss of air.

## Design Procedures

To provide a plumbing system that conforms to specific acoustic standards, the designer requires the following information:

1. The maximum noise levels allowable in each habitable room.
2. The data on the acoustic performance of the building materials and the method of construction.

3. The acoustic ratings of the plumbing appliances and fixtures, piping and valves.

4. The acoustic performance of the noise-isolation devices that can be incorporated in the plumbing installation (i.e., vibration mountings and rubber spacers).

5. The data on the effects of the background noises to screen out the effects of the plumbing noises.

6. Supervision of the plumbing installation in order to ensure adherence to the acoustic details.

Specific acoustic performance guarantees for plumbing installations should be avoided where sufficient research has not been carried out. On any critical projects, the retention of an acoustical consultant may be essential. In the end, the final results are as much dependent on the quality of the workmanship and supervision as they are on the design details. Many well-designed projects fail to achieve the required performance because of inadequate supervision, which is necessary in order to pinpoint and correct substandard details.

## Noise and Vibration Control

All noise-control problems can be reduced to three basic elements: source, path, and receiver.

Noise-control problems frequently involve consideration of several sources of noise, several paths for the transmission of noise, and several different receivers. The relationship among these elements defines the seriousness of the problem. In order to solve a noise problem, the source strength can be reduced, the path can be made less effective in transmitting sound, or the receiver can be made more tolerant of disturbance. However, most practical solutions involve a trade-off, so concentration on only a single aspect of the problem may result in overdesign or an unsatisfactory solution.

For sources that not only produce noise and vibration problems but also have the potential to lead to damage or decrease the useful channel space for liquids, it is desirable to reduce the source strength. Cavitation is a typical example of this kind of problem. The solution to this problem hinges on the pump suction (i.e., net positive suction head, NPSH). One may consider placing the pump at a lower elevation, if practical, or one may improve the suction piping and raise or pressurize the supply tank. In recent years, efficient suction-assisting devices (such as booster pumps) have become commonly used where low NPSH must be handled at low cost. Several manufacturers supply add-on or built-in inducers for end-suction pumps.

For noise and vibration sources that do not influence systems operating conditions or reliability, control of noise transmission (i.e., the path) from the source to the noise-sensitive area may be the most economical solution. Noise may be transmitted through structure-borne, airborne, and fluid-borne paths. Structure-borne noise travels in the form of high-frequency structural vibrations; airborne noise travels in the form of sound waves; and fluid-borne noise travels in the form of pressure fluctuations. The structure-borne path usually plays an important role because the noise source within the pump, or piping, often can communicate with the surroundings only by setting the enclosure into vibration. These vibrations may radiate sound directly or may be transmitted through the supporting structure, to be converted to airborne sound elsewhere.

Vibration isolators, such as resilient mounts and resilient pipe hangers, are commonly used to reduce structure-borne vibrations. Theoretically, in order to design an adequate isolation system, the engineer must realize how much vibratory force is generated by the equipment and the maximum permissible force transmission to the building. Since these design parameters cannot readily be obtained, some practical guidelines have been formulated to provide effective isolation at a reasonable cost. These are generally adequate for all but the most critical or special applications (such as very light or flexible structures or equipment installed above adjoining very quiet spaces).

To ensure that the desired noise isolation is achieved, a detailed vibration-control specification and its stringent enforcement are required. With increased public awareness of noise, government agencies such as the General Services Administration (GSA Guide Specification Number 4-1515-71, *Public Building Service*) and Federal Housing Administration (FHA *A Guide to Airborne, Impact and Structure-Borne Noise Control in Multifamily Dwellings*) have established recommended guidelines on noise and vibration control.

# Chapter 10 — Acoustics in Plumbing Systems

However, it is not enough merely to have noise-control specifications. Adequate detailing techniques are most essential for communications between the design engineers and the contractors. For most situations, acoustical details are well developed and are available for most applications. From a practical point of view, most plumbing fixtures cannot be effectively isolated, although they can be installed to minimize vibration.

When pipes are connected to vibrating equipment installed on vibration isolators, sufficient flexibility must be built into the piping systems to match the equipment vibration isolators. In addition, adequate flexibility is required in order to protect the equipment from any strains imposed by misalignment and by thermal movement of the piping. Piping flexibility can be achieved by the inherent resilience of the pipe in simple bend-and-loop configurations (if there is sufficient length) or by the use of flexible pipe connectors, which also attenuate the transmission of noise and vibration along the piping system. However, their use as vibration-isolation devices should be considered very carefully for the following reasons:

- They are the weakest component in the piping system. (Without proper specification of the material, installation, and maintenance, they may fail and cause severe water damage.)
- In many instances, sound energy may flank the flexible pipe connectors so that pipe-wall noise is exited downstream of the resilient break. Indeed, the various restraints added by the manufacturers to reduce the possibility of failure make the flexible pipe connectors almost as rigid as the pipe itself.

Studies have found that flexible pipe connectors are most effective in the case of cavitation. Flexible pipe connectors have also been found to be effective in the attenuation of the tonal components at the impeller passage frequency of a pump.

**Equipment design** Quiet operation of pumps begins with proper design. Although today's state-of-the-art design and development of pumps and plumbing fixtures has a long heritage, noise is still seldom considered by the manufacturer. This is perhaps because of the designer's lack of awareness and experience, but even knowledgeable designers yield to economic pressures for cost reduction.

It is obvious that the primary purpose of a pump or plumbing fixture is to move liquids and to perform the necessary plumbing functions. These considerations must come first. However, noise and vibration controls should be integrated in the design and may then be expected to lead to improved performance with little or no cost penalty. Quite often, the cost of modification is negligible. The key to effective noise control is a complete understanding of the noise-generating mechanisms.

By simply changing the cut-water clearance of a pump, a major reduction in the blade-passage-frequency noise is achieved. Similarly, water faucets can easily be designed for quiet operation. For a particular value of pressure drop, a valve can be designed to minimize cavitation and its resulting noise within the water pressure design range.

Some water-closet manufacturers indicate that, like dishwashers and food-waste disposers, economy models are noisier than more expensive ones. Nevertheless, quietness in water closets is a marketable attribute. One of the problems with flush-valve-operated water closets is the high initial noise impulse that is associated with the opening of the flush valve. However, if the valve discharges against a properly selected resistance, the noise impulse can be substantially reduced. There is no doubt that a cost-saving, quiet fixture could be achieved with more research.

## SYSTEM DESIGN

**Equipment selection** To select a quiet unit, the engineer must have an understanding of the noise characteristics of plumbing fixtures and appliances. Good matching between machine characteristics and system requirements is essential for performance, as well as for noise control. For example, in a system operating over a narrow load range, a pump of single-volume design (selected for near-peak efficiency operation) is acceptable because the unbalanced radial load on the impeller is the least at optimum delivery.

Adequate criteria should be established for equipment vibration to ensure that there are no excessive forces that must be isolated or will adversely affect the performance or the life of the equipment. There are many ways to develop equipment-vibration criteria. A simple but satisfactory approach would be to use the criteria that have been developed on the basis of the experience of persons and firms involved with vibration testing of mechanical equipment in the building construction industry.

**Pressure** Most model plumbing codes have established the rate of flow desirable for many common types of fixtures as well as the average pressure necessary to provide this rate of flow. Although the pressure varies with the design of the fixture, a pressure of 5 to 8 psi (34.5 to 55.2 kPa) at the entrance to the fixture is generally the minimum required for good service at lavatory faucets and tank-type water closets. A pressure of 15 psi (103.4 kPa) may be ample for most of the manufacturer's requirements. Some fixtures, especially wall-hung water closets, require a pressure up to 25 psi (172.4 kPa).

Water pressure in many mains is typically 50 to 80 psi (334.7 to 551.6 k Pa). As the water flows through a pipe, the pressure continually decreases along the pipe, due to the loss of energy from friction and the difference in elevation between the water main and the fixture.

From a noise-control point of view, it is desirable to keep the fixture inlet water pressure as low as possible. This condition usually can be achieved by installing pressure-regulating devices in order to balance the pressure gradient in the water system. Many cities experience large pressure fluctuations in the hydraulic gradient of their water systems due to demand changes (such as after working hours). The inlet water pressure must be kept higher than the required minimum pressure to ensure good service. The alternative, if it were practical, would be to have continuous adjustments of the system's inlet pressure as the demand changes. The system pressure also has a great effect on the occurrence of cavitation. The plumbing system must be operated at a pressure level high enough to prevent cavitation.

**Speed** Changes in the operating conditions of the pump have a significant effect upon the level of pressure fluctuations, particularly for plumbing systems in which resonance exists. It is possible that a 5% change in the pump speed may result in a 70% change in the pressure fluctuations. Also, a valve's pressure-flow characteristics and structural elasticity may be such that, at some operation point, it will oscillate (perhaps in resonance with parts of the piping system) so as to produce excessive noise or even physical damage. A change in the operating conditions or details of the valve geometry may then result in significant noise reduction.

**Pipe sleeves** One very important consideration for piping system noise control that seldom receives any attention is the detailing of the piping sleeves at the wall and floor penetrations. Each type of piping sleeve has a specific application and its acoustical treatment cannot easily be generalized. For example, the acoustical requirements for a piping sleeve used on water piping that passes through a foundation wall will be different than those for a piping sleeve used for sprinkler pipes that pass through a double-wall construction enclosing a concert hall. However, each case should be treated so that the pipe penetration will match the acoustical value of the wall and provide the proper separation between the piping and the building construction. This requirement must be made clear to the contractor, which entails showing the construction details on the contract drawings.

Most plumbing systems contain many points at which the piping must penetrate floors, walls, and ceilings. If such penetrations are not properly treated, they provide a path for noise transmission which can destroy the acoustical integrity of the occupied space. Accepted practice is to seal the openings with fibrous material and caulking in a manner similar to that illustrated in Figures 10-1 and 10-2. Some penetration seals, as shown in Figure 10-3, are also commercially available.

**Water hammer** A common method of controlling water hammer noise is to install a shock absorber or air chamber where the water hammer is most likely to originate, such as at a faucet or a control valve. In many residential systems, it is common to install one similar to that shown in Figure 10-4.

**Pipe wrapping** The noise from a pipe may be reduced by applying a wrapping (lagging) to the pipe. Such a wrapping normally consists of a layer of porous material placed between the pipe surface and an external, impervious cover. The cover must be supported by the blanket with no structural ties between the outer cover and the pipe. Structural connections reduce the effectiveness of the pipe wrapping. The porous material serves three purposes:

# Chapter 10 — Acoustics in Plumbing Systems

- It keeps the external, impervious cover separated from the surface.
- It attenuates sound (particularly, at high frequencies).
- It reduces the amplitude at the resonant frequency defined by the mass of the cover and the stiffness of the layer of porous material.

The typical noise reduction from a pipe wrapping is in the range of 0 to 5 dB at low frequency and 15 to 25 dB at high frequency.

**System layout** A system that is undersized (or that contains a section of undersized piping) will usually generate excessive noise. It is good engineering practice to use simple-design pipe layout (i.e., long straight runs with a minimum of elbows and tees) and long radius elbows and connectors. The straight run can be estimated as being 12 times the diameter of the pipe. Piping layout near pumps and valves is also of great importance. Figure 10-5 illustrates some examples of suction-piping installations.

**Vibration isolation** The sources most commonly responsible for the generation of noise in plumbing systems are discussed in this section. However, most plumbing noise problems are not caused directly by the noise radiated to the air from these sources. Usually, the plumbing system transmits the sounds so that the mechanical vibration follows its support system to the surface and is eventually radiated as noise.

A complete discussion of vibration-isolation theory is beyond the scope of this chapter. Only the methods of vibration control that are readily applied and broadly useful in practical problems are considered here. This chapter does not

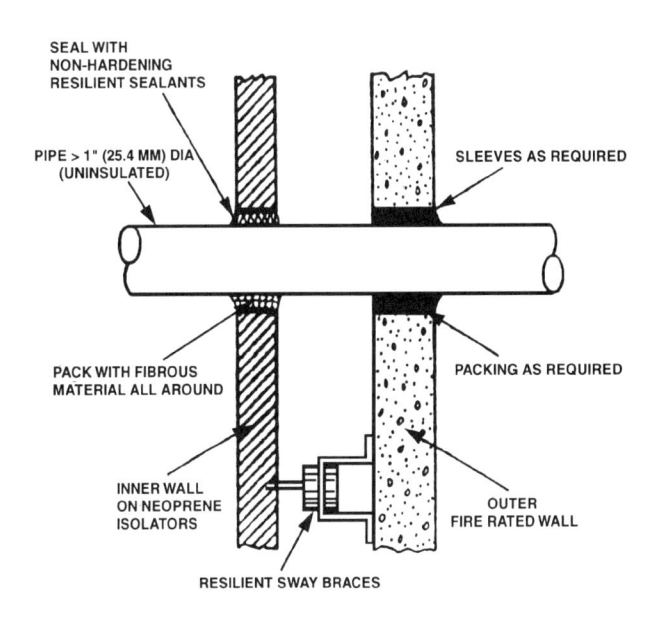

**Figure 10-2 Acoustical Treatment for Pipe-Sleeve Penetration at Spaces with Inner Wall on Neoprene Isolators**

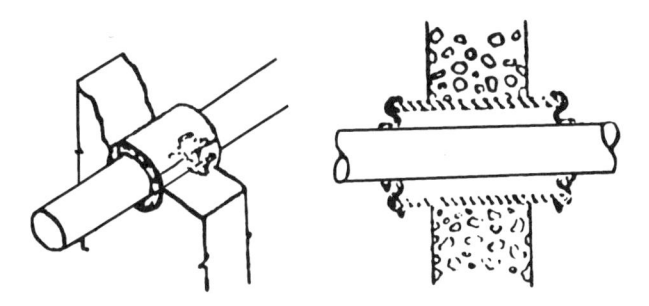

**Figure 10-3 Acoustical Pipe-Penetration Seals**

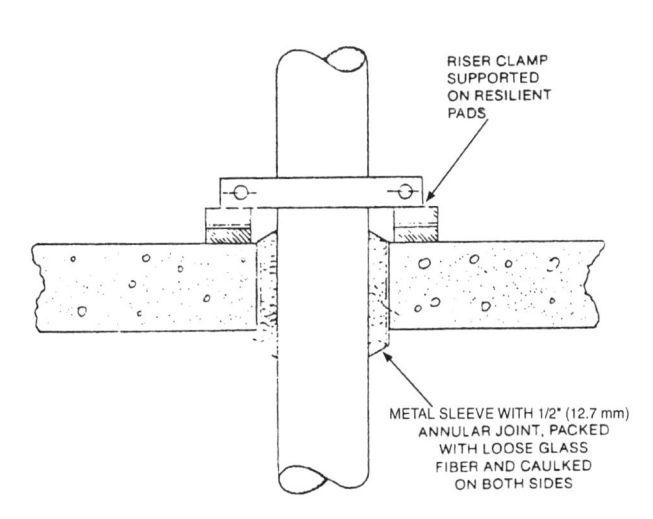

**Figure 10-1 Pipe-Sleeve Floor Penetration**

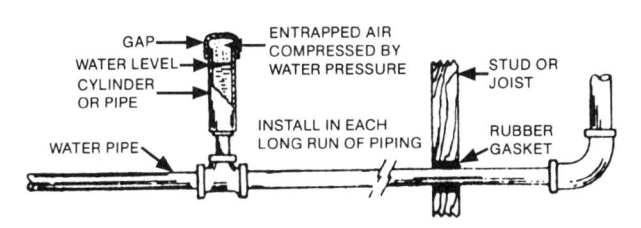

**Figure 10-4 Installation of an Air Lock in a Residential Plumbing System**

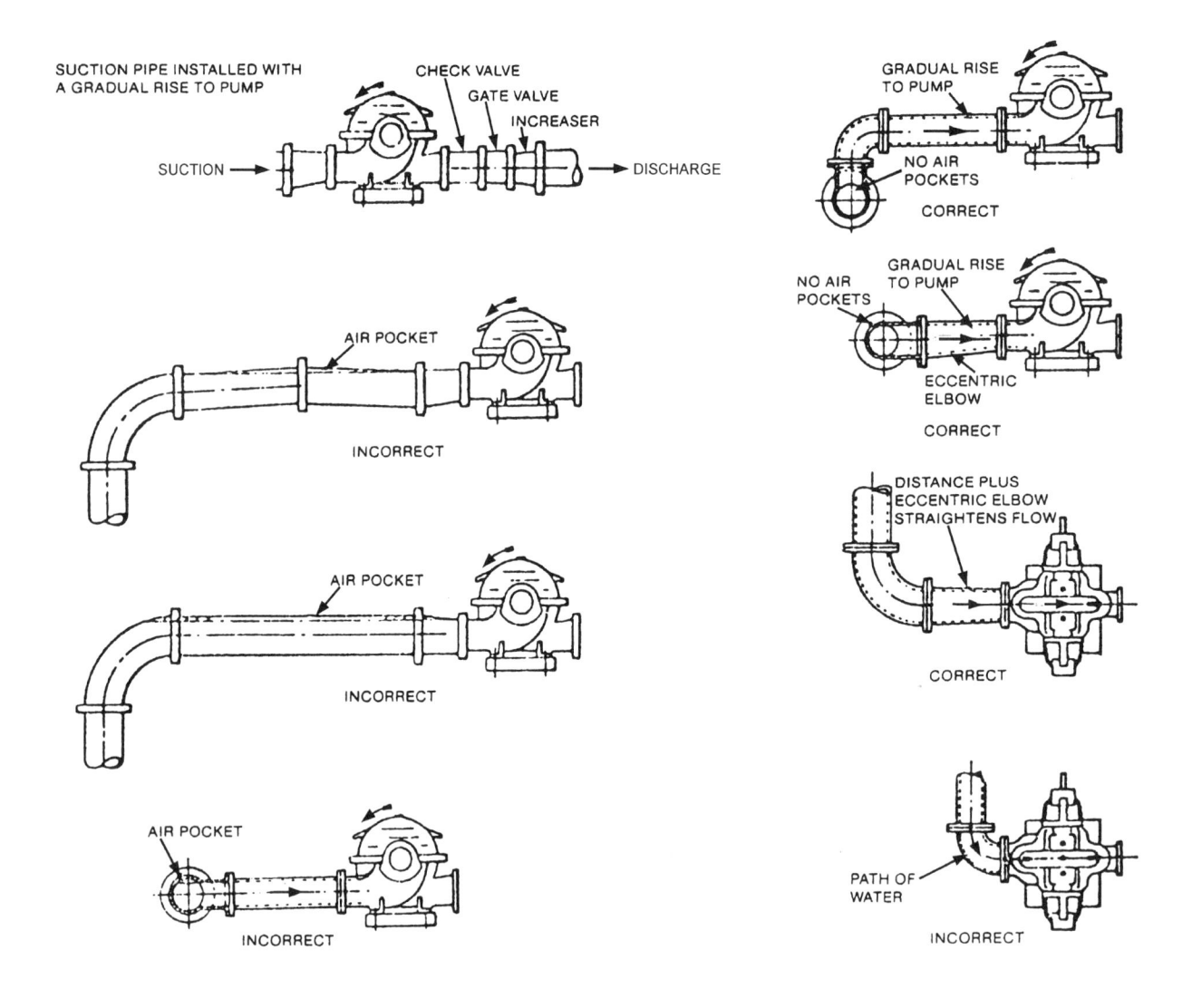

Figure 10-5 Examples of Suction-Piping Installations

address the various, specific, vibration-control techniques that are useful only in the hands of a specialist or that require detailed measurements and analyses.

***Selection criteria*** A vibration-isolating device should be selected using the following criteria:

1. It must be soft enough to provide the desired isolation effect and have a stiffness that is less than the local stiffness of each of the items it connects.

2. It must provide a natural frequency that is considerably lower than the lowest excitation frequency of concern.

3. It must be capable of carrying the loads imposed on it.

4. It must be able to withstand the environment to which it will be exposed.

***Vibration control devices*** In plumbing systems, vibration-control devices generally consist of steel springs, air springs, rubber isolators, pads or slabs of fibrous (or other resilient) materials, isolation hangers, flexible pipe connectors, concrete bases, or any combination of these items. Some of the most common vibration-isolation devices are illustrated in Figures 10-6 and 10-7.

# Chapter 10 — Acoustics in Plumbing Systems

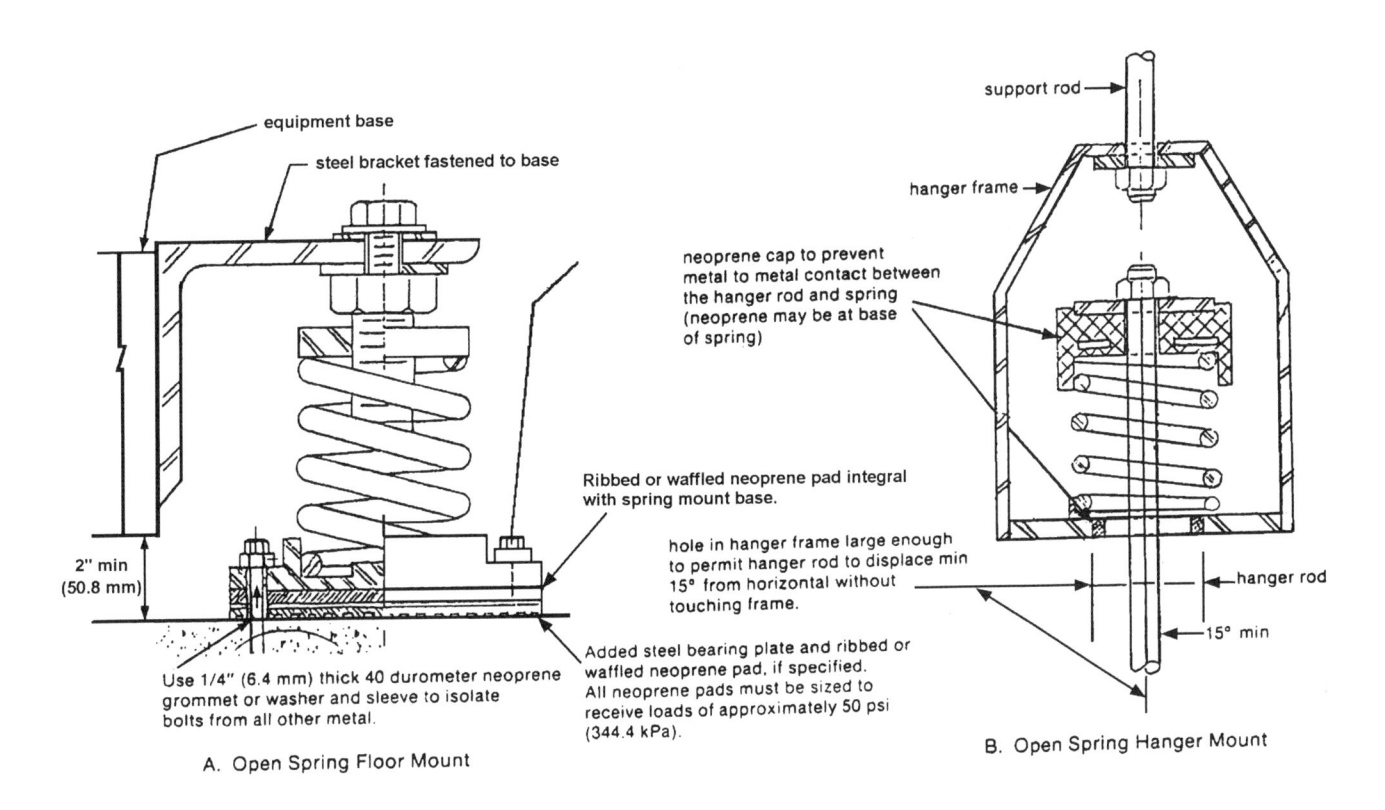

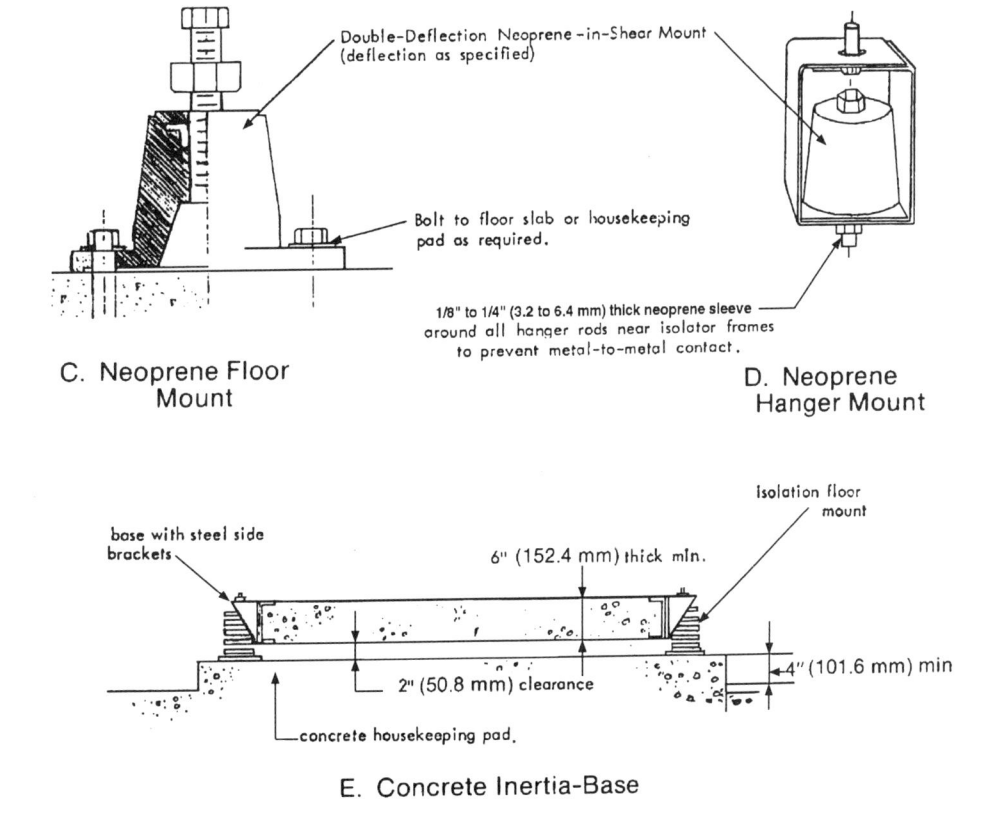

**Figure 10-6  Typical Vibration-Isolation Devices**

*Steel springs* Steel springs are available for almost any desired deflection. These devices are generally used as vibration isolators that must carry heavy loads or where the environmental conditions make other materials unsuitable. The basic types of steel spring mountings are as follows:

1. Housed-spring mountings.
2. Open-spring mountings.
3. Restrained-spring mountings.

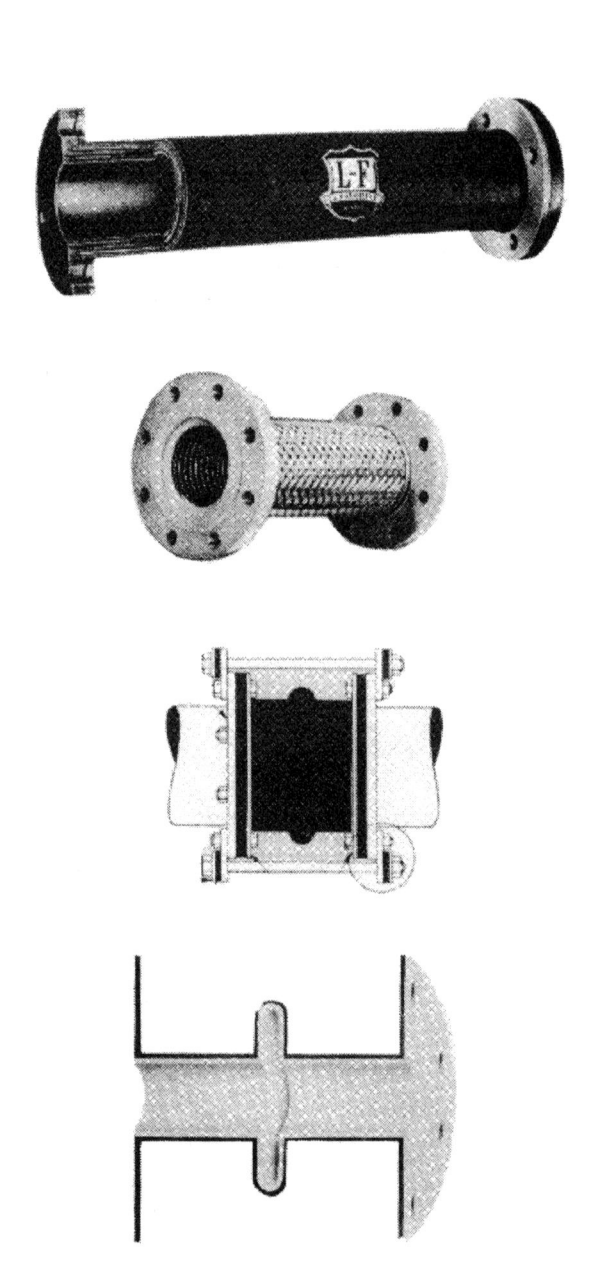

**Figure 10-7 Typical Flexible Connectors**

Because steel springs have little inherent damping and can increase their resonance in the audio-frequency range, all steel-spring mountings should be used in series with pads of fibrous or other resilient materials to interrupt any possible vibration-transmission paths.

*Air springs* Air springs, as steel springs, are available for almost any desired deflection where 6 in. (152.4 mm) or more is required. Air springs have the advantage of virtually no transmission of high-frequency noise.

*Rubber isolators* Rubber isolators are generally used where deflections of 0.3 in. (7.6 mm) or less are required. These devices can be molded in a wide variety of forms designed for several combinations of stiffness in the various directions. The stiffness of a rubber isolator depends on many factors, including the elastic modules of the material used. The elastic modules of the material vary with the temperature and frequency and are usually a characteristic of a durometer number, measured at room temperature. Materials in excess of 70 durometers are usually ineffective as vibration isolators. Rubber isolating devices can be relatively light, strong, and inexpensive; however, their stiffness can vary considerably with the temperature.

*Precompressed, glass-fiber pads* These devices are generally used where deflections of 0.25 in. (6.4 mm) or less are required. Precompressed, glass-fiber pads are available in a variety of densities and fiber diameters. Although glass-fiber pads are usually specified in terms of their densities, the stiffness of the pads supplied by different manufacturers may differ greatly, even for pads of the same density.

*Sponge rubber* Sponge-rubber vibration isolators are commercially available in many variations and degrees of stiffness. The stiffness of such a material usually increases rapidly with increasing load and increasing frequency.

*Concrete base* Concrete-base devices are usually masses of concrete, poured with steel channel, weld-in reinforcing bars and other inserts for equipment hold-down and vibration-isolator brackets. These devices perform the following functions:

1. Maintain the alignment of the component parts.
2. Minimize the effects of unequal weight distribution.

# Chapter 10 — Acoustics in Plumbing Systems

3. Reduce the effects of the reaction forces, such as when a vibration-isolating device is applied to a pump.
4. Lower the center of gravity of the isolated system, thereby increasing its stability.
5. Reduce motion.

Concrete bases can be employed with spring or rubber vibration isolators. Usually, industrial practice is to make the base in a rectangular configuration approximately 6 in. (152.4 mm) larger in each dimension than the equipment being supported. The base depth needs not to exceed 12 in. (0.3 m) unless specifically required for mass, rigidity, or component alignment. A concrete base should weigh at least as much as the items being isolated (preferably, the base should weigh twice as much as the items). The plumbing designer should utilize the services of a structural engineer when designing the concrete base.

*Flexible connectors* When providing vibration isolation for any plumbing system or component, the engineer must consider and treat all possible vibration-transmission paths that may bypass (short-circuit or bridge) the primary vibration isolator. Flexible connectors are commonly used in pipe connecting between isolated and unisolated plumbing components. Flexible pipe connectors are usually used for the following reasons:

### Table 10-1 Recommended Static Deflection for Pump Vibration-Isolation Devices

| Equipment Location | Power Range, HP (kW) | Speed, RPM | Indicated Floor Span, in. (mm) 30 ft (9.1 m) | 40 ft (12.2 m) | 50 ft (15.2 m) |
|---|---|---|---|---|---|
| Slab on grade | Up to 7.5 (5.6) | 1800 | ¾ (19.1) | ¾ (19.1) | ¾ (19.1) |
| | | 3600 | ¼ (6.4) | ¼ (6.4) | ¼ (6.4) |
| | Over 7.5 (5.6) | 1800 | 1 (25.4) | 1 (25.4) | 1 (25.4) |
| | | 3600 | ¾ (19.1) | ¾ (19.1) | ¾ (19.1) |
| | 50-125 (37.3-93.2) | 1800 | 1½ (38.1) | 1½ (38.1) | 1½ (38.1) |
| | | 3600 | 1 (25.4) | 1 (25.4) | 1 (25.4) |
| Upper floor above noncritical areas | Up to 7.5 (5.6) | 1800 | ¾ (19.1) | ¾ (19.1) | 1½ (38.1) |
| | | 3600 | ¾ (19.1) | ¾ (19.1) | 1 (25.4) |
| | Over 7.5 (5.6) | 1800 | 1 (25.4) | 1½ (38.1) | 2 (50.8) |
| | | 3600 | ¾ (19.1) | 1 (25.4) | 1½ (38.1) |
| | 50-125 (37.3-93.2) | 1800 | 1½ (38.1) | 2 (50.8) | 2½ (63.5) |
| | | 3600 | 1 (25.4) | 1½ (38.1) | 2 (50.8) |
| Upper floor above critical areas | Up to 7.5 (5.6) | 1800 | 1 (25.4) | 1½ (38.1) | 2 (50.8) |
| | | 3600 | ¾ (19.1) | 1 (25.4) | 1½ (38.1) |
| | Over 7.5 (5.6) | 1800 | 1½ (38.1) | 2 (50.8) | 3 (76.2) |
| | | 3600 | 1 (25.4) | 1½ (38.1) | 2 (50.8) |
| | 50-125 (37.3-93.2) | 1800 | 2 (50.8) | 3 (76.2) | 4 (101.6) |
| | | 3600 | 1½ (38.1) | 2 (50.8) | 3 (76.2) |

1. To provide flexibility of the pipe and permit the vibration isolators to function properly.
2. To protect the plumbing equipment from strains due to the misalignment and expansion or contraction of the piping.
3. To attenuate the transmission of the noise and vibration along the piping system.

For plumbing systems, the flexible pipe connectors usually consist of hose connectors and expansion joints.

Most commercially available flexible pipe connectors are designed for objectives (1) and (2) denoted above and not primarily for noise reduction. For noise control, resilient pipe isolators should be utilized.

***Vibration isolation of plumbing fixtures*** From a practical point of view, most plumbing fixtures cannot be effectively isolated, although these components can be installed in a manner to minimize vibration transmission. Figures 10-8, 10-9, 10-10 and 10-15 illustrate some examples of resiliently mounted plumbing fixtures.

***Vibration isolation of pumps*** Concrete bases with spring isolators or neoprene pads are preferred for all floor-mounted pumps. It is common practice to isolate a pump in a manner similar to that illustrated in Figure 10-11. Figure 10-12 shows some of the most common errors found in pump-isolation systems. Table 10-1 contains the recommended static deflection for the selection of pump vibration-isolation devices.

For critical system applications, sump pumps and roof drains should also be isolated. See Figure 10-13 for a typical installation.

***Vibration isolation of piping*** All hot water piping, including the heat exchanger and the hot-water storage tank, should be isolated in addition to the following:

1. All piping in the equipment room.
2. All piping outside of the equipment room, within 50 ft (15.2 m) of the connected pump.
3. All piping over 2 in. in diameter (nominal size) and any piping suspended below or near a noise-sensitive area.
4. The first three (3) supports provide the same deflection as the pump vibration isolators. They should be a precompressed type in order to prevent a load transfer to the equipment when the piping systems are filled.

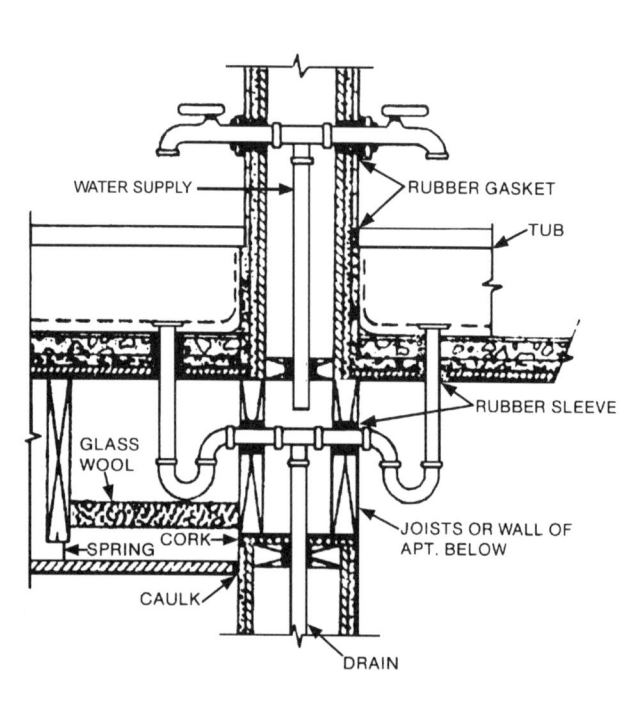

**Figure 10-8** Bathtub and/or Shower Installation

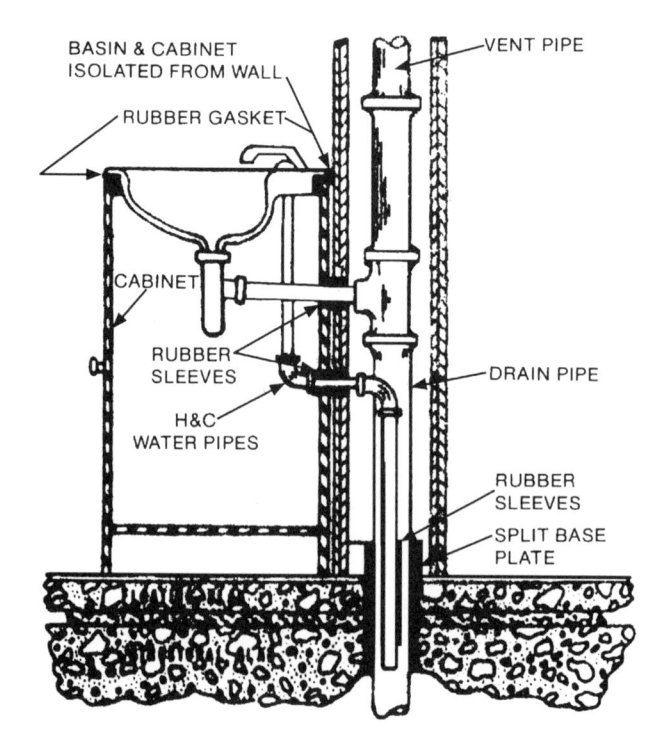

**Figure 10-9** Suggested Mounting of Piping and Plumbing Fixture

# Chapter 10 — Acoustics in Plumbing Systems

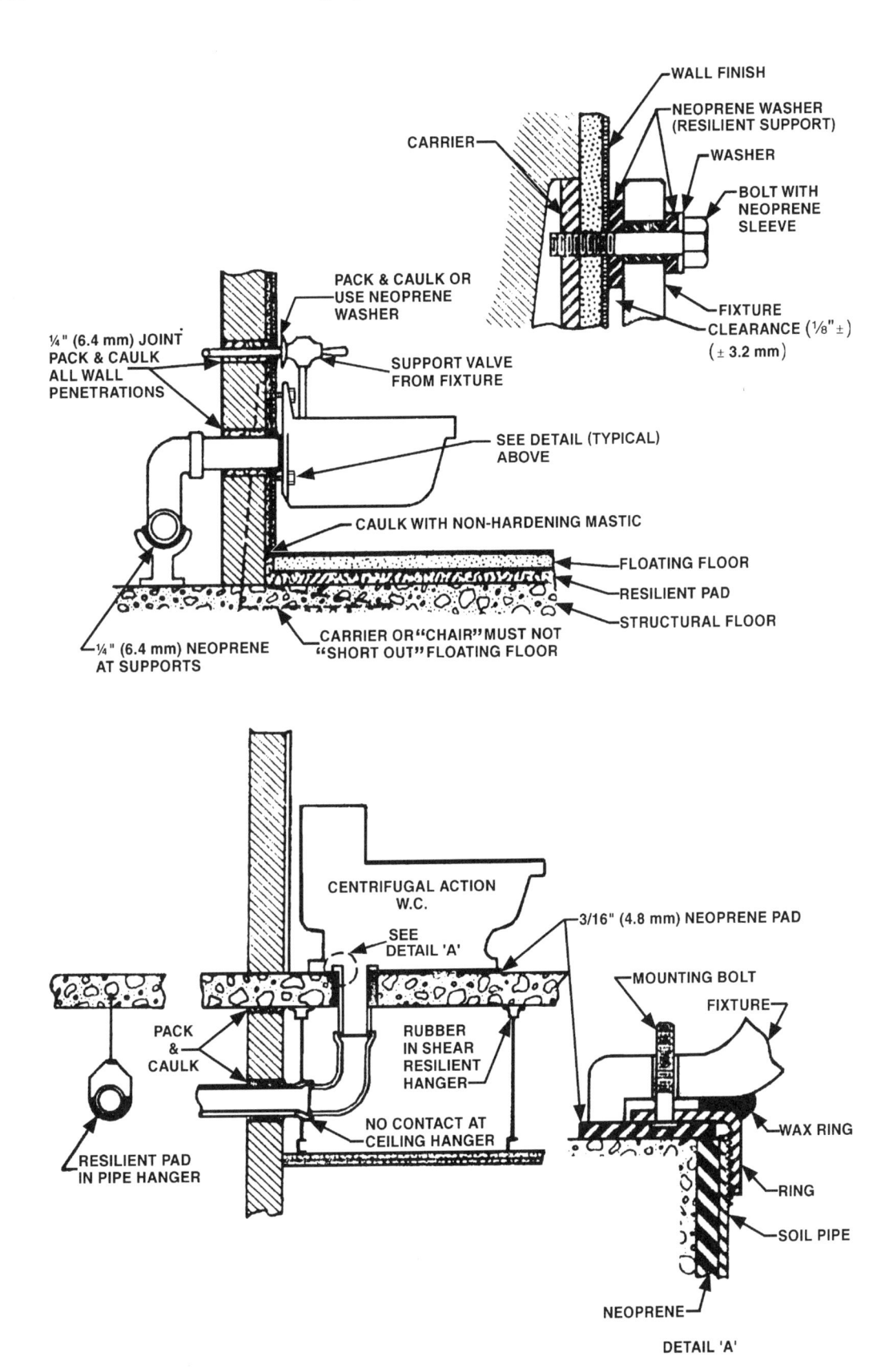

**Figure 10-10  Suggested Installation of Plumbing Fixtures**

5. The remaining vibration isolators should provide half (½) the deflection of the pump isolators or 0.75 in. (19.1 mm) deflection, whichever is larger.

All piping connected to plumbing equipment should be resiliently supported or connected. See Figures 10-14 and 10-15 for typical installations.

**Seismic protection** The seismic protection of resiliently mounted systems presents a unique problem for vibration isolation selection and application. Since resiliently mounted systems are much more susceptible to earthquake damage (due to resonances inherent in the vibration isolators), a seismic specialist should be consulted if seismic protection of such a system is desired. (Refer to the *Data Book* chapter "Seismic Protection of Plumbing Equipment" for more information on this topic.)

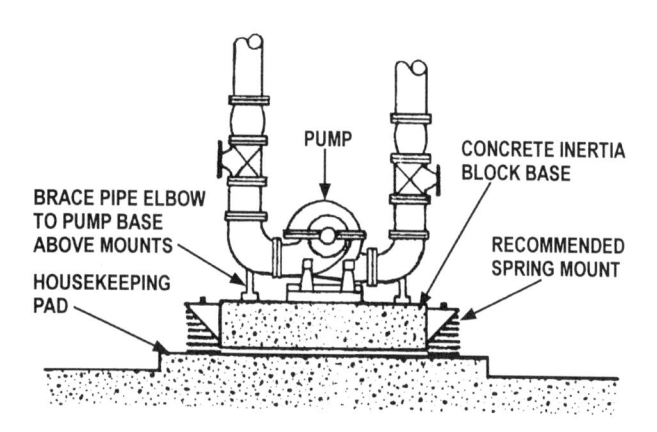

**Figure 10-11** Vibration Isolation of Flexible-Coupled, Horizontally Split, Centrifugal Pumps

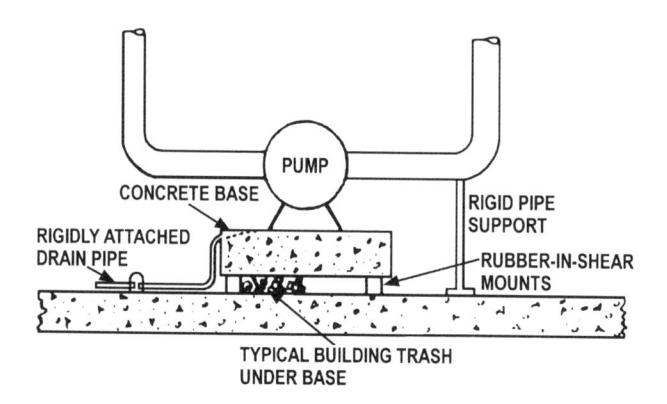

**Figure 10-12** Common Errors Found in Installation of Vibration-Isolated Pumps

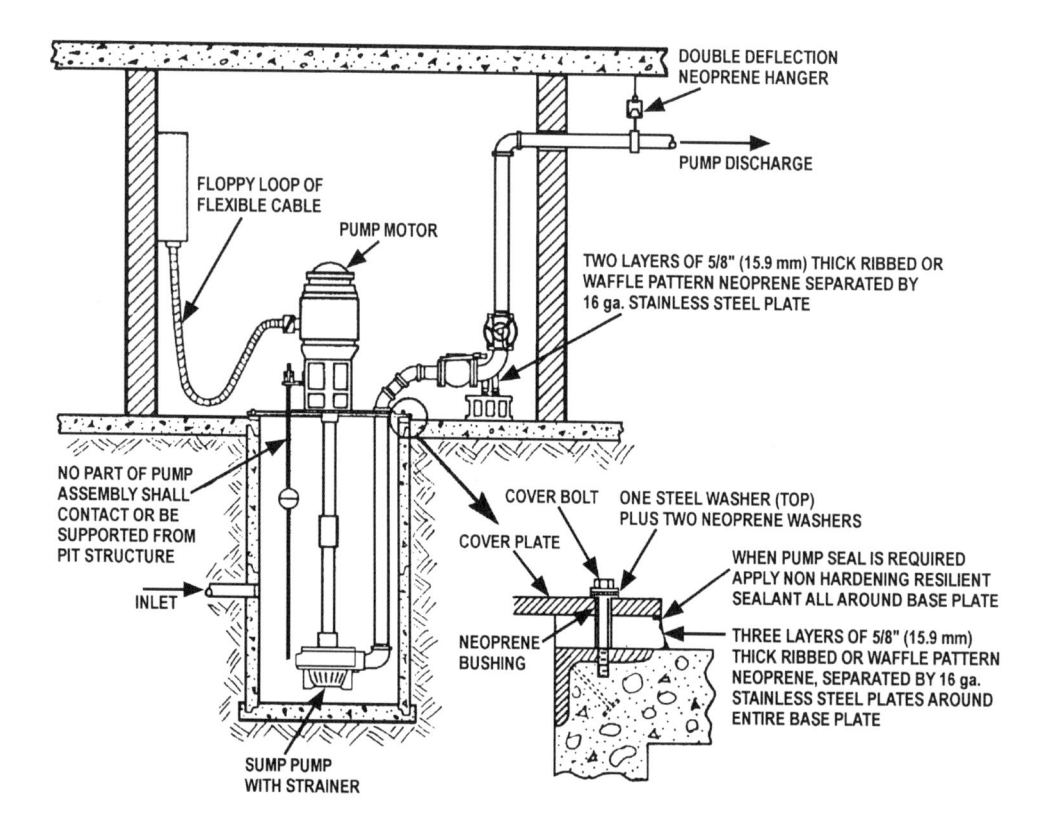

**Figure 10-13** Vibration Isolation of a Sump Pump

# Chapter 10 — Acoustics in Plumbing Systems

## GLOSSARY

***Acoustics*** The study of airborne sound and structural vibration propagation over the frequency range 2 to 20 kHz.

***Decibel*** The unit used to qualify the level of sound (or loudness) relative to an arbitrary reference point [zero is equal to 20 Pa (20 x 10$^{-6}$ pascals)]. These units are also employed to quantify sound power. This unit is the smallest increment of change in sound intensity that a normal human being can detect, while a 10-decibel change of increasing or decreasing sound is commonly regarded as a subjective doubling or halving of loudness, respectively. Abbreviated "dB."

***Decibel (A) scale*** A frequency-modified sound level in which low-frequency and high-frequency sounds are attenuated in a similar manner to that in which the human ear responds to wide-range sounds. It is the most common unit used for sound measurement to relate sound intensity to normalized subjective loudness. Abbreviated "dB(A)."

***Hertz*** The unit of frequency internationally accepted to be equivalent to cycles per second of sound. 1 Hertz is equal to 1 cycle/second. Abbreviated "Hz."

***Net positive suction head (NPSH)*** Actual fluid energy available or required at the inlet of a pump.

***Noise criteria (NC) curves*** Employed to assess loudness or annoyance on an octave band basis. These noise criteria (NC) curves have been partially superseded by the "preferred noise criteria (PNC) curves."

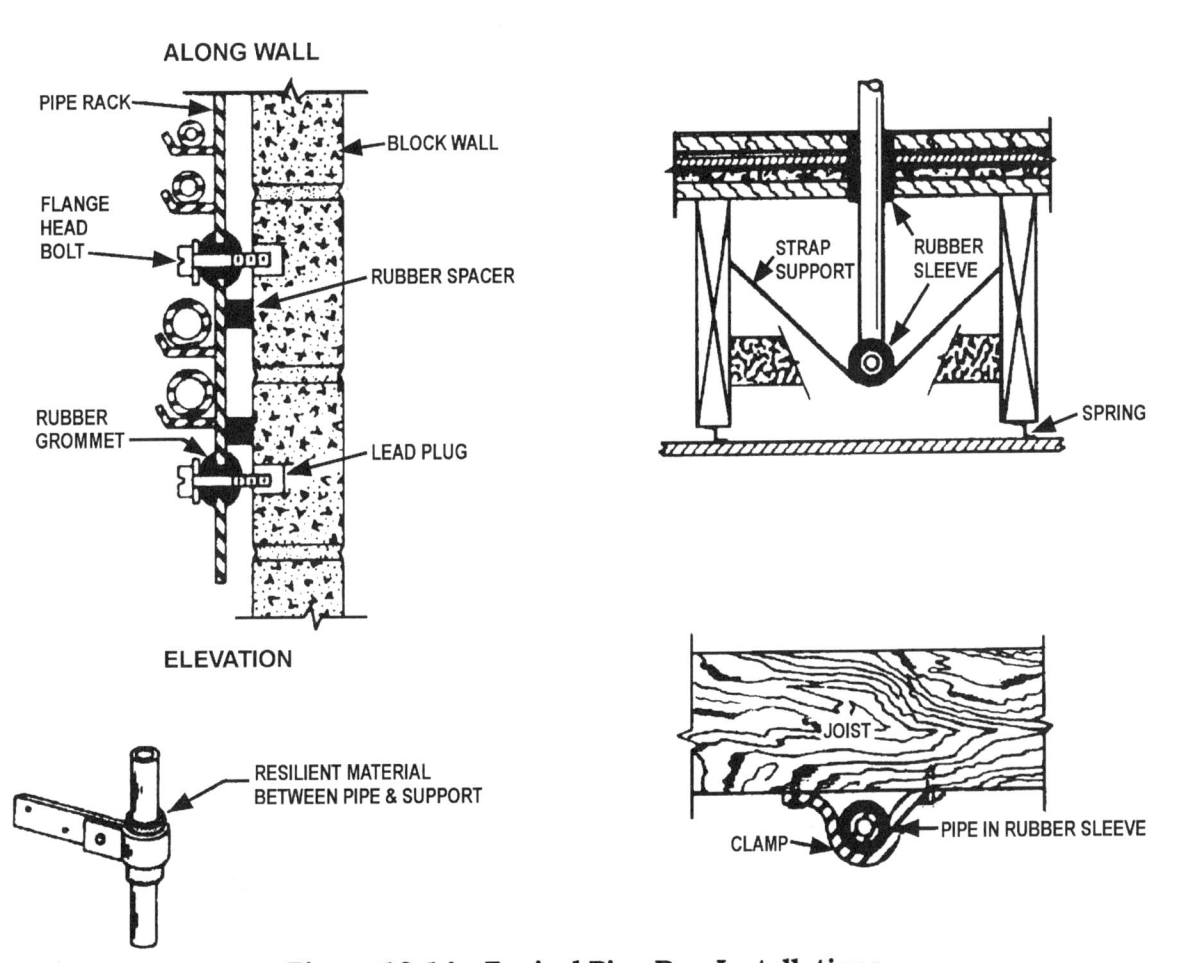

**Figure 10-14 Typical Pipe Run Installations**

**Octave** A doubling of the frequency. Also used as the most common frequency division for the specification of filters employed for acoustical analyses.

**Pure tones** Detectable and generally audible frequency components with characteristics similar to whistles or shrieks generally regarded as being more obtrusive and more likely to give rise to annoyance than other broadband sounds devoid of such components.

**Sound power** The total acoustical energy radiated by a device or fitting operating under normal working conditions.

**Sound-power level** The acoustical output, in decibels, radiated by a device or fitting with reference to a sound power of watts and normally determined in octave bands and, typically, at octave bands center frequencies in the range between 63 Hz and 8 kHz.

**Sound pressure** The oscillation in pressure that gives rise to a sound field in a gas or a liquid.

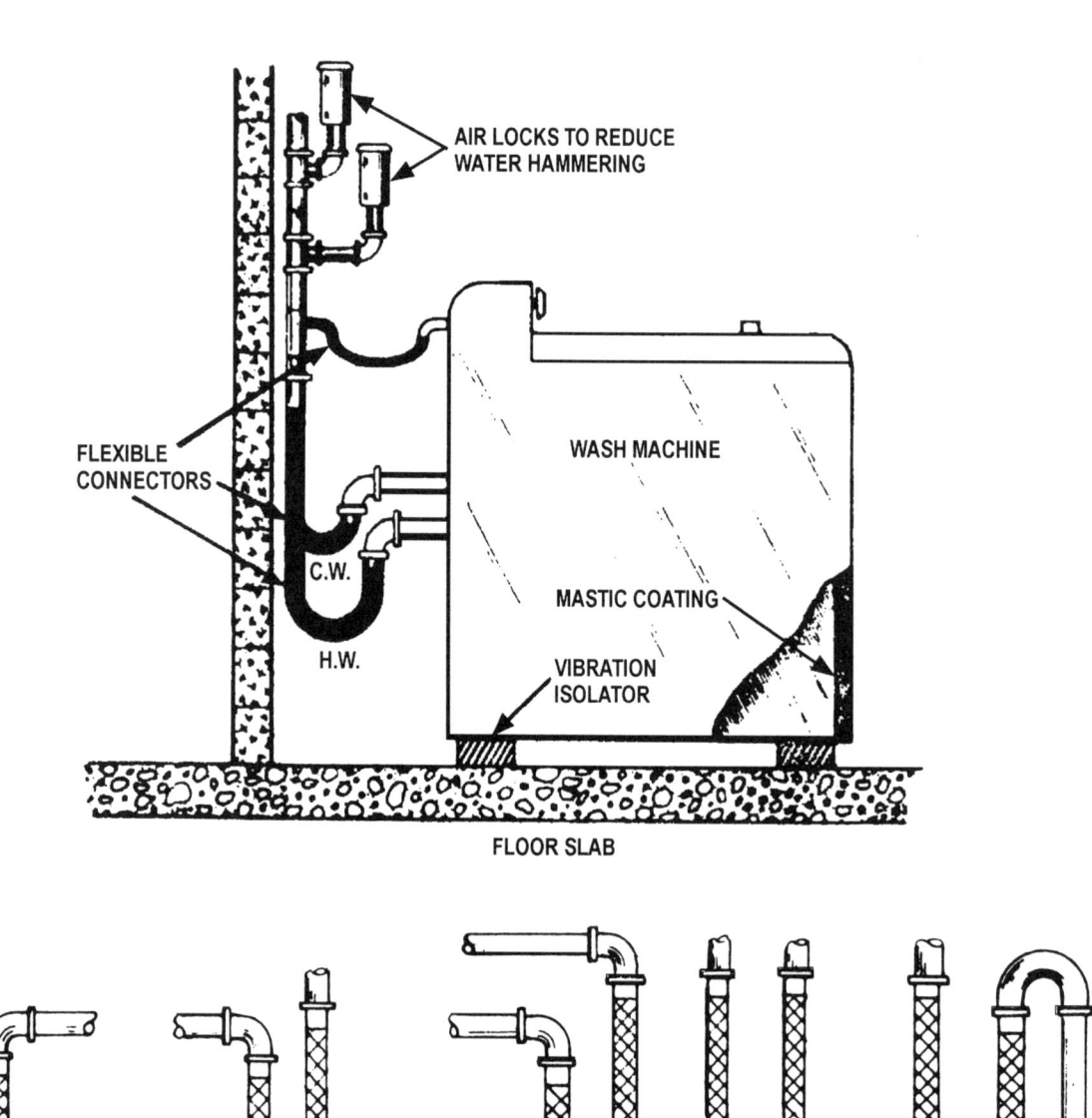

Figure 10-15 **Typical Flexible Pipe-Connector Installations**

# Chapter 10 — Acoustics in Plumbing Systems

***Sound-pressure level*** The logarithmic value of sound pressure referenced to a point of absolute zero (usually 20 µPa) in order to provide a convenient numerical value, in decibels, typically occurring within the range 0 to 120 dB. Typical sound levels are denoted in Table 10-2.

### Table 10-2  Typical Sound Levels

| Sound Level (dB) | Type of Operation |
|---|---|
| 100 | Hammering on pipes |
| 70 | Normal speech levels |
| 50 | Background sound in office |
| 30 | Background sound in urban bedroom |
| 10 | Threshold of hearing for normal adult |

***Transmission loss*** A measure of the sound insulation of a partition or wall, in decibels. It is equal to the number of decibels by which the sound energy passing through it is reduced. The value of the transmission loss is independent of the acoustical properties of the two spaces separated by the partition.

***Vibration*** The generation of cyclic or pulsating forces through a physical medium other than air that converts to sound energy at the boundary between a solid and a gaseous medium. Sound energy may be converted to vibration at the interface between a gaseous medium and a solid medium.

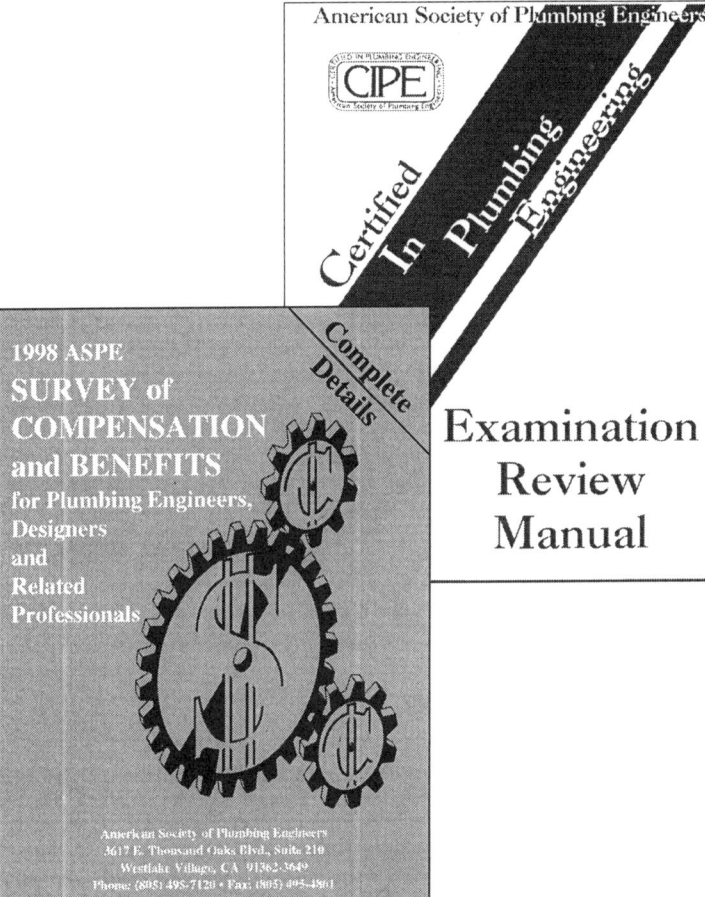

# American Society of Plumbing Engineers
## The Plumbing Engineer's and Designer's Resource
Professional Materials • Engineering Resources • Technical Data • Professional Development
Plumbing Design • Plumbing Specifications • Piping Design • Engineering Information

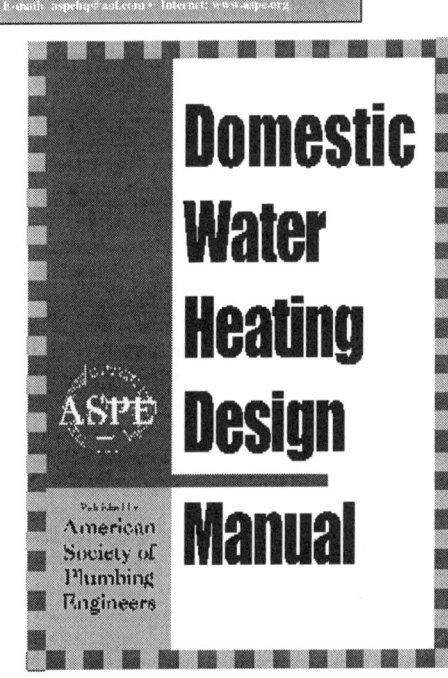

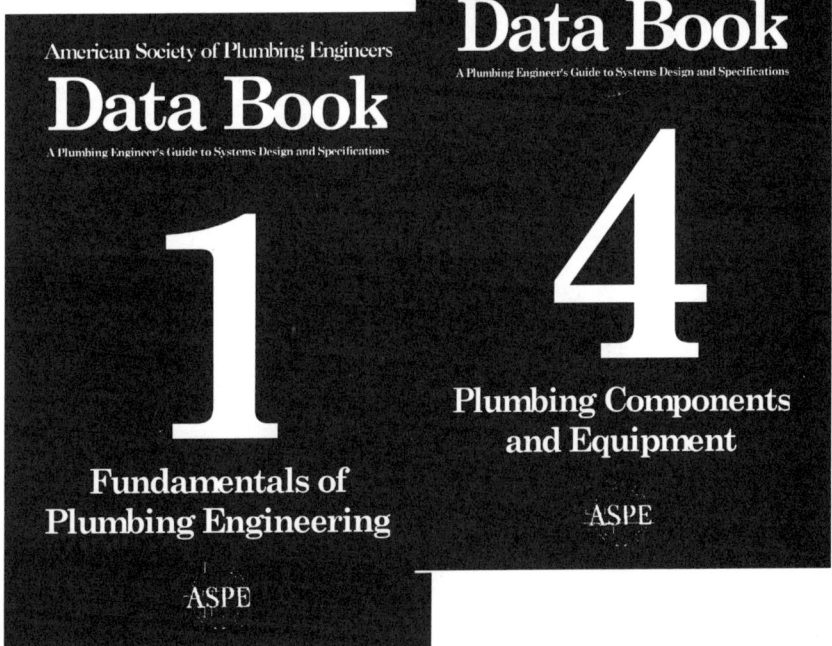

American Society of Plumbing Engineers
3617 E. Thousand Oaks Blvd, Suite 210 • Westlake Village, CA 91362
(805) 495-7120 • Fax: (805) 495-4861
E-Mail: aspehq@aol.com • Internet: www.aspe.org

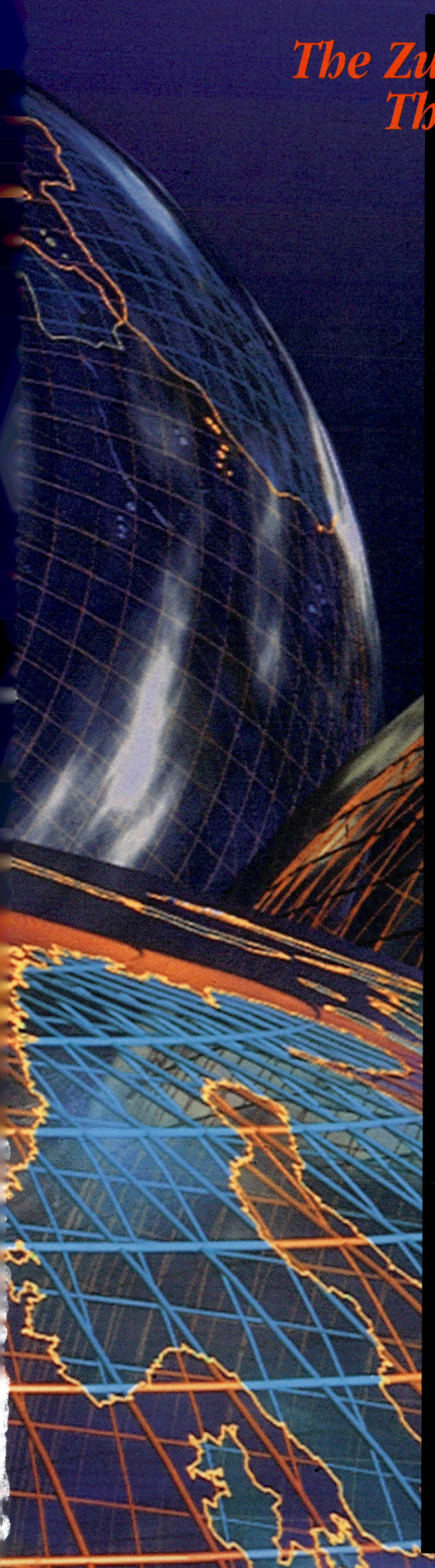

# The Zurn Plumbing Products Group.
# The Ultimate Plumbing Package.

If you're looking for the *Ultimate Plumbing Package*, then turn to ZURN. Zurn has literally thousands of roof-to-basement water and water drainage controls *packaged* for the commercial, industrial and institutional building markets. Products are in stock and ready to ship from our many warehouses for immediate door-to-door delivery!

Call your local Zurn representative or link up with us through our network of specialty companies listed below for quality engineered plumbing products and the *ultimate in Plumbing Packages*.

Zurn Industries, Inc., Plumbing Products Group
1801 Pittsburgh Ave., P.O. Box 13801, Erie, PA 16514-3801

## Roof To Basement Drainage Products

Zurn/Specification Drainage Operation
Phone: 814/455-0921
Fax: 814/875-1402

## Manual and Sensor Flush Valves, Sensor Faucets; Commercial and Institutional Faucets

Zurn/Commercial Brass Operation
Phone: 919/775-2255
Fax: 919/775-3541

## Trench Drain Systems

Zurn/Flo-Thru Operation
Phone: 716/665-1132
Fax: 716/665-1135

## Backflow Preventers, Pressure Regulators, Tub and Shower Valves

Zurn/Wilkins Operation
Phone: 805/238-7100
Fax: 805/238-5766

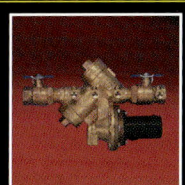

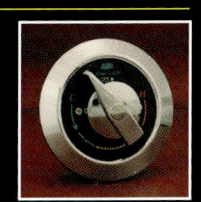

## Tubular Brass and Specialty Products

Sanitary-Dash Operation
Phone: 860/923-9533
Fax: 860/923-2780
A Zurn Company

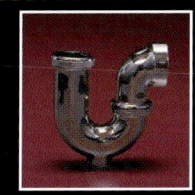

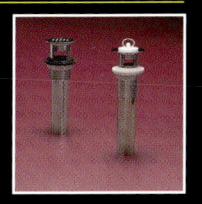

## Plastic Drainage Products and Hydrants

Jonespec Specialty
Plumbing Products
Phone: 716/665-1131
Fax: 716/665-3126
A Zurn Company

**ZURN**

Zurn International Plumbing Products Group
Phone: 814/455-0921  Fax: 814/875-1240

In Canada:
Phone: 905/795-8844  Fax: 905/795-8850

# The 100-Year Difference
## Zurn Specification Drainage Will Be Celebrating Its 100th Year In 1999.

### How Have We Set Ourselves Apart From The Rest?

- **Domestically Owned Foundry Dedicated To Quality Drainage Products.**
- **The Versatility Of 7,000 Patterns.**
- **Most Experienced Sales Representatives In The Industry.**
- **Dynamic Distribution Network With 70 Service Centers For Unmatched Service.**
- **Made-To-Order Design, In-House Pattern Making and Production Capabilities.**
- **Installation and Labor Saving Designs.**
- **New and Innovative Products.**

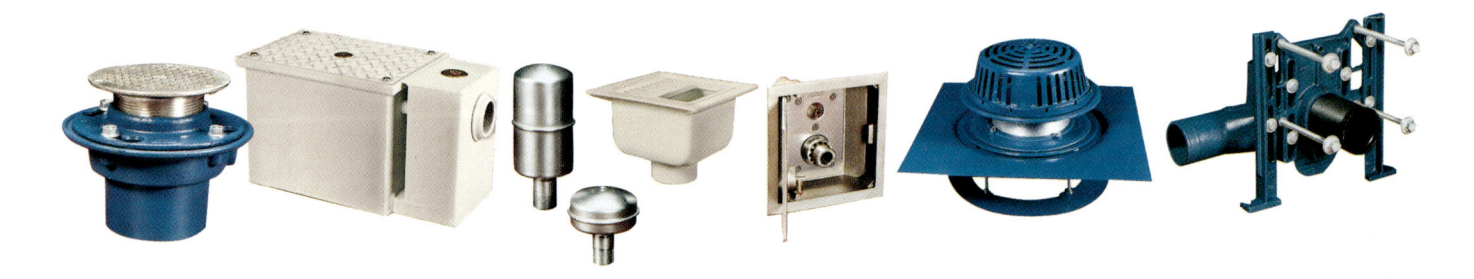

**ZURN**

**PLUMBING PRODUCTS GROUP**
SPECIFICATION DRAINAGE OPERATION
1801 PITTSBURGH AVE.
ERIE, PA, U.S.A. 16514
PHONE: 814/455-0921
FAX: 814/454-7929